Lecture Notes in Computer Science   1259

Edited by G. Goos, J. Hartmanis and J. van Leeuwen

Advisory Board: W. Brauer   D. Gries   J. Stoer

Springer
Berlin
Heidelberg
New York
Barcelona
Budapest
Hong Kong
London
Milan
Paris
Santa Clara
Singapore
Tokyo

Tetsuya Higuchi   Masaya Iwata
Weixin Liu   (Eds.)

Evolvable Systems:
From Biology to Hardware

First International Conference, ICES96
Tsukuba, Japan, October 7-8, 1996
Proceedings

 Springer

Series Editors

Gerhard Goos, Karlsruhe University, Germany

Juris Hartmanis, Cornell University, NY, USA

Jan van Leeuwen, Utrecht University, The Netherlands

Volume Editors

Tetsuya Higuchi
Masaya Iwata
Weixin Liu
Electrotechnical Laboratory, Computer Science Division
1-1-4 Umezono, 305 Tsukuba, Ibaraki, Japan
E-mail: higuchi@etl.go.jp
        miwata@etl.go.jp
        w-liu@etl.go.jp

Cataloging-in-Publication data applied for

Die Deutsche Bibliothek - CIP-Einheitsaufnahme

Evolvable systems : from biology to hardware ; first international
conference ; proceedings / ICES '96, Tsukuba, Japan, October 7 - 8,
1996. Tetsuya Higuchi ... (ed.). - Berlin ; Heidelberg ; New York ;
Barcelona ; Budapest ; Hong Kong ; London ; Milan ; Paris ; Santa
Clara ; Singapore ; Tokyo : Springer, 1997
  (Lecture notes in computer science ; Vol. 1259)
  ISBN 3-540-63173-9

CR Subject Classification (1991): B.6, B.7,F.1, I.6, I.2.9, J.2, J.3

ISSN 0302-9743
ISBN 3-540-63173-9 Springer-Verlag Berlin Heidelberg New York

© Springer-Verlag Berlin Heidelberg 1997
Printed in Germany

Typesetting: Camera-ready by author
SPIN 10549828   06/3142 – 5 4 3 2 1 0   Printed on acid-free paper

Preface

This book describes the newest developments in the area of evolvable hardware systems, as presented at the First International Conference on Evolvable Systems (ICES96). Interest in evolvable systems is largely the result of recent technological developments that have led to reconfigurable hardware: that is, hardware with an architecture and —— as a result —— a functionality that can be changed online.

A number of years ago, researchers in Japan, Switzerland, and the UK recognized the potential of such developments for building adaptive hardware, i.e., hardware that can function properly in a continuously changing environment and can even continue to perform when some of its components malfunction. The key to the success of such an approach is to make the adaptive changes fast enough to enable the hardware to be used in real-time applications. In principle, any adaptation technique, be it reinforcement learning, evolutionary learning, or neural networks can be used to reconfigure the hardware. So far, however, evolutionary computation has been the main focus of research because of the simplicity of regarding the configuration bits of programmable logic devices as a genotype. Thus, this new field is now commonly called evolvable (rather than adaptive) hardware systems, or simply evolvable hardware (EHW), and attracts researchers and engineers from such diverse areas as hardware design, evolutionary computation, and neural networks.

ICES96 was held at the Electrotechnical Laboratory (MITI), Tsukuba, Japan in October 1996, and was actually the sequel of a workshop held in Lausanne a year earlier. The proceedings of the Lausanne workshop are also available as a volume in the Lecture Notes in Computer Science series (LNCS 1062). While the workshop mainly assembled researchers who were working with actual hardware, ICES96 also included researchers working on software and theoretical aspects, as well as representatives from industry, and interested researchers from related fields. In total, there were almost 80 participants, from 13 different countries. Of these, about 40% were university staff, another 40% were researchers from companies or government research laboratories, and the remainder were postgraduate students. Just over a third of participants came from outside Japan.

The conference itself included 34 oral presentations (of which 6 were invited talks), separated into sessions such as Cellular Systems, Engineering Applications of EHW, Evolutionary Robotics, Innovative Architectures, and Genetic Programming. An overview of the whole conference can be found in the paper 'Recent advances in evolvable systems —— ICES96' by Ian Frank, Bernard Manderick, and Tetsuya

Higuchi, which is scheduled to appear in the journal 'Evolutionary Computation'.

Due to the growing interest in the evolvable systems, it has been decided that ICES will be held every two years. The next conference, ICES98, will be hosted by the EPFL (Swiss Federal Institute of Technology) in Lausanne, October 1998. Information on ICES98 is available at http://lslwww.epfl.ch/ices98/.

We gratefully acknowledge the sponsorship and the support of the Science and Technology Agency (STA), the New Technology Development Foundation, the New Energy and Industrial Technology Development Organization (NEDO), the Japanese Society of Artificial Intelligence, and the Electrotechnical Laboratory.

April 1997

Tetsuya Higuchi
Masaya Iwata
Weixin Liu

Contents

Cellular Systems

Engineering Applications of EHW

Evolutionary Robotics

Innovative Architectures

Evolvable Systems

Evolvable Hardware

Genetic Programming

Invited Talks

Iconic Learning in Networks of Logical Neurons

Igor Aleksander F.Eng.
Neural Systems Engineering Group
Imperial College, London SW7 2BT, UK
i.aleksander@ic.ac.uk

Abstract

A development of dynamic neural networks has been to view them as learning state machines (Aleksander and Morton 1993,1995 and Aleksander 1996). This paper formalises ideas previously presented as empirical possibilities in order to show how particular modes of learning influence the capacity of such systems to model their environments which too have been assumed to have the character of a state machine.

1. Introduction

This paper reviews recent work done with logical or 'weightless' neurons summarised descriptively elsewhere (Aleksander and Morton 1993, 1995). A particular feature of these dynamic networks is that the state assignment is executed through a form of learning called 'iconic'. This attempts to retain the character of the symbols which drive the input so as to achieve internal representations in the language of the environment in which the system is doing its learning. Here a formal description is first given of the logical neuron, which is then made the state variable of a neural state machine which also is described formally together with the iconic learning algorithm. The novel task introduced in this paper to learn the state structure of an environment which in itself behaves like a state machine. Various examples of operation are investigated so as to provide a design guide. These are presented as progressively more comprehensive forms of learning ranging from inferring the rules of a language purely from its strings to the anticipation of output states of an environment resulting from actions generated by the learning organism itself.

2. The logical neuron

As pointed out in Aleksander and Morton (1995) logical neurons have been investigated since 1965 in various guises and forms. Here one of the most recent forms of the device is described, the G-RAM neuron (which stands for Generalising Random Access Memory).

The logical neuron processes a binary n-vector input:
$$\mathbf{x} = (x_1, x_2, \ldots x_n).\qquad [1]$$
It performs a function selected from the set of all possible binary functions of n binary variables:
$$\Phi(\mathbf{x}) = \{f_1(\mathbf{x}), f_2(\mathbf{x}) \ldots f_m(\mathbf{x})\} \text{ where } m = 2^{2^n}. \quad [2]$$

A function from $\Phi(\mathbf{x})$ is selected by a process of training which consists of training set

$$T = \{x_{10}, x_{20} ..., x_{11}, x_{21} ...\} = \{X_0, X_1\}$$

where x_{j0} is an input vector from X_0 which requires to be mapped into a 0 and x_{k1} is an input vector from X_1 which requires to be mapped into a 1.

The neuron *generalises* according to the G-algorithm:

> **G-algorithm:**
>
> Any element $x_j \in X$ the set of all input vectors
>
> and $\notin \{X_0 \cup X_1\}$,
>
> is assigned to X_0 or X_1 depending on whether it is closer in Hamming distance to an element in X_0 or X_1 respectively.
>
> If there there are any $x_u \in \{X_0 \cap X_1\}$
>
> or if there are x_u which are equidistant from their nearest neighbours in X_0 and X_1,
>
> **then** the neuron will map x_u randomly into 0 or 1,
>
> the choice being made for each system clock period.

Therefore the G-Ram behaves like a bit-organised random access memory up to the point at which there is a clash between attempting to set an address to both 0 and 1, or an address cannot be set according to the Hamming distance rule. In such cases the address remains undefined and behaves as a random binary string generator. This form of generalisation is sometimes described as a process of **spreading**, where information set in the RAM through the training set spreads to other locations. In some applications **spreading limited to a radius** is employed in the sense that Hamming distances greater than a set limit are ignored.

3. A neural state machine

The definition of a neural state machine follows the classical form of a state machine for which a 5-tuple is defined as:

$$< I, Q, Z, \partial, \beta > \ ;$$

where I is the set of input messages:

$$I = \{i_1, i_2, ... , i_d\} \ ,$$

each input vector being formed on a input variables:

$$i_j = (i_1, i_2, ... i_a)_j, \qquad\qquad [3]$$

so that $\qquad d = 2^a.$

Similarly, Q is the set of internal states:

$$Q = \{q_1, q_2, ... , q_e\} \ ,$$

each state vector being formed on b input variables:

$$q_j = (q_1, q_2, ... q_b)_j, \qquad\qquad [4]$$

so that $\qquad e = 2^b.$

Also, Z is the set of outputs:

$$Z = \{z_1, z_2, ... , z_g\} \ ,$$

each output vector being formed on c input variables:

$$z_j = (z_1, z_2, ... z_c)_j,$$

so that $\qquad g = 2^c.$

The next-state function ∂ is defined as $\partial(IxQ) \rightarrow Q$, while for the output function β, the classical Moore model is assumed $\beta(Q) \rightarrow Z$.

What makes this a **neural** system is that each $q_j \in \{q_1, q_2, \dots q_b\}$ and $z_j \in (z_1, z_2, \dots z_c)$ are the output of a logical neuron and perform a neural function according to equations [1] and [2].
That is,

$$q_j = f_r(\mathbf{x_j}), \; z_k = f_s(\mathbf{xk})$$

selected by training as described earlier. The constituents of $\mathbf{x_j}$ depend on the **connectedness** of the neural state machine. That is, if the net is **fully connected**,

$$\mathbf{x_j} = (i_1, i_2, \dots i_a, q_1, q_2, \dots q_b),$$

and is the same for all j, whereas if the net is **partly connected** $\mathbf{x_j}$ is drawn from an arbitrary selection from $(i_1, i_2, \dots i_a, q_1, q_2, \dots q_b)$ and no two $\mathbf{x_j}$ vectors are the same. The balance of connection between input and state variables may be controlled by the designer. In this paper the effect of such variations is not crucial as the focus of the paper is on the state machine characteristics of these systems, and, unless otherwise stated, a full connection is assumed.

4. State assignment

Having created a neural state machine with initially undefined functions for its neurons (i.e. state variables), the process of training is intended to create some useful sequential function for the system. This means that the state assignment problem arises in this case as it does in traditional state machine design. In the neural systems domain, this is normally handled in two ways. The first is unsupervised learning with localised representations created through competition between topologically differentiated neurons (the schemes of Teuvo Kohonen, 1988 and Steve Grossberg,1987 are dominant in this area). Here representations emerge where states are made meaningful according to the topological area in which activity is taking place. That is, $(q_1, q_2, \dots q_b)$ is organised in some topologically meaningful fashion (e.g. a two-dimensional or a three-dimensional array). The second method relies on arbitrary representations (as in the work of Elman, 1990 and Servan-Schreiber et. al. (1991) where some generalisation is obtained through weighted structures and error back-propagation through a layer of internal state neurons. Also the work of Giles et. al. (1996) is notable for generating usable state machines where efficiency of computation is a quality factor.

The character of the work in this paper which is different from that quoted above is that in work on 'artificial consciousness' briefly discussed at the end of this paper, it is paramount that the state assignments be such as to retain the features of the inputs which require them (this allows the formation of mental imagery, planning and prediction to be modelled in these machines). A scheme called 'iconic training' has been devised to achieve this, and it is the stated purpose of this paper to examine the theoretical effects of this in state-machine terms.

5. Iconic training

First it is assumed that for $(i_1, i_2, \dots i_a)$ in [3] and $(q_1, q_2, \dots q_b)$ in [4],

$$a \geq b$$

An arbitrary one-to-one mapping is defined between (some) elements of $(i_1, i_2, \dots i_a)$ and all the elements of $(q_1, q_2, \dots q_b)$. We then say that each q_j is an 'iconic reflection' of one and only one i_k and indicate this as:

$$q_j : \eta (i_k).$$

Iconic training is then a process of applying pattern i_j to the input of the state machine, which may be in state q_k and training according to the G-algorithm so that $\eta (i_k)$ in q_l is trained to the value of i_k in i_j (within the constraints of the G-algorithm). The so created state q_j is either a bit-for-bit copy of i_j (if $a = b$) or a sampled version of it (if $a > b$). Whichever the case, the state structure so created is a reflection of the time structure of the input. An example might help. To summarise this it is said that :

$$q_j = \eta(i_j) \tag{5}$$

6. Example 1: Learning a sequential grammar

To illustrate the operation of an iconically trained neural state machine we consider learning a *sequential* grammar which is a subset of a finite-state grammar and which may be defined in contrast to the latter as follows. We recall that a finite state grammar is defined by production rules such as:

$$U_j := v_k.U_l / v_m ,$$

where U_j and U_l are non-terminal symbols, v_k and v_m are the terminal symbols which appear in the language, (/) signifies 'or', and (.) is the concatenation sign.

The sequential grammar rule
A *sequential* grammar is defined here as one for which there are no two non-terminal symbols U_2 and U_4 and terminal symbol v_4 such that the following is part of the set of production rules:
$$U_1 := v_k.U_2,$$
$$U_3 := v_k.U_4 .$$

In other words, in a sequential grammar, a terminal symbol cannot occur several times in a grammatical string separated by other terminal symbols. This may be illustrated with two examples in fig 1. In this figure generators of sequences according to defined grammars are shown. A grammatical sequence is defined as one which takes the machine from the beginning state (U_{be}) to the end state (U_{en}). We note that in machine (a) there are several instances of breaking the sequential grammar rule:

e.g.
$$U_0 := v_p.U_2,$$
$$U_4 := v_p.U_5 \dots \text{ etc.}$$

Strings such as $v_b.v_p.v_v.v_p.v_s.v_e$ are possible, but illegal for sequential grammars. The machine in (b) however does not break the sequential grammar rule. It can be noted in passing that given any finite state grammar, this can be transformed into a sequential grammar by enhancing the set of terminal symbols so that if
$$U_1 := v_k.U_2,$$
$$U_3 := v_k.U_4 \text{ occurs,}$$

the second rule is changed to
$$U_3 := v_k'.U_4,$$

where v_k' is an alternative binary coding for the terminal symbol v_k. Indeed, the grammar in fig. 1(b) was generated precisely in this way, by letting:

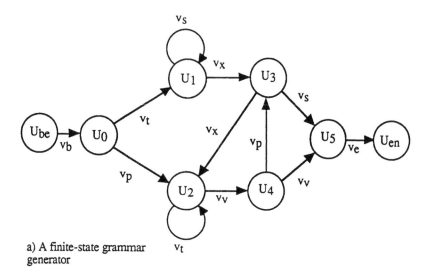

a) A finite-state grammar generator

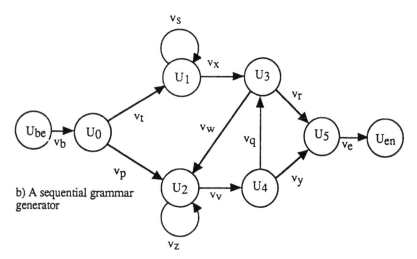

b) A sequential grammar generator

Figure 1. Examples of a finite state grammar generator and a sequential grammar generator.

v_r be an alternative coding for v_s;
v_w be an alternative coding for v_x;
v_y be an alternative coding for v_v and
v_z be an alternative coding for v_t.

Finally, it is necessary to define what "learning a sequential grammar" means in terms of the neural state machine. Assume that the learning system is, as part of its iconic training, fed all the possible paths that take the generator machine from the beginning state (U_{be}) to the end state (U_{en}). Whenever U_{en} is reached a special resetting terminal symbol r_r is emitted by the generator which takes it back to U_{be} (not shown on any of the diagrams). On the completion of training, the input to the neural state machine is replaced by noise. The learning machine will have learned the sequential grammar of the generator if, with noise at its input it is capable of generating all the legitimate sequences within the grammar. This can be expressed as a result.

Result 1. *A fully connected neural state machine can learn, by means of iconic training, a sequential grammar from positive examples only provided by a generator .*

Justification 1.

1.1   For a pair of terminal symbols v_j and v_k successively provided by the generator, the learning machine will create two states U_j and U_k with a transition under v_k between them, that is every generator rule of the form ,

$$U_1 := v_k.U_2$$

is represented in the neural state machine as

$$U_j := v_k.U_k ;$$

where $U_k = \eta(v_k)$

and $U_j = \eta(v_j)$ according to [5].

1.2   As rules of the form

$$U_1 := v_j.U_2$$

and $U_3 := v_k.U_2$

are allowed in sequential grammars the learning automaton will create several states for one state in the generating automaton, one for each transition into a generator state.

1.3   Assuming that the system is fully connected, if, when learning is complete, the input is replaced by noise, say it is in state U_j having learned the transition $U_j := v_k.U_k$, the total input to the neurons will be ($\emptyset$, U_j) [$\in$ (IxQ), as in section 3], where $\emptyset$ is a noise signal. It is assumed that all noise signals are roughly orthogonal to each other and to any training patterns. Then, according to the G-algorithm the next state will be U_k as its binary elements will, on average, be at a Hamming distance of $(a+b)$ $(1/2 + 0)$ whereas those states which are not 'next states' to U_j will be at a Hamming distance of $(a+b)$ $(1/2 + 1/2)$. Should there be more than one transition from U_j, then the last part of the G-algorithm ensures that entry into each of these states is equiprobable.

1.4   If the output of the neural state machine is made the state itself,then assuming $\eta=1$ in $\eta(v_j)$, the learning system will generate the exact codes of the terminal symbols of the language in sequences which do not violate the learned sequential grammar. If $\eta<1$ then the coding of the terminal symbols will be sampled, and the assumption of sufficient sparseness has to be made to show that the correct language is generated by the learning system.

This concludes justification 1.

Comment ary.
While it could be said that all that the iconic neural state machine has done is to have stored all the sequences of the sequential grammar, it needs to be pointed out that the extra which has been achieved is that the function of the generator itself has been learned.

It is also important to note that the learning has been achieved on the basis of positive samples alone, which, in studies of language acquisition is critical in debates about nativism. The result is an existence justification that positive sequences are sufficient to learn a grammar, albeit of a very simplified kind. The nature of the learned grammar and the resulting state splitting is illustrated in fig. 2.

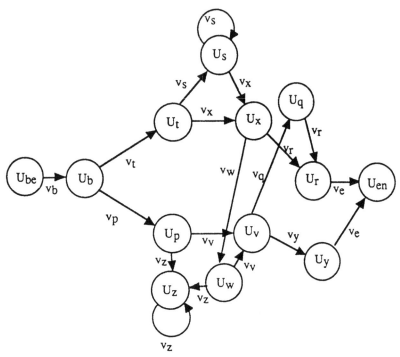

Figure 2. The state structure learned by the neural state machine.

Although this appears to be a rather clumsy structure, it represents the grammar faithfully. The proliferation of states is not of great consequence as redundant, sparse coding is generally assumed in the design of the neural state machine. Of course, when run with noise at the input, the structure becomes an autonomous probabilistic state machine, the states of which have a coding of the desired grammar or a sampled version of it from which the full codes can be retrieved.

7. Example 2: Auxiliary codes and learning a non-reentrant finite-state grammar

A non-reentrant grammar is one which has no reentrant loops of states in its generator (but one which contains the offending rule combination excluded in sequential grammars).

Were the environment to behave as a generator of such a grammar, this can be learned by an iconic neural state machine provided that iconic state codes are enhanced by auxiliary ones. Such codes are arbitrarily generated as a series of noise-like patterns by untrained neurons:
$$\emptyset_0, \emptyset_1, \emptyset_2, \emptyset_3, \ldots \quad .$$

During training, for every input v_k the iconic state U_k ($= \eta(v_k)$) is generated coupled with

one of the of the arbitrary states $\emptyset_j$ say $< \eta(v_k)\emptyset_j >$. This is given the symbol U_{kj} where j is incremented for each new occurrence of U_k. Clearly, the arbitrary codes act as a counter.

However, as the sequences are generated one by one, it is important to recognise that a code has been assigned to some transitions previously. This is easily done through the fact that a previous transition is recognised as the new input generates its iconic representation without learning, and learning should not take place. We trace this learning behaviour in the example shown in fig.3 which in (a) shows the generator grammar and in (b) the way it is learned in the iconic neural state machine. This gives a clue as to the justification of the next result.

Result 2. *A fully connected neural state machine with provision for auxiliary, arbitrary state codes can learn, by means of iconic training, a non-reentrant grammar from positive examples only provided by a generator .*

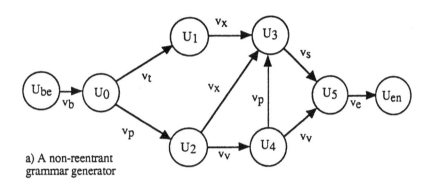

a) A non-reentrant
grammar generator

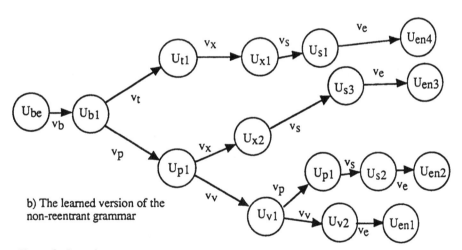

b) The learned version of the
non-reentrant grammar

Figure 3. Learning a non-reentrant grammar.

Justification 2.

2.1   For a pair of terminal symbols v_j and v_k successively provided by the generator, the learning machine will create two states U_{ja} and U_{kb} with a transition under v_k between them, that is every generator rule of the form ,
$$U_1 := v_k.U_2$$
is represented in the neural state machine as
$$U_{ja} := v_k.U_{kb} ;$$

where $U_{kb} = < \eta(v_k)\emptyset_b >$

and $\qquad$ $U_{ja} = <\eta(v_k)\emptyset_a>$ $\qquad$ according to [5] and the definition of the auxiliary code.

2.2 $\quad$ If $\qquad$ $U_1 := v_k.U_2,$

$\qquad\qquad\qquad$ $U_3 := v_k.U_4$ occurs in the generator,

the provision of auxiliary codes in the neural state machine will create states

such as $<\eta(v_k)\emptyset_a>$ for $<\eta(v_k)\emptyset_b>$ for U2 and U4, providing the distinction which, without auxiliary codes, would have caused ambiguities.

2.3 $\quad$ Therefore every rule of the form $U_1 := v_k.U_2,$ in the generator is represented in the neural state machine.

2.4 $\quad$ When a new sequence is learned, learning only takes place when the sequence departs from a previously learned one. Therefore the learned structure is that of an expanding tree, with the creation of many end states of the form U_{enj}.

2.5 $\quad$ When the input is replaced by noise, a branching structure such as

$$U_1 := v_j.<\eta(v_j)\emptyset_a>,$$

$$U_1 := v_k.<\eta(v_k)\emptyset_b>$$

becomes $\qquad$ $U_1 := \emptyset_p.<\eta(v_j)\emptyset_a>,$

$\qquad\qquad\qquad$ $U_1 := \emptyset_q.<\eta(v_k)\emptyset_b>$

$\emptyset_p$ and $\emptyset_q$ being orthogonal noise inputs, the system becomes probabilistic with

equiprobable transitions into the appropriate states $<\eta(v_j)\emptyset_a>$ and$<\eta(v_k)\emptyset_b>$.

2.6 $\quad$ If the output mapping of the neural state machine is such as to transmit the $\eta(v_j)$ part of

the state $<\eta(v_j)\emptyset_a>$ only, the exact ($\eta{=}1$) or sampled ($\eta{<}1$) version of the appropriate terminal symbols will be generated when noise is applied, satisfying the definition for having learned the non-reentrant grammar.

This completes the justification of result 2.

Comment ary.
The essence of learning a non-reentrant grammar is that the learning system can create a tree-like version of the structure of the generator. This explains why the non-reentrant structure is important: were there loops in the structure of the generator, the learning system would attempt to implement them by infinitely expanding branches of the tree. However, this allows extensions of the algorithm using auxiliary states which is briefly discussed below. Again, the fact that this grammar is learned from positive examples only should be noted.

8. Extension of auxiliary code iconic learning to some reentrance

Although we have said that during training the generator needs to produce each valid sequence just once, this becomes problematic when there are reentrant loops in the generator structure as the set of legitimate sequences becomes infinite. The generator can then indicate its structure either through an agreed limit on the number of times a loop is used in the generation of training sequences or by generating sequences probabilistically: by treating each exit from a state as an equiprobable event. This would result in a diminished probability of generating sequences with many repetitions of the reentrant loops. In either case, the iconic machine would be able to learn that a finite number of loops is possible although it would represent the repetitions as new iconic states. A more subtle way of dealing with reentrant environments is first to restrict the transversal of each loop to just one, and then to ensure that the same iconic code is always held for at least two learning steps. The reader may wish to work out that this would result in the representation of single loops in the state structure of the learning machine. Protocols can be created which cope with longer loops.

9. Example 3: Learning the structure and means of control of observable environments

Where we have sought to infer grammatical environments only from the terminal symbols of a language, observable environments are those for which the states are measurable. The object for the learning machine is still to absorb the state structure of the environment, but this time through exploration by means of a predefined list of terminal symbols. The arrangement is shown in fig. 4.

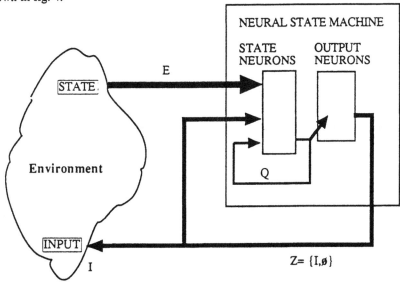

Figure 4. Learning from observable environments.

The environment is a state machine with the state set
$$E = \{e_1, e_2, \dots, e_e\},$$
and an input set $\quad I = \{i_1, i_2, \dots, i_d\}$.
The environment is fully observable as its output set is congruent with the state set E.
The neural state machine has an output or 'action' set which is an augmentation of set I:
$$Z = \{I, \emptyset\} \text{ where } \emptyset \text{ is the 'no action signal' added to the I set which has}$$
no effect on the environment-automaton.
The overall input set to the neural state machine is:
$$\Gamma = (E \times Z).$$
As shown in fig. 4, the neural state machine has also its own set of states:
$$Q = \{q_1, q_2, \dots, q_f\}$$

The state machine starts in a condition where the none of the neurons have learned anything. This *exploratory mode* starts in $\emptyset$, selects an element from I at random and returns to $\emptyset$. (This could be a natural response within the definition of x_u states in the **G-algorithm** where a large number of codes at the Z terminals can have the interpretation of $\emptyset$).

Result 3. *By iconic learning of the behaviour of an observable environment under actions randomly provided by the output of the neural machine itself, the neural state machine can, on disconnection from the environment recall, with noise at its input, how action leads to changes in state. That is, the learning machine builds an inner model of the effect of its own action on the environment.*

Justification 3.

3.1   Every transition in the environment is of the form:

$$(e_i, i_j) \rightarrow e_k.$$

3.2   In the first phase of learning where there is no learning between state and output neurons, the sequence of random actions from the neural state machine in its exploratory mode is i_j, ø, i_k, ø,... where j and k take up arbitrary values. We assume that the output is generated and that the environment reacts immediately. This means that the sequence of inputs seen by the learning system for $(e_i, i_j) \rightarrow e_k$ is $(e_i, ø), (e_i, ø),..(e_i, ø), (e_k, i_j),(e_k, ø), (e_k, ø),...$

3.3   Iconic learning creates the following form of state transitions:

$$[\eta(e_i, ø) \in Q_n], [(e_i, ø) \in I_n] \rightarrow [\eta(e_i, ø) \in Q_n],$$

(where the suffix n indicates the state machine sets for the neural machine)

$$[\eta(e_i, ø) \in Q_n], [(e_k, i_j) \in I_n] \rightarrow [\eta(e_k, i_j) \in Q_n].$$

$$[\eta(e_k, i_j) \in Q_n], [(e_k, ø) \in I_n] \rightarrow [\eta(e_k, ø) \in Q_n]$$

The first of these is a reentrant state while the latter two indicate the transition between two reentrant ones.

3.4   As in previous examples, when the input is replaced by noise (Ø), through the generalisation of the neurons, the learning system will take autonomous probabilistic transitions such as:

$$[\eta(e_i, ø) \in Q_n], [Ø \in I_n] \rightarrow [\eta(e_i, ø) \in Q_n],$$

$$[\eta(e_i, ø) \in Q_n], [Ø \in I_n] \rightarrow [\eta(e_k, i_j) \in Q_n]$$

$$[\eta(e_k, i_j) \in Q_n], [Ø \in I_n] \rightarrow [\eta(e_k, ø) \in Q_n]$$

This fully represents the state transitions in the environment and their causal inputs.

This concludes the justification of result 3.

Commentary.
While the learning state machine has learned to represent the reactions of the environment to arbitrarily chosen inputs, unfortunately it has no representation that it itself is generating the moves. This is done in a more elaborate form of learning which is formulated as result 4.

Result 4: *Having learned the structure according to result 3, a link can be learned between the state and the action of the output neurons. This provides the neural state machine with a power of generating sequences and anticipating the states of the environment.*

Justification of Result 4:

4.1   For state $[\eta(e_i, ø) \in Q_n]$ if the output changes from ø to i_j allowing the link between the state neurons and the output neurons simply to associate the state with the output, the linked representations are formed as follows before the environment has time (say) to react:

$$[<\eta(e_i, ø)/ø> \in (Q/Z)_n] [(e_i, ø) \in I_n] \rightarrow [<\eta(e_i, ø)/ i_j> \in (Q/Z)_n].$$

The notation $(Q/Z)_n$ and $<\eta(e_i, \emptyset)/\emptyset>$ indicates the link between output and state which is the focus of this result (see commentary, below).

4.2 Once the environment has reacted and become e_k, the output is noticed at the input, and the output has returned to $\emptyset$, the following new transition is learned:

$$[<\eta(e_i, \emptyset)/i_j> \in (Q/Z)_n] \ [(e_k, i_j) \in I_n] \rightarrow [<\eta(e_k, i_j)/ \emptyset> \in (Q/Z)_n].$$

4.3 Clearly, when the input is turned to noise, the automaton will not only recall the transitions of the environment but actually act on it while anticipating its states.

This completes the justification of Result 4 .

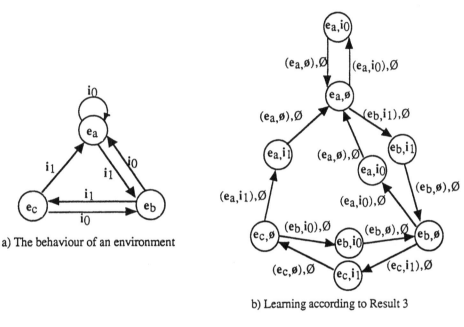

a) The behaviour of an environment

b) Learning according to Result 3

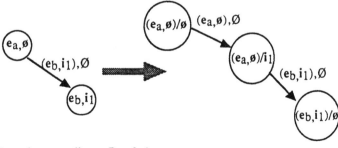

c)Learning according to Result 4.

Figure 5. Learning according to results 3 and 4.

Comment ary.
A simple example might help at this point . This is shown in fig. 5. In (a) the behaviour of a simple environment is shown with $E = \{e_a, e_b, e_c\}$ and $I = \{i_0, i_1\}$. In (b) is shown the result of learning according to result 3. At this point the knowledge of the machine could be expressed in anthropomorphic terms as "*given being at rest with the environment in some particular state, the neural state machine knows , through being able to explore its state space (Ø transitions) what action/state combinations may be expected next leading to corresponding rest/state combinations*".

In (c) the situation is slightly different. Here a modified Moore formulation worth noting is assumed. The the link to the output variables means that they are treated just as part of the statevariables. Then a change of only the output state variables becomes equivalent to a change of state. The effect of this is for the sytem to "know" that the environmental state changes occur as a result of its own state changes.

10. Current work: habituation and independence

While in the last example above the neural state automaton is capable of predicting the state of the environment even when disconnected from it, in no sense can it be said to be deciding in which way to control the environment. This is a matter of current research concern and has been sketched out in Aleksander, 1996 (Ch.5). Without going into details here, it may be indicated that the neural state machine can become capable of traversing its state space even in the presence of an input, by means of a habituation and arousal control. Habituation causes the input to be replaced by noise even if the input signal persists. This frees the state machine to explore its state space if the input is not changing. In a sense this allows the learning system to explore its options and become primed to take action according to some superimposed need. The methods used in this paper are being applied to give a rigorous description of this.

11. Conclusions and summary

While iconic training may not necessarily be the best way of creating state machines that mimic or control state-machine environments , it forms an important class of learning algorithms as it leads to representations of terminal symbol codes (linguistic environments) and states (observable environments). The technique has been used in an empirical way in systems which are meant to act as cognitive models and this paper contains the formal implication of some of these methods.

Four major results have been discussed. In all cases the test for whether learning has been accomplished involves supplying noise to the input of the learning machine which should then generate sequences which mimic the behaviour of the environment. The first result depends on a definition of 'sequential grammars' which do not generate sequences where identical terminal symbols in remote parts of a string. The result shows that such grammars may be learned directly through iconic learning and, were the grammar not sequential (and non-reentrant) it could still be learned but the learned model would have the structure of a sequential grammar with more than one code per terminal symbol. A key feature of this and the next result is that the grammars are learned from positive examples only. This impinges on the broader question of language acquisition in children. Result 2 relates to a broader class of systems which can learn non re-entrant grammars without recoding terminal symbols. The key to this is a proper use of auxiliary (non-iconic) symbols which are generated by the learning system to disambiguate repeated occurrences of the terminal symbols. Ways of dealing with some reentrance have also been mentioned.

Results 3 and 4 relate to observable environments the former showing how, if the output of the learning system is not linked to the state, the machine builds for itself a model of the environment in terms of a probabilistic automaton of all input/state transitions which the environment can deliver. The latter looks at the situation in which the output action of the learning automaton is linked to its states which enables the automaton to model and anticipate the effect of its own actions on the environment.

This is by no means a complete theory of iconic learning but merely a beginning of its rigorous characterisation. As indicated in Aleksander (1996) iconic learning has a place in the cognitive modelling which is part of a broad programme in 'artificial consciousness'. It has been hinted in the last part of this paper that through a process of habituation the system can form representations of its own actions which can form the basis for planning and models of 'will' within the above programme of work.

References

Aleksander, I and Morton, H.B., *Neurons and symbols: the stuff that mind is made of.* London, Chapman and Hall, 1993.

Aleksander, I and Morton, H.B., *Introduction to neural computing. (2nd Ed.)* London, International Thompson, 1995.

Aleksander, I, *Impossible minds: my neurons, my consciousness.* London, Imperial College Press, 1996 (Ch.5).

Elman, J.L., Finding structure in time, *Cognitive science* , 14, 179-211, 1990.

Giles, C.L., Omlin, C.W. and Horne, W.G., Recurrent neural networks: representation and synthesis of automata and sequential machines. *Proc. ICNN 96.* Washington, 1996.

Grossberg, S., Competitive learning: from adaptive interaction to adaptive resonance, *Cognitive Science* 11, 23-63, 1987.

Kohonen, T. *Self-organisation and associative memory.* 2nd Ed. Heidelberg: Springer, 1988.

Servan-Schreiber, D., Cleermans, A. and McClelland, J.L. *Machine Learning.* 7, 2/3, 162-193, 1991.

Hardware Requirements for Fast Evaluation of Functions Learned by Adaptive Logic Networks

William W. Armstrong[1,2]

[1] Department of Computing Science
University of Alberta, Edmonton, Canada T6G 2H1
[2] Dendronic Decisions Limited, 3624 - 108 Street
Edmonton, Alberta, Canada T6J 1B4

Abstract. Adaptive Logic Networks (ALNs) represent real-valued, continuous functions by means of piecewise linear approximants. ALNs have been used successfully in many applications requiring supervised learning, including control, data mining and rehabilitation. The hardware needed to train an ALN, or to evaluate the function it represents, can be quite modest, since ALN algorithms achieve high speed and efficiency by controlling the flow of execution: omitting computations involving linear pieces that are not needed for the current state change or output. Efficient flow of control can be combined with programmable hardware, as is illustrated by a field programmable gate array processor built for evaluating functions learned by ALNs.

1 Adaptive Logic Networks

The adaptive logic network (ALN) [1] is a special kind of multilayer perceptron (MLP) [8, 10]. The goal of training an ALN is to obtain a representation of a function

$$x_n = f(x_1, \ldots, x_{n-1}) \ , \tag{1}$$

where a training set of samples $(x_1, \ldots, x_n)$ is given together with some user-imposed constraints on the partial derivatives and/or convexity of f (perhaps involving different constraints on different subregions of the function's input space). An ALN represents a function as a relation involving the input variables and the output variable of f: $x_1, \ldots, x_n$ are related if and only if $x_n \leq f(x_1, \ldots, x_{n-1})$. In order to actually compute the function, or an inverse of the function, other structures called ALN decision trees are used. They are derived from the ALN and are highly optimized for speed of evaluation. ALNs have been used on applications in control [3] [3], data mining [4] and rehabilitation of persons with incomplete spinal cord injury [5, 6, 7, 9]

Architecturally, an ALN is formed by interconnection of three kinds of units: input units, linear threshold units, and AND and OR logic gates.

---

[3] Contract W7702-2-R328/01-XSG for the Defence Research Establishment Suffield led to the development of the ALN algorithms and FPGA card described in this article.

Input units These hold the inputs of the function $x_1, \ldots, x_{n-1}$, and a value x_n which is either a desired output of f during training, or, after training, a value to be compared with the value of the learned function f. The inputs to the ALN can be Boolean, integer, floating point or (in theory) real.

Linear threshold units (LTUs) These accept values from the input units and produce a Boolean 1 output if and only if

$$w_0 + w_1 x_1 + \ldots + w_{n-1} x_{n-1} - x_n \geq 0 \ . \tag{2}$$

The LTU produces a Boolean 0 otherwise. The weights w_i of each LTU (we omit an index for the unit number) are adjusted during training. The LTU represents the linear function

$$x_n = w_0 + w_1 x_1 + \ldots + w_{n-1} x_{n-1} \tag{3}$$

by producing a 1 output at points on and under its graph (the "1-class" in n-dimensional space).

Logic gates AND and OR Their inputs (two or more) are connected to the outputs of LTUs or other logic gates to form a tree. The set of points in the 1-class of a logic gate of type AND (OR) is the intersection (union) of the 1-classes of its inputs.

One of the earliest types of MLP had LTUs for all units and was not necessarily a tree. AND and OR gates are special cases of LTUs, so an ALN is a special case of that type of MLP. The "relational" way we use the net, though, is unusual. It offers significant advantages over using a net to compute functions directly: more efficient computation, more control over the learned function during training, and enhanced possibilities for checking the result. One major advantage of the relational approach is the ability to obtain the inverse of a function represented by an ALN, if it exists, without additional training.

2 Training Algorithms for ALNs

Training algorithms for ALNs take advantage of the real values of functions represented by parts of the tree, and thus use more information than is contained in the Boolean values flowing through the ALN. Perhaps this explains why ALNs succeed where early attempts at MLP training failed [8].

To understand the training algorithms for ALNs, it is convenient to look at a hypothetical functional network that parallels the ALN and is composed of units that compute linear functions (3) for each LTU, and units that compute the minimum (or maximum) of functions for each AND (or OR, respectively). When an input is presented during training, then, except for rare ties, only one linear function is active, i.e. forms the piece of the graph that defines the output value. At a minimum node, it is the child with the lesser value, and at a maximum node, it is the child with the greater value. Hence instead of diffusing the assignment of credit (or blame) for the error of the synthesized function over many linear function units, the training algorithm picks out one linear piece for adjustment.

Even in the case of ties, one responsible linear function unit is picked, based on the order of the children of the logic units. Once a linear function has been picked as responsible by the above criterion, its weights can be adapted by some technique for incremental linear regression. An ALN training algorithm based on the above ideas is implemented in an educational kit that can be downloaded [2].

There is a close connection between ALN architecture and qualitative properties of the unknown function represented by the training data. Namely, if the function f represented by an ALN is a minimum (maximum) of several linear functions, then f is convex-up, dome-shaped, (or convex-down, bowl-shaped, respectively).

Achieving a good fit to an unknown function depends on having a good correspondence between its convex parts and the convex parts of the synthesized function. To illustrate this, consider how easy it is to represent a convex-up function as a minimum of linear functions. Now contrast that with the problem of representing a convex-down function using a maximum of several minima, corresponding to an ALN with two layers of gates. This is analogous to representing a function as a Riemann sum as is done in elementary integration, except that the pillars forming the Riemann sum have to have slanted sides so the minima of the sides and tops become convex-up functions. In general, particularly in high dimensional spaces, many minima are required and even then the synthesized function is only roughly convex-down.

We note that this construction can be turned into a proof that any continuous function can be approximately represented by an ALN with just two layers of logic gates. This can be done to arbitrary precision on any compact set. However the number of linear pieces required in such an approximation could be much higher than in a logic tree of greater depth but appropriate shape to follow the convexities of the data. Results with ALN training algorithms which let the ALN structure grow during training to better follow the patterns in the data have shown marked improvement in speed and generalizing ability over algorithms that fix the tree first.

3   Rapid evaluation of learned functions

The piecewise linear function f represented by an ALN can be computed by a network of linear function units and maximum and minimum units as outlined above. However for any given input, this would be very wasteful as only one linear function is responsible for the value of the synthesized function. Unfortunately, to find out which one, the ALN would have to compute the values of many of its linear functions. There is a better way that reduces the number of multiplications and additions.

After training, a decision tree is built as follows. The linear pieces forming the function are examined and a threshold on some variable is found such that about half of the linear pieces are active on one side of the threshold and about half on the other side. Successive divisions of parts of the input space are carried out, forming a decision tree at whose leaves are small blocks of the input space. In each block, all but a small number of linear pieces can be omitted from the computation of the function value. During function evaluation, only the simple min-max expressions and the coefficients of the linear functions are retrieved once the block is known. We call such a decision tree an "ALN decision tree". It is important to emphasize that despite the amount of computation omitted, the function synthesized remains unchanged and is still continuous.

In general, it is possible to reduce the number of linear pieces that have to be evaluated in each block to a number at most equal to the dimension of the space n. Sometimes it is necessary to alter the function slightly to do this, but the needed change can be arbitrarily small. Thus if the function $x_3 = f(x_1, x_2)$ is synthesized by 300 linear pieces, only three pieces will have to be evaluated in any block of the decision tree. This amounts to omitting 99% of the parameters for each evaluation.

4 ALN Evaluation Using Programmable Hardware

During tests to see if ALNs could control a vehicle active suspension system in real time, a card for use in a personal computer was built around a field-programmable gate array (FPGA) [3]. The sole function of the card was, given a vector of digitized values from the test apparatus, to determine the value of an index indicating which block of the ALN decision tree was involved. The FPGA contained hardware units that implemented multiplexers. A multiplexer with three inputs realizes the function $MUX(x, y, z)$ whose value is y if x is true and z if x is false. This is an appropriate circuit component to perform the index computation.

Suppose the inputs of the function f, namely $x_1, \ldots, x_{n-1}$ are loaded into registers and compared with predetermined thresholds, with perhaps several thresholds on each variable. The bits representing all results of threshold computations are, of course, adequate to distinguish the leaves of the decision tree, but not all bit combinations are used. In general this vector of bits could be too long to be useful. In order to get an index into a small array, the results of comparisons have to be manipulated into a shorter vector.

To illustrate the process, let the blocks of ALN decision trees be numbered arbitrarily. We can structure the FPGA according to the following algorithm. The proof that we can do the construction proceeds by induction on the depth of the ALN decision tree. If there is only one test threshold c in the ALN decision tree, then $MUX(c, 1, 0)$ or $MUX(c, 0, 1)$ provides one bit in any possible k-bit numbers that could be assigned to index the two blocks. Suppose now an ALN decision tree is of depth d, and that the blocks at the leaves are numbered arbitrarily with k-bit numbers. Suppose our algorithm can generate hardware to compute the k-bit numbers for all decision trees of depth $d - 1$ or less. Then we execute the algorithm twice, once for each numbering obtained by fixing the topmost decision of the tree to 0 or, respectively, to 1. The hardware that computes these two indexes $(a_1, \ldots, a_k)$, $(b_1, \ldots, b_k)$, can be extended with k multiplexers controlled by the topmost comparison c of the decision tree to obtain hardware for the d-level decision tree which computes the k-bit index $(MUX(c, a_1, b_1), MUX(c, a_2, b_2), \ldots, MUX(c, a_k, b_k))$. This completes the construction. Allowing some flexibility in numbering can reduce the hardware requirements.

We see that by using the multiplexers of the FPGA, it is feasible to address memory that compactly stores min-max expressions, which in turn, refer to another memory that stores the coefficients of the linear functions.

Suppose now an ALN decision tree must be executed at extremely high speed, say for use in processing data coming from an accelerator to recognize when an interesting particle has been produced. Only the data on such particles needs to be collected. Because of the tight time requirements, where nanoseconds count, not only the computation of the block index, but also the computation of the expressions at the leaves may have to be parallelized to some extent. If there are n variables in the problem, then an ALN decision tree can be constructed to compute a discriminant function, no matter how complex it is, such that at the leaves of the decision tree, the expressions contain at most n linear functions combined by a tree of fewer than $n - 1$ maximum and minimum operations.

The number of multiplications required is $n - 1$ for each linear function. All multiplications could be done in parallel. The additions to combine the products can be done in parallel to a great extent, one level of additions cutting the number left to do in half; so the process is of order $log(n)$. The min-max expressions can be evaluated in time of order n. Note that the computation needed for any block is bounded independently of the complexity of the entire function; it depends only on the dimension n of the problem! A more complex function has a deeper decision tree or circuit to find the right block and a RAM circuit with more address lines to store data about the blocks. That takes time that grows logarithmically in the number of linear pieces involved in the synthesis.

Since the expression to be computed in each block of the ALN decision tree is different, programmable hardware could be useful in achieving high parallelism. There arise questions of loading the min-max expression, loading the parameters of the linear functions involved, and doing this with programmable hardware which is fast yet not too expensive. This may be an application for evolvable hardware which could learn to set itself up to execute efficiently the variety of tasks involved in different ALN decision trees and different blocks of the same tree.

The extremely high efficiency and speed of ALNs may encourage researchers in the area of evolvable hardware to try this as an example of a system that must judiciously exploit data-driven flow of control and complement that with parallel computation when appropriate. In our opinion, combining such flow of control with programmable parallelism (or, dare we say it: "flow of hardware"?) is the most promising way to make practical systems that rapidly solve large problems.

5 Conclusions

We have discussed methods for fast evaluation of the piecewise linear functions synthesized by ALN training. Comparisons of the inputs of the function to fixed thresholds in an ALN decision tree are used to allow large parts of the computation to be omitted for any given input. Programmable FPGA hardware can accelerate the determination of an index of the block of the input space where the current input lies, thus omitting all but the evaluation of a few linear functions and several maximum and minimum operations. That part itself could be accelerated by parallelizing multiplication, addition, and minimum and maximum operations in a way that exploits programmable hardware.

References

1. Armstrong, W. W., Thomas, M. M., Adaptive Logic Networks Section C1.8 in The Handbook of Neural Computation (HONC), Fiesler, E., Beale, R., eds., Institute of Physics Publishing and Oxford University Press USA, ISBN 0-7503-03123, (ca. 1000 A4 pp., looseleaf), 1996.
2. Armstrong, W. W., Thomas, M. M., Atree 3.0 Educational Kit (software for Windows) and User's Guide (HTML format), 1996.
   ftp://ftp.cs.ualberta.ca/pub/atree/atree3/atree3ek.exe.
3. Armstrong, W. W., Thomas, M. M., Control of a vehicle active suspension model using adaptive logic networks Section G2.1 in HONC (see reference 1).
4. Armstrong, W. W., Chu, C., Thomas, M. M., Using Adaptive Logic Networks to Predict Machine Failure, Proc. World Congress on Neural Networks, Washington, D. C. , July 17 - 21, 1995 vol. II, pp. 80-83.
5. Armstrong, W. W., Kostov, A., Stein, R. B., Thomas, M. M., Adaptive Logic Networks in Rehabilitation of Persons with Incomplete Spinal Cord Injury, Chap. 18 in Proc. Battelle Pacific Northwest Labs. Workshop on Environmental and Energy Applications of Neural Networks, Richland WA, USA, March 30-31, 1995, World Scientific Publishing Co., Progress in Neural Processing Series. Also available from: ftp://www.cs.ualberta.ca/pub/atree/battelle/alnwalk.ps.Z
6. Kostov, A., Andrews, B. J., Popovic, D. B., Stein, R. B., Armstrong, W. W., Machine Learning in Control of Functional Electrical Stimulation Systems for Locomotion, IEEE Trans. Biomed. Eng. vol. 42 no. 6, 1995, pp. 541 - 551.
7. Kostov, A., Armstrong, W. W., Thomas, M. M., Stein, R. B. Adaptive logic networks in rehabilitation of persons with incomplete spinal cord injury Section G5.1 in HONC (see reference 1.)
8. Minsky, M. L., Papert, S., Perceptrons: An Introduction to Computational Geometry, Expanded edition, MIT Press 1987, ISBN 0-262-63111-3.
9. Popovic, D., Stein, R. B., Jovanovic, K., Dai R., Kostov, A., Armstrong, W., Sensory Nerve Recording for Closed-loop Control to Restore Motor Functions, Trans. Biomed. Eng., vol. 40 no. 10, 1993, pp. 1024 - 1031.
10. Werbos, P. J., Beyond regression: New tools for prediction and analysis in the behavioral sciences. PhD dissertation, Harvard Univ., Cambridge, Mass. 1974.

FPGA as a Key Component for Reconfigurable System

Shinichi Shiratsuchi

Engineering   XILINX KK
2-8-5 No.2 Nagaoka Bldg. Tokyo, Japan   E-mail   shira@xilinx.com

Abstract. This paper describes several key point to be considered when the system   designer start devising the reconfigurable system.   At first to give a clear image on reconfiguration, style of that is summarized in "Static Reconfiguration" and "Dynamic Reconfiguration."   Then several approaches to design reconfigurable system is discussed. Finally, XC6200 partially reconfigurable FPGA is described to give a clear image on realistic dynamically   reconfigurable system.

1   Introduction

At the period of introduction phase of SRAM based FPGA(Field Programmable Gate Array), it had been considered just as the replacement of logic such as SSI/MSI. However, recent progress of IC fabrication technology provides good performances in both density and speed which are comparable with a   low end MPGA(Mask Programmable Gate Array) now. Xilinx estimates FPGA performance as 100K gates with speed of 100Mhz system clock 1997.

In addition to these performance enhancement, the nature of SRAM (i.e. reprogrammability) offers the reconfigurability to a system . These performance enhancements and reprogrammbility provide different approach for designing sophisticated system to designers as CPU changed the manner of designing digital processing unit.

The   reconfigurability has being considered as one of key technology to adding feature sets into a system such as reconfigurable computer and parallel processor recently. Also in the field of commercial type application, reconfigurability is having attentions to enhance the system   performance. In these system, some of the processing elements change its logic contents or its figure of communication path at a time the system needs it. To accomplish this type of design, SRAM based FPGAs are not only necessary but also main part of such systems.

By utilizing the reprogrammability of the FPGA, there are several products already on the market. For example, Giga operations, Virtual Computer, Annapolis Micro Systems and Time Logic are providing the reconfigurable computing systems. The   reconfigurability   provide   significant   performance enhancement to these systems which employ advantages of processing speed in   hardware   and   benefits   of   flexibility   in software.

The key is considering how well system can use hardware advantages and software benefits without   exhausting the reprogramming   time.

XC6200 which described in other section is one of the best candidates for the dynamically reconfigurable system.

2   The style of   Reconfiguration

Generally   the   style   of   reconfiguration   falls   into   two   groups   i.e.   Static Reconfiguration and Dynamic Reconfiguration.

The Static Reconfiguration means that a system loads the programming data to FPGA while the system is not actually operating. On the contrary, dynamic reconfiguration means that the processing module in FPGA is swapped during intermediate time of system operation dynamically. Also in the dynamic reconfiguration, there are two types which are mostly depend on the FPGA configuration circuitry. One is "Full Chip Reconfiguration" in which the system reprograms the entire FPGA even if a most of circuit information of FPGA is the same. The other is "Partial Reconfiguration" in which the system reprograms only a portion of FPGA to update circuit information. The rest of circuit stay resident. Table-1 shows the summarized form of reconfiguration styles.

	Static Reconfiguration (Full Chip)	Dynamic Reconfiguration	
		Full Chip	Partial
When	Compile time	During system operation	
Time to Reprogram	Don't Care	millisecond	nano to micro seconds
Reprogram Area	Entire FPGA	Entire FPGA	Portion of FPGA
Purpose	-System alternation -Field hardware upgrade	-Logic swap -performance Enhancement	-Logic swap -performance Enhancement

Table-1 Category of Reconfiguration

There is several discussion on categorizing the reconfiguration style(Chean 1990, Lysaght and Dunlop 1994), in some cases the dynamic reconfiguration is under the "Partial Reconfiguration", here however, dynamic reconfiguration is compared with static reconfiguration since they are both in time domain. This is a point of view from device engineering side. In the group of dynamic reconfiguration, the full chip and partial reconfiguration is compared with each other giving a contrast in domain of area or logic gates.

It is important that system designer has to think about both in density and in time domain for a job scheduling, while the old style logic replacement consider about in a domain of logic density.

As a note for designers, pin locking capability is the critical issues to successfully complete the projects. XC5200 and XC4000EX have special routing resources called VersaRing to accomplish pin locked design. This capability is needed for all the reconfigurable system.

2.1 Static Reconfiguration

The static reconfiguration is referred as a compile-time reconfiguration. In this style, system loads one of programming data from the set of data to FPGA in advance to system begins actual operation. Because the reconfiguration time can be ignored in this style, the reconfiguration circuit is simple. Some samples are described in the section "Approach to Reconfigurable System".

2.2 Dynamic Reconfiguration

The dynamic reconfiguration can swap the logic contents of a FPGA during system operation without disturbing the operation of the system. Full chip and Partial reconfigurations belong to this group. However if the system employs multiple FPGAs, then the system itself can be partially reconfigurable even they use full chip reprogrammable devices. Most of reconfigurable applications employ multiple FPGA solution. Recently this type of computer, coprocessor and logic are sometimes called as RADD (Reconfigurable Architectures on DemanD.)

Device	PROM Size Data Bits	Time to Reprogram	Typical Gates *1	Reprogram mode *2
XC3120A	14,819	1.5 msec	1K-1.5K	Serial
XC3195A	94,984	9.5 msec	6.5K-7.5K	Serial
XC4003E	53,976	5.4 msec	2K-5K	Serial
XC4013E	247,960	24.8 msec	10K-30K	Serial
XC4028EX	668,167	8.4 msec	18K-50K	Express
XC4062EX	1,433,847	17.9 msec	40K-130K	Express
XC5202	42,416	0.5 msec	2K-3K	Express
XC5206	106,288	1.3 msec	6K-10K	Express
XC5215	237,744	3.0 msec	15K-23K	Express
XC6216	Depend on logic size	~200 µsec/chip 40 nsec/cell	16K-24K	Parallel/Partial

*1: Typical gates include logic and RAM(if available)
*2: 10MHz configuration clock is assumed.

Table-2 Time to Reprogram

2.2.1 Full Chip Reconfiguration

Table-2 indicates the reconfiguration time consumed to reprogram entire FPGA. In the most of case reprogramming time is in order of milliseconds generally. So, system designer must consider a trade off between the reprogramming time and actual processing time.

The frequency of the reconfiguration in a task also affects to the total through put of the system. Here task means a set of process of job. As a result, some kind of job scheduling is necessary to manage the efficiency. To estimate efficiency of reconfiguration, following point should be considered by the system designers.

- Reprogramming time vs. processing time
- Frequency of reconfiguration vs. Quantity of output in a "Task"
- Comparison of processing time with software only solution

Figure-1 shows a reconfigurable system's job transition. Here the job means a module of signal processing. In case of a full chip reconfiguration, transition of job J2 to J4 should wait for finishing the job J1 and J3, even if job J2 finished its process very shortly. Otherwise intermediate processing data of job J1 and J3 is destroyed. Regarding to Job J5 and J6, similar constraints exists. For this reason, the system which utilizes full chip reconfigurable FPGA had better to employ multiple FPGA solution to avoid conflicts of job scheduling.

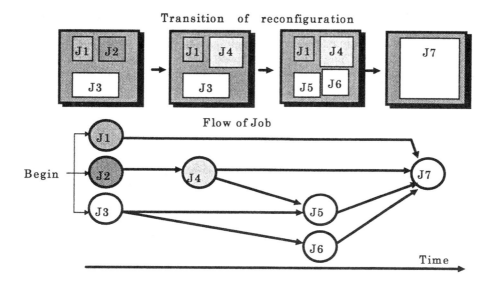

Figure-1 Job transition for reconfiguration

2.2.2   Partial Reconfiguration

Partially reconfigurable FPGA is considered as the   best   device to   construct the dynamically reconfigurable system. Figure-1 also indicates the concept of the partial reconfiguration. Partially reconfigurable FPGA can swap any job whenever the system need. As a result, the system can swap from job J2 to J4 immediately when J2 finished its process. Job J5 start operation just after the finishing of job J3 and J4. Also Job J6 can start its operation after finishing the Job J3.   So there is no wait state for the job if there is the available logic space in FPGA. XC6200 offers an ability of partial reconfiguration and very short reprogramming time from in order of nanoseconds to microseconds depending on the circuit size. This type of FPGA creates different type of applications from the conventional one.

3   Approach to Reconfigurable system

Concerning to the static reconfigurable system, programming time is not a issue, so any kind of reconfiguration is acceptable. Figure-2 indicates sample loader circuits. Fig-2)-a) is a PROM programmable. The programming data sets are stored in the PROM, then system loads one of them   to FPGA at the power on phase. Fig-2-b) is software programmable. This also programs the FPGA at the boot up stage.
Type a) is preferred for the embedded system because of circuit simplicity while type b) is preferred for the dedicated computational application because of flexibility to add new module of libraries.

Figure-3 indicates sample loading circuit by employing the full chip reconfigurable FPGA. This is for XC4000EX and XC5200 with an "Express" mode. Concerning to dynamic reconfiguration with full chip programmable FPGA, the reprogramming circuit can be similar to obtain a quick reconfiguration. Main CPU loads the reprogramming data sets for FPGA from system storage to the SRAM, which is detachable from the main bus, at any time before reconfiguration when the bus is idle, then loading control logic manages the reconfiguration, so that the main CPU does not lose milliseconds of time for reconfiguration. In this case, the main CPU only pass a start vector of the adequate programming data to the loading controller.

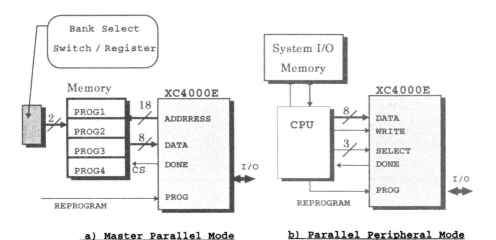

Figure-2 Program loading circuit for static reconfiguration

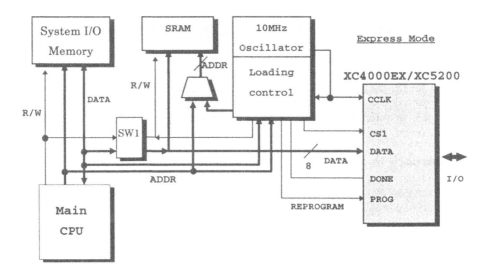

Figure-3   Program Loader for XC4000EX/XC5200

3.1 Over-Head Factor

To talking about efficiency of reconfiguration, over-head factor concerning to reconfiguration time should be considered. The over-head factor is the simply the ratio of total reconfiguration time and processing time. This simple parameter indicates what amount of time will be consumed just for reconfiguration compared with its processing time in a ratio.

Over-head factor (OHF) of a task which has job J1 through Jn is as follows.

$$OHF = (\Sigma Tr_i)/(\Sigma Tp_i) , \ i=1 \ to \ n$$

Tr_i: Reconfiguration time for job Ji
Tp_i: Processing time for job Ji

If OHF is close to zero, system can safely ignore the reprogramming time, even if it uses full chip reconfigurable FPGAs.

However if it is near one or greater, the system designer must consider to choose processing speed or flexibility. A system having larger OHF is suitable as a static reconfigurable system. Because the reprogramming is performed only at the starting point of the processing.

To obtain small OHF, the system needs to process fairly large amount of data for a specific configuration. As a result, system becomes somewhat like in Figure-4. In this figure the input buffer stores the large quantity of data to be processed, then FPGA processes its data and stores that results into output buffer storage. This is a case of utilizing full chip reconfiguration.

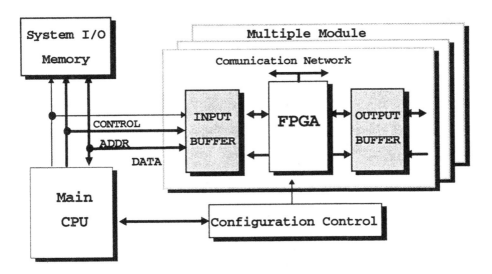

Figure-4   Sample block diagram for small OHF

There are some sort of over-head for the full chip reconfigurable FPGA as described. The system designer must consider the structure so that the over-head is minimized. The Partially configurable FPGA, however, offers a very short reconfiguration time and ease of job scheduling. Next section describes partial reconfiguration with XC6200 as sample.

4   XC6200 Partially Reconfigurable FPGA

In advance to discuss the dynamic reconfiguration with a partially reconfigurable FPGA, the unique architecture and feature sets of XC6200 are described.

4.1 XC6200 Architecture and Features

Figure-5 and Figure-6 indicate the overview   and the basic architecture of XC6200 respectively. This device is intended to be utilized in a dynamically reconfigurable system. Followings are the architectural features.

- Built in CPU Interface (FastMap)
  @ 16 bits address line
  @ 8/16/32 bits data line
  @ Read/write control
  @ Chip selects

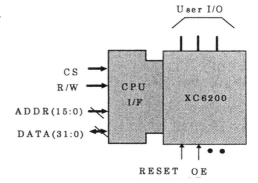

Figure-5 Overview of XC6200

- All the internal registers can   be accessible through CPU I/F   by utilizing dedicated routing   resources without consuming the   local resources.
- Hierarchical array structure   which allows relocatable module design.

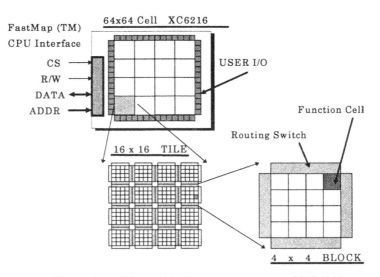

Figure-6   Hierarchical array structure of XC6200

Figure-6 shows hierarchical array structure of XC6200. The fundamental logic cell is called "Function Cell" as shown in Figure-7. The function cell consists of routing elements and function unit. Function unit is indicated in Fig-7-a). By utilizing the function unit, any type of two input function can be constructed in a cell.

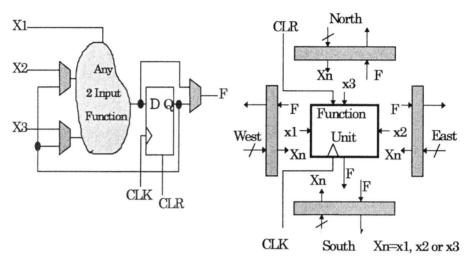

a) **Function Unit**   b) **Function Cell Connection**

Figure-7   Function Cell structure of XC6200

Each cell has routing resources to connect with its nearest neighboring cell as shown in Fig-7-a). 4x4 cell array shown in Figure-6 called "Block" also has special routing resources to connect with its adjacent blocks. Furthermore 4x4 Block e.g. 16x16 cell array forms "Tile". 4x4 Tile e.g. 64x64 cell array forms entire XC6216 device. The special routing resources are laid out in each hierarchy of cell, block, tile, and 64x64 cell array. These routing resources connect each hierarchy of the array and are called FastLane4, FastLane16 and FastLane64 respectively. The hierarchical and symmetrical structure of XC6200 provides relocatability to the logic module so that system can place it anywhere in the device with the identical topology.

Each register in a cell shown in Fig-7-a) has an address, so that the system can access though the built in CPU interface. This means that the processing modules do not necessarily route the signal line from function cell to I/O block. It is already on chip of XC6200 as a dedicated CPU interface. By employing this feature, some module might be designed as a visible I/O less module which communicates with system controller through the CPU interface directly. Xilinx called this type of I/O as "Wireless I/O". This ability enhances the relocatability of the module since such kind of modules does not need to have routings for I/O block connection.

Figure-8 shows the partial reconfiguration with using wireless I/O capability. This capability enhances the ease of module design and also ease of interfacing to the main processing unit.

Figure-9 and Figure-10 shows a sample of the data processing flow. It looks like a kind of a data flow machine, which has Tag for each data sets as a job identifier. The tag is interpreted by CPU or DSP then replaced with the actual programming data for XC6200. The device is reconfigured then accept the data followed. Data stream shown in Figure-10 can be applied through the CPU bus or different I/O on XC6200 devices to the modules.   This is a sort of Dynamic Link Hardware.

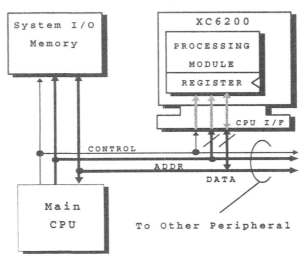

Figure-8   Wireless I/O concept

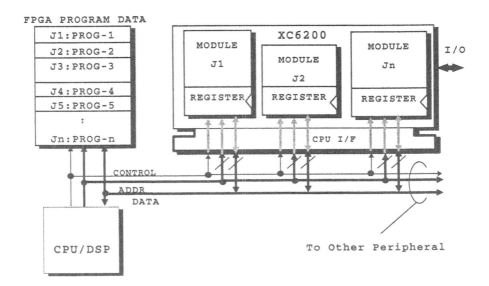

Figure-9   Partial reconfiguration with XC6200

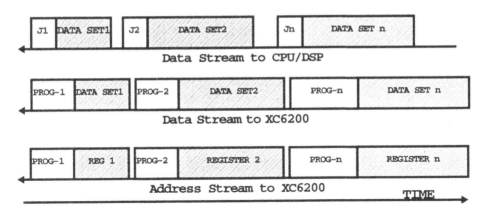

Figure-10 Data Flow type application with XC6200

As shown in these figures, partial reconfiguration offers a lot of different approaches to consider complex systems from a different point of views. By employing the partially reconfigurable FPGA, a concept of the Dynamic Link Hardware is realistic with high speed, flexible and less over-head.

5 Conclusion

The reconfigurable system is one of the solution with high speed , flexible and low cost. The several approaches already exist and new approach with XC6200 was described. The most of this paper spent on the hardware structure and its efficiency. However, recent research type of work for a Reconfigurable Logic ,indicates the need of sophisticated compiler for that type of system. Such compiler generates the both objects for CPU and the reconfigurable logic from the source code such as C, C++ or VHDL. To generate object code for reconfigurable system, compiler needs to understand the software structure and to divide it into FPGA portion and true software portion then execute the Place and Route program which creates the reprogramming data. This is fairly difficult task but need to be solved for future of the reconfigurable system.

Acknowledgment
I acknowledge XILINX KK engineers for their help to accomplish this paper.

Reference
Chean,M. and Fortes,J .,"A Taxonomy of reconfiguration Techniques for
        Fault-Tolerant Processor Arrays".
        IEEE Computer, vol.23, no.1   January 1990
Lysaght,P., " Dynamically reconfigurable Logic in Undergraduate
        Projects" in FPGAs,
        W.Moore and W.Luk,Eds., Abingdon EE&CS Books, England,1991
Fawcett,B.,"Applications of Reconfigurable Logic" in More FPGAs,
        W.Moore and W.Luk,Eds., Abingdon EE&CS Books, England,1993
Kanai,T.,"FPGAs Aid PCB Inspection", ASIC & EDA, Jan,1992
        Conner,D.,"Reconfigurable Logic" EDN,Mar,1996 ,
        Xilinx,"The programmable Logic Data Book 1996" ,July,1996

Overview

.

Phylogeny, Ontogeny, and Epigenesis: Three Sources of Biological Inspiration for Softening Hardware

Eduardo Sanchez, Daniel Mange, Moshe Sipper, Marco Tomassini, Andres Perez-Uribe, and André Stauffer

Logic Systems Laboratory, Swiss Federal Institute of Technology, IN-Ecublens, CH-1015 Lausanne, Switzerland. E-mail: {Name.Surname}@di.epfl.ch

Abstract. Living beings are complex systems exhibiting a range of desirable qualifications that have eluded realization by traditional engineering methodologies. In recent years we are witness to a growing interest in Nature exhibited by engineers, wishing to imitate the observed processes, thereby creating powerful problem-solving methodologies. If one considers Life on earth since its very beginning, three levels of organization can be distinguished: the *phylogenetic* level concerns the temporal evolution of the genetic programs within individuals and species, the *ontogenetic* level concerns the developmental process of a single multicellular organism, and the *epigenetic* level concerns the learning processes during an individual organism's lifetime. In analogy to Nature, the space of *bio-inspired* systems can be partitioned along these three axes, phylogeny, ontogeny, and epigenesis, giving rise to the *POE* model. This paper is an exposition and examination of bio-inspired systems within the POE framework. We first discuss each of the three axes separately, considering the systems created to date and plotting directions for continued progress along the axis in question. We end our exposition by a discussion of possible research directions, involving the construction of bio-inspired systems that are situated along two, and ultimately all three axes. This presents a vision for the future which will see the advent of novel systems, inspired by the powerful examples provided by Nature.

1 Introduction: Biological inspiration as a bridge from the natural sciences to engineering

Traditionally, the development of the engineering disciplines (civil, electrical, computer engineering, etc') and that of the natural sciences (physics, chemistry, biology, etc') have proceeded along separate tracks. The natural scientist is a *detective*: faced with the mysteries of nature, such as meteorological phenomena, chemical reactions, and the development of living beings, he seeks to *analyze* existing processes, to *explain* their operation, to *model* them, and to *predict* their future behavior. The engineer, on the other hand, is a *builder*: faced with social and economic needs, he tries to *create* artificial systems (bridges, cars, electronic devices) based on a set of *specifications* (a description) and a set of *primitives* (elementary components such as bricks, beams, wires, motors, transistors).

These two major branches of human endeavor have been drawing closer together during the past decades. It is nowadays common for scientists to use tools created by engineers; to cite one example of many, we are witness to the systematic use of electronics in the medical world for such tasks as decoding the human genome, visually representing highly complex chemical molecules, computerized tomography, and so on.

More recently, engineers have been allured by certain natural processes, giving birth to such thriving domains as artificial neural networks and evolutionary algorithms. Living beings are complex systems exhibiting a range of desirable qualifications, such as evolution, adaptation, and fault tolerance, that have proved difficult to realize using traditional engineering methodologies. Such systems are characterized by a genetic program, the genome, that defines their development, their functioning and their extinction. If one considers Life on earth since its very beginning, then the following three levels of organization can be distinguished [9, 10]:

Phylogeny The first level concerns the temporal evolution of the genetic program, the hallmark of which is the evolution of species, or *phylogeny*. The "multiplication" of living beings is based upon the reproduction of the program, subject to an extremely low error rate at the individual level, so as to insure that the identity of the offspring remains practically unchanged; this error rate is higher at the group or species level [43]. It is precisely these copying errors, due to mutation (asexual reproduction) or mutation along with recombination (sexual reproduction), that gives rise to the emergence of novel species or new organisms. The phylogenetic mechanisms are fundamentally non-deterministic, with the mutation and recombination rate providing a major source of diversity; this diversity is indispensable for the survival of living species, for their continuous adaptation to a changing environment, and for the appearance of new species.

Ontogeny Upon the appearance of multicellular organisms, a second level of biological organization manifests itself. The successive divisions of the mother cell, the zygote, with each newly formed cell possessing a copy of the original genome, is followed by a specialization of the daughter cells in accordance with their environment, i.e., their position within the ensemble; this latter phase is known as cellular differentiation. *Ontogeny* is therefore the developmental process of a multicellular organism; this process is essentially deterministic: an error in a single base within the genome can provoke an ontogenetic sequence which results in notable, possibly lethal, malformations.

Epigenesis The ontogenetic program is limited in the amount of information that can be stored, thereby rendering the complete specification of the organism impossible. A well-known example is that of the human brain with some 10^{10} neurons and 10^{14} connections, far too large a number to be completely specified in the four-character genome of length 3×10^9. Therefore, upon reaching a certain level of complexity, there must emerge a different process that permits the individual organism to integrate the vast quantity of interactions with the outside world. This process is known as *epigenesis*, and primarily includes the nervous system, the immune system and the endocrine system. These systems are characterized by the possession of a basic structure that is entirely defined by the genome (the innate part), which is then subjected to modification through interactions of the individual with the environment (the acquired part). The epigenetic processes can be loosely grouped under the heading of *learning* systems.

In analogy to Nature, the space of *bio-inspired* systems can be partitioned along these three axes: phylogeny, ontogeny, and epigenesis; we refer to this as the *POE* model (Figure 1). As an example, consider the following three paradigms, each of which is positioned along one axis: (P) evolutionary algorithms are the (simplified) artificial counterpart of phylogeny in Nature; (O) self-reproducing automata are based on the concept of ontogeny, where a single mother cell gives rise, through multiple divisions, to a multicellular organism; (E) artificial neural networks embody the epigenetic process, where the system's synaptic weights and perhaps topological structure change through interactions with the environment. Within the domains collectively referred to as *soft computing* [67], characterized by ill-defined problems coupled with the need for continual adaptation or evolution, the above paradigms yield impressive results, rivaling those of traditional methods.

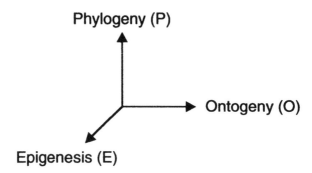

Fig. 1. The POE model. Partitioning the space of bio-inspired systems along three axes: phylogeny, ontogeny, and epigenesis.

Our goal in this paper is to introduce the basics of bio-inspired systems along each of the three axes; due to space restrictions, and in following the main theme of the conference, we shall concentrate on the phylogenetic axis (Section 2). In Section 3 we present a brief account of the ontogenetic axis, considering the role of self-reproduction within the scheme of bio-inspired systems. Section 4 considers the third axis, namely epigenesis. Our paper ends in Section 5 with our conclusions and directions for future research, based on the POE model; specifically, we shall consider the possibilities of combining two axes, along with the ultimate goal of combining all three. This presents a vision for the future which will see the construction of novel systems, inspired by the powerful examples provided by Nature.

2 The phylogenetic axis: Evolvable hardware

2.1 Artificial evolution

The idea of applying the biological principle of natural evolution to artificial systems, introduced more than three decades ago, has seen an impressive growth in the past few years. Usually grouped under the term *evolutionary algorithms* or *evolutionary computation*, we find the domains of genetic algorithms, evolution strategies, evolutionary programming, and genetic programming [2, 17, 22, 28, 32, 41, 42, 56]. As a generic example of artificial evolution, we consider genetic algorithms, originally invented by John Holland in the 1960s [28].[1]

A genetic algorithm is an iterative procedure that consists of a constant-size population of individuals, each one represented by a finite string of symbols, known as the *genome*, encoding a possible solution in a given problem space. This space, referred to as the *search space*, comprises all possible solutions to the problem at hand; generally speaking, the genetic algorithm is applied to spaces which are too large to be exhaustively searched. The symbol alphabet used is often binary due to certain computational advantages purported in [28] (see also [22]); this has been extended in recent years to include character-based encodings, real-valued encodings, and tree representations.

The standard genetic algorithm proceeds as follows [63]: an initial population of individuals is generated at random or heuristically. Every evolutionary step, known as a *generation*, the individuals in the current population are *decoded* and *evaluated* according to some predefined quality criterion, referred to as the *fitness*, or *fitness function*. To form a new population (the next generation), individuals are *selected* according to their fitness, by using a given selection procedure. Selection alone cannot introduce any new individuals into the population, i.e., it cannot find new points in the search space; these are generated by genetically-inspired operators, of which the most well-known are *crossover* and *mutation*. Crossover is performed with probability p_{cross} (the "crossover probability" or "crossover rate") between two selected individuals, called *parents*, by exchanging parts of their genomes (i.e., encodings) to form two new individuals, called *offspring*; in its simplest form, substrings are exchanged after a randomly selected crossover point. This operator enables the evolutionary process to move toward "promising" regions of the search space. The mutation operator is introduced to prevent premature convergence to local optima by randomly sampling new points in the search space. It is carried out by flipping bits at random, with some (small) probability p_{mut}. Genetic algorithms are stochastic iterative processes that are not guaranteed to converge; the termination condition may be

---

[1] For the purposes of this presentation, the differences with other evolutionary algorithms are inconsequential.

specified as some fixed, maximal number of generations or as the attainment of an acceptable fitness level.

Evolutionary algorithms are ubiquitous nowadays, having been successfully applied to numerous problems from different domains, including optimization, automatic programming, machine learning, economics, immune systems, ecology, population genetics, studies of evolution and learning, and social systems [42]. For recent reviews of the current state of the art, the reader is referred to [62, 63].

2.2 Large scale programmable circuits

An integrated circuit is called programmable when the user can configure its function by programming. The circuit is delivered after manufacturing in a generic state and the user can adapt it by programming a particular function; the programmed function is coded as a string of bits representing the configuration of the circuit. In this paper we consider solely programmable *logic* circuits, where the programmable function is a logic one, ranging from simple boolean functions to complex state machines.

The first programmable circuits allowed the implementation of logic circuits that were expressed as a logic sum of products; these are the PLDs (Programmable Logic Devices), whose most popular version is the PAL (Programmable Array Logic). More recently a new player has appeared on the scene, affording higher flexibility and more complex functionality: the Field-Programmable Gate Array (FPGA) [54]. An FPGA is an array of logic cells placed in an infrastructure of interconnections (Figure 2), which can be programmed at three distinct levels: (1) the function of the logic cell; (2) the interconnections between cells; (3) the input and outputs. All three levels are programmed via a string of bits that is loaded from an external source, either once or several times; in the latter case the FPGA is considered *reconfigurable*.

FPGAs are highly versatile devices that offer the designer a wide range of design choices. However, this potential power necessitates a plethora of tools in order to design a system; essentially, these generate the configuration bit string upon given such inputs as a logic schema or a high-level functional description.

2.3 Evolvable hardware: The present

If one carefully examines the work carried out to date under the heading 'evolvable hardware', it becomes evident that this mostly involves the application of evolutionary algorithms to the synthesis of digital systems [55] (recently, Koza has studied analog systems as well [33]). Taking this point of view, evolvable hardware is simply a sub-domain of artificial evolution, where the final goal is the synthesis of an electronic circuit. The work of [32], where genetic programming was applied to the evolution of a three-variable multiplexer, may be considered an early precursor along this line; it should be noted that at the time the main goal was that of demonstrating the capabilities of the genetic programming methodology, rather than designing actual circuits. We argue that the term *Evolutionary Circuit Design* would be more descriptive of such work than that

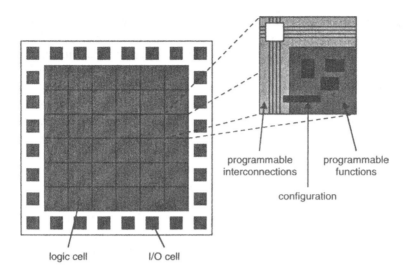

programmable
interconnections

programmable
functions

configuration

logic cell   I/O cell

Fig. 2. A schematic diagram of a Field-Programmable Gate Array (FPGA).

of 'evolvable hardware'. For now, we shall remain with the latter (popular) term, however, we shall return to the issue of clarifying definitions in Section 2.5.

Taken as a design methodology, evolvable hardware offers a major advantage over classical methods; the designer's job is reduced to that of specifying the circuit requirements and the basic elements, whereupon evolution "takes over" to "design" the circuit. Currently, most evolved digital designs are sub-optimal with respect to traditional methodologies, however, improved results are continuously attained. Examining work carried out to date, one can derive a rough classification of current evolvable hardware, in accordance with the genome encoding (i.e., the circuit description), and the calculation of a circuit's fitness.

Genome encoding

- *High-level languages.* The first works carried out used a high-level functional language to encode the circuits in question, a representation far-removed from the structural (schematic) description. In [32], the evolved solution is a program describing the (desired) multiplexer rather than an interconnection schema of logic elements (the actual hardware representation). The work of [25] uses a high-level hardware language to represent the genomes. [31, 33] used the rewriting operation, in addition to crossover and mutation, in order to enable the formation of a hierarchical structure; this is still within the framework of a high-level language.
- *Low-level languages.* The idea of directly incorporating within the genome the bit string representing the configuration of a programmable circuit was

expressed early on by [13], though without demonstrating its actual implementation. As a first step one must choose the basic logic gates (e.g., AND, OR, and NOT), and suitably codify them, along with the interconnections between gates, to produce the genome encoding. An example of this approach is the work of [61]. [26] used a low-level bit string representation of the system's logic schema to describe small-scale PALs, where the circuit is restricted to a logic sum of products. The limitations of the PAL circuits have been overcome to a large extent by the introduction of FPGAs, as used, e.g., by [60].

The use of a low-level circuit description that requires no further transformation is an important step forward since this potentially enables placing the genome directly in the actual circuit, thus paving the way toward truly evolvable hardware. However, up until recently, FPGAs had introduced their own share of problems: (1) the genome's length was on the order of tens of thousands of bits, rendering evolution practically impossible using current technology; (2) one still had to extend the genome into a logic schema, a phase for which automatic methods do not exist; (3) within the circuit "space", consisting of all representable circuits, a large number were invalid (e.g., containing short circuits).

With the introduction of the new family of FPGAs, the Xilinx 6200, these problems have been attenuated [60]. As with previous FPGA families, there is a direct correspondence between the bit string of a cell and the actual logic circuit, however, this now always leads to a viable system. Moreover, as opposed to previous FPGAs where one had to configure the entire system, the new family permits the separate configuration of each cell, a markedly faster and more flexible process. [60] has employed this latter characteristic to reduce the genome's size, without, however, introducing real-time, partial system reconfigurations.

Fitness calculation

- *Offline evolvable hardware.* The use of a high-level language for the genome representation means that one has to transform the encoded system in order to evaluate its fitness. This is carried out by simulation, with only the final solution found by evolution actually implemented in hardware. This form of simulated evolution is known as *offline* evolvable hardware [55].
- *Online evolvable hardware.* As noted above, the low-level genome representation enables a direct configuration (and reconfiguration) of the circuit, thus entailing the possibility of using real hardware during the evolutionary process; this has been called *online* evolution by some of the works found in [55].

2.4   Common features of current phylogenetic hardware

Examining work carried out to date we find a number of common characteristics that span both online and offline systems, often differing from biological evolution:[2]

---

[2] This is not necessarily disparaging, as discussed in Section 5.

- Evolution pursues a predefined goal: the design of an electronic circuit, subject to precise specifications; upon finding the desired circuit, the evolutionary process terminates.

- The population has no material existence; at best, in what has been called online evolution, there is one circuit available, onto which individuals from the (offline) population are loaded *one at a time*, in order to evaluate their fitness.

- The absence of a real population in which individuals coexist simultaneously entails notable difficulties in the realization of interactions between "organisms". This results in a completely local fitness calculation, contrary to nature which exhibits a co-evolutionary scenario.

- If one attempts to resolve a well-defined problem, involving the search for a specific combinatorial or sequential logic system, there are no intermediate approximations; fitness calculation is carried out by consulting a lookup table which is a complete description of the circuit in question. This casts some doubts into the utility of using an evolutionary process, since one can directly implement the lookup table in a memory device, a solution which may often be faster and cheaper.

- The evolutionary mechanisms are carried out outside the resulting circuit. This includes the genetic operators (selection, crossover, mutation) as well as fitness calculation. As for the latter, while what is currently advanced as online evolution uses a real circuit for fitness evaluation, the fitness values themselves are stored elsewhere.

- The different phases of evolution are carried out sequentially, controlled by a central software unit.

2.5 Evolvable hardware: A look ahead along the phylogenetic axis

The phylogenetic axis admits a number of qualitative sub-divisions (Figure 3):

- At the bottom of this axis, we find what is in essence *evolutionary circuit design*, where all operations are carried out in software, with the resulting solution possibly loaded onto a real circuit. Though a potentially useful design methodology, this falls completely within the realm of traditional evolutionary techniques. As examples one can cite the works of [25, 26, 31, 33].

- Moving upward along the axis, one finds works in which a real circuit is used during the evolutionary process, though most operations are still carried out offline, in software. An example is the work of [60], where fitness calculation is carried out on a real circuit. It is important to note that while this has been referred to as online evolution, it would probably be more appropriate to reserve this term for the next sub-division.

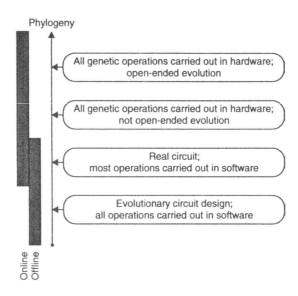

Fig. 3. The phylogenetic axis.

- Still further along the phylogenetic axis, one finds systems in which *all* genetic operations (selection, crossover, mutation, fitness evaluation) are carried out *online*, in hardware. The major aspect missing concerns the fact that evolution is not open-ended, i.e., there is a predefined goal and no dynamic environment so to speak of. An example is the work of [21].
- This represents the top of the phylogenetic axis, where a *population* of hardware entities evolves in an *open-ended* environment.

We argue that only the last category can be truly considered evolvable hardware, a goal which still eludes us at present. A natural application area for such systems is within the field of autonomous robots, which involves machines capable of operating in unknown environments without human intervention [4]. Another interesting example would be what we call "Hard-Tierra"; this involves the hardware implementation of the Tierra "world" [51], which consists of an open-ended environment of evolving computer programs. A small-scale experiment along this line was undertaken in [18]. The idea of Hard-Tierra is important since it demonstrates that 'open-endedness' does not necessarily imply a real, biological environment.

3 Ontogeny and self-reproducing hardware

The fundamental principle of embryology in real life is illustrated in Figure 4 (based on [11, 12]) which covers a period of two generations preceded and followed by an indefinite number of generations. The first condition is that there must be replicators, entities capable, like DNA molecules, of self-replication. The second condition is our main concern: there must be an embryonic process. The genome should influence the development of the external characteristics of the being, the phenotype; and the replicators must be able to wield some phenotypic power over their world, such that some of them are more successful at replicating themselves than others (this point is crucial for the phylogenetic process). It is important to understand that genes, the basic constituents of the genome, act on two quite different levels: they participate in the embryonic process, influencing the development of the phenotype in a given generation, and they participate in genetics, having themselves copied down the generations (reproduction). This is epitomized by an empirical separation between the disciplines of genetics and embryology; genetics is the study of the vertical arrows in Figure 4, i.e., the relationship between genotypes in successive generations, while embryology is the study of the horizontal arrows, i.e., the relationship between genotype and phenotype in any one generation [12].

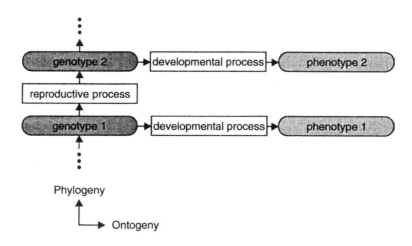

Fig. 4. The embryonic process in Nature: An interplay between phylogeny and ontogeny.

Research into self-reproducing machines, inspired by the ontogeny of living beings, began with von Neumann in the late 1940s. This line of research can be divided into five stages, placed along the ontogenetic axis (Figure 5):

1. Von Neumann [64] and his successors Banks [3], Burks [5], and Codd [8] developed self-reproducing automata capable of universal computation (i.e., able to simulate a universal Turing machine [29]) and of universal construction (i.e., able to construct any automaton described by an artificial genome). Unfortunately, the complexity of these automata is such that no physical im-

plementation has yet been possible, and only partial simulations have been carried out to date.

2. Langton [34] and his successors Byl [6], Reggia *et al.* [52], and Morita *et al.* [45] developed self-reproducing automata which are much simpler and which have been simulated in their entirety. These machines, however, lack any computing and constructing capabilities, their sole functionality being that of self-reproduction.

3. Tempesti [59], and Perrier *et al.* [49] developed self-reproducing automata inspired by Langton's work, yet endowed with finite ([59]) or universal ([49]) computational capabilities.

In biological terms, a cell can be defined as the smallest part of a living being which carries the complete plan of the being, that is its genome [39]. In this respect, the above self-reproducing automata are unicellular organisms: there is a single genome describing (and contained within) the entire machine; their reproduction is then analogous to the *asexual* reproduction of *unicellular* living beings.

4. Mange *et al.* [36, 38, 39] and Marchal *et al.* [40] proposed a new architecture called *embryonics*, or embryonic electronics. Based on three features usually associated with the ontogenetic process in living organisms, namely multicellular organization, cellular differentiation, and cellular division, they introduced a new cellular automaton, complex enough for universal computation, yet simple enough for a physical implementation through the use of commercially available digital circuits; in addition to self-reproduction, this multicellular "organism" also exhibits self-repair capabilities, another biologically-inspired phenomenon. In order to embed universal construction, they are designing the basic cell with a molecular organization, similar to that of the transcription-translation mechanism (ribosome) [37]. These self-reproducing machines are clearly multicellular artificial organisms, in the sense that each of the several cells comprising the organism contains one copy of the complete genome; their reproduction is analogous to that of asexual multicellular living beings (as in the budding process of the hydra, described by [65]).

All the above machines are characterized by an *asexual* reproductive process; the genome is therefore haploid.

5. The use of a diploid genome was discussed by [22], and more recently by [27]. This idea, coupled with the recombination of genetic material from two parents, could be introduced within the embryonics framework, representing an ultimate phase with respect to reproducing machines.

4 Epigenesis: Learning through interactions with an environment

To the best of our knowledge, there exist three major epigenetic systems in living multicellular organisms: the nervous system, the immune system and the

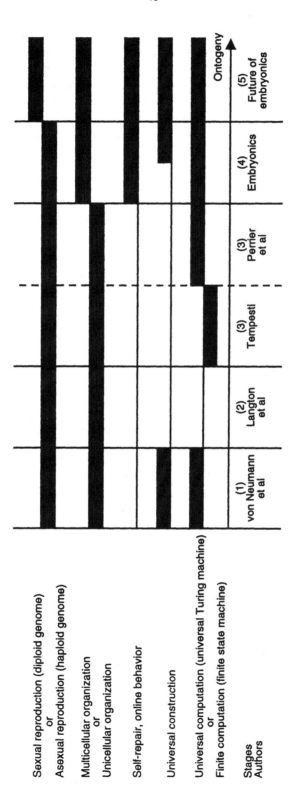

Fig. 5. The ontogenetic axis.

endocrine system, the first two having already served as inspiration for engineers. The nervous system has received the most attention, giving rise to the field of artificial neural networks; this will be the focus of our discussion below. The immune system has inspired systems for detecting software errors [66], as well as immune systems for computers [30]. Immunity of living organisms is a major domain of biology; it has been demonstrated that the immune system is capable of learning, recognizing, and, above all, eliminating foreign bodies which continuously invade the organism. Moreover, when viewed from the engineering standpoint, it is most interesting that immunity is maintained when faced with a dynamically changing environment. This feature leads us to surmise that the immune system, if implemented as an engineering model, can provide a new tool suitable for confronting dynamic problems, involving unknown, possibly hostile, environments.

The nervous system remains the most popular epigenetic model used by engineers. From a biological point of view, it has been determined that the genome contains the formation rules that specify the outline of the nervous system [9, 10]. It is primarily the synapses, the zones of contact between two neurons, where learning takes place, through interactions with the environment during the organism's lifetime. The nervous system of living beings thus represents a mixture of the innate and the acquired, the latter precluding the possibility of its hereditary transmission.

Artificial neural networks have been implemented many times over, mostly in software rather than in hardware, though only the latter concerns us here. Online learning is essential if one wishes to obtain learning systems as opposed to merely learned ones; such systems must learn rapidly from examples with no external guidance. Thus, while neural-network hardware had appeared already in the 1980s [1, 23], only today are we seeing the birth of the technology that enables true online learning. From a hardware point of view, it seems that the possibilities of using memory to retain training examples [53], and the use of adaptable connectivity structures have been widely under-explored; this can be due to the lack of appropriate technology. FPGAs have been used to implement neural networks with a dynamically reconfigurable structure, where neurons and connections may be added or removed, in accordance with environmental changes; this increases the online learning capabilities of the network, coupled with high-speed, parallel operation [47, 48]. The work of [44] has also investigated the possibility of restructuring the network, online. Another recent system of interest is that of Field-Programmable Interconnection Circuits (FPICs), which can be used in conjunction with FPGAs to further improve the network's online capabilities.

Other interesting paths are those that combine two or three axes of the POE model, as discussed in Section 5.

5 Conclusions: Softening hardware by combining phylogeny, ontogeny, and epigenesis

We presented the POE model for classifying soft hardware, based on three axes found in nature: phylogeny, ontogeny, and epigenesis (Figure 1). It is relatively straightforward to place the works presented at ICES'96 along these axes (Figure 6). Taking a look at the results obtained to date reveals a particular emphasis on the phylogenetic axis (fifteen papers, among which six concern offline evolution and nine concern partially or fully online evolution). This conforms with the prime theme of the conference, namely evolvable hardware. The epigenetic axis exhibits seven works on neuronal hardware, which can be grouped into three groups: brainware, learning and control processors for autonomous robots, and artificial neural networks. Finally, the ontogenetic axis is represented by two works concerning uni and multicellular, self-reproducing hardware. In addition, there are a number of overview papers, as well as works concerning specialized hardware systems.

A natural extension which suggests itself is the combination of two, and ultimately all three axes, in order to attain novel bio-inspired hardware (Figure 7). As examples we propose:

- *The PO plane.* This involves self-reproducing, evolving hardware, situated in the phylogenetic-ontogenetic plane. For example, [57, 58] have co-evolved non-uniform cellular automata to act as random number generators; [36] have shown that such evolved generators can be implemented by a multi-cellular automaton that exhibits self-reproduction and self-repair. Thus, the eventual combination of these two projects can be considered to be in the phylogenetic-ontogenetic plane.

- *The PE plane.* The architecture of the brain is the result of a long evolutionary process, during which a large set of specialized subsystems interactively emerged, carrying out tasks necessary for survival and reproduction [19]. Learning (epigenesis) in biological neural systems can be considered to serve as a mechanism for fine-tuning these broadly laid out neural circuits [24]. Although it is impossible that the genes code all structural information about the brain (Section 1), they may be the ultimate determinant of what it can and cannot learn [7].
  The idea of evolutionary, artificial neural networks, situated in the PE plane, has received attention in recent years; this involves a population of neural networks, where evolution takes place at the global (population) level, with learning taking place at the individual (neural network) level. Examples are the works of [35, 46, 68], though they are currently completely offline. The work of [14, 20] can also be situated in the PE plane, with a partially-online implementation; epigenetic learning takes place online, with the phylogenetic (population) existing offline.

- *The OE plane.* According to selectionism (e.g., [15]), selective pressures operate on epigenetic variation during the ontogeny of the individual (in "somatic" time), not on a phylogenetic time scale [50]. This suggests the pos-

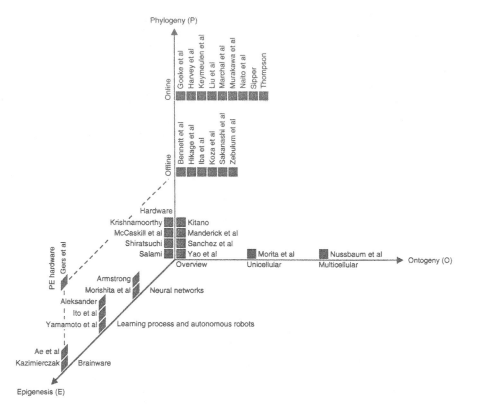

Fig. 6. Placing the works presented at the ICES'96 conference within the POE framework.

sibility of combining the ontogenetic mechanisms discussed above, with the epigenetic (neural network) learning algorithms.

Inductive learning can be interpreted as the capability to infer a response to an unknown situation, achieved through generalizing from previously-encountered, known situations. Engineers are perpetually confronted with a trade-off between generalization and robustness; while adding a multitude of neurons increases the system's fault tolerance, there is a risk of "learning nothing", if we do not attempt to generalize [53]. Implementing neurons in hardware is generally quite expensive, so it is imperative that cost-effectiveness be considered, trying to obtain the smallest possible network. Once a good generalization is obtained (with respect to a certain problem or situation), fault tolerance can be achieved through other self-repair mechanisms, e.g., those used by the embryonics system.

- *The POE space.* The development of an artificial neural network (epigenetic axis), implemented on a self-reproducing multicellular automaton (ontoge-

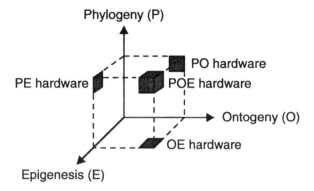

Fig. 7. Combining POE axes in order to create novel bio-inspired systems.

netic axis), whose genome is subject to evolution (phylogenetic axis), constitutes an ultimate example situated in the POE space (Figure 7).

As a final remark we note that the systems considered in this paper are bio-*inspired*; this means that, while motivated by our observations of Nature, we do not have to strictly adhere to its solutions. As an example, consider the issue of Lamarckian evolution, which involves the direct inheritance of acquired characteristics. While the biological theory of evolution has shifted from Lamarckism to Darwinism, this does not preclude the use of artificial Lamarckian evolution [16]. Another example concerns the time scales of natural processes, where phylogenetic changes occur at much slower rates than either ontogenetic or epigenetic ones, a characteristic which need not necessarily hold in our case. Thus, "deviations" from what is strictly natural may definitely be of use in our bio-inspired systems.

Looking (and dreaming) toward the future, one can imagine nano-scale (bioware) systems becoming a reality, which will be endowed with evolutionary, self-reproducing, self-repairing, and neural capabilities; such systems could give rise to novel *species* which will coexist along with carbon-based living beings.

This constitutes, perhaps, our ultimate challenge.

Acknowledgment

We are grateful to Antoine Danchin of the Pasteur Institute, Paris for his careful reading of this manuscript and his many helpful remarks and suggestions.

References

1. L. E. Atlas and Y. Suzuki. Digital systems for artificial neural networks. *IEEE Circuits and Devices magazine*, pages 20–24, November 1989.

2. T. Bäck. *Evolutionary algorithms in theory and practice: evolution strategies, evolutionary programming, genetic algorithms.* Oxford University Press, New York, 1996.

3. E. R. Banks. Universality in cellular automata. In *IEEE 11th Annual Symposium on Switching and Automata Theory*, pages 194–215, Santa Monica, California, October 1970.

4. R. A. Brooks. New approaches to robotics. *Science*, 253(5025):1227–1232, September 1991.

5. A. Burks, editor. *Essays on cellular automata.* University of Illinois Press, Urbana, Illinois, 1970.

6. J. Byl. Self-reproduction in small cellular automata. *Physica D*, 34:295–299, 1989.

7. J. Changeux and A. Danchin. Selective stabilisation of developing synapses as a mechanism for the specification of neural networks. *Nature*, 264:705–712, 1976.

8. E. F. Codd. *Cellular Automata.* Academic Press, New York, 1968.

9. A. Danchin. A selective theory for the epigenetic specification of the monospecific antibody production in single cell lines. *Ann. Immunol. (Institut Pasteur)*, 127C:787–804, 1976.

10. A. Danchin. Stabilisation fonctionnelle et épigénèse: une approche biologique de la genèse de l'identité individuelle. In J. -M. Benoist, editor, *L'identité*, pages 185–221. Grasset, 1977.

11. R. Dawkins. *The Blind Watchmaker.* W.W. Norton and Company, 1986.

12. R. Dawkins. The evolution of evolvability. In C. G. Langton, editor, *Artificial Life*, volume VI of *SFI Studies in the Sciences of Complexity*, pages 201–220. Addison-Wesley, 1989.

13. H. de Garis. Evolvable hardware: Genetic programming of a Darwin machine. In R. F. Albrecht, C. R. Reeves, , and N. C. Steele, editors, *Artificial Neural Nets and Genetic Algorithms*, pages 441–449, Berlin, 1993. Springer-Verlag.

14. H. de Garis. "Cam-Brain" ATR's billion neuron artificial brain project: A three year progress report. In *Proceedings of IEEE Third International Conference on Evolutionary Computation (ICEC'96)*, pages 886–891, 1996.

15. G. M. Edelman. *Neural Darwinism: The Theory of Neuronal Group Selection.* Basic Books, New York, 1987.

16. J. D. Farmer and A. d'A. Belin. Artificial life: The coming evolution. In C. G. Langton, C. Taylor, J. D. Farmer, and S. Rasmussen, editors, *Artificial Life II*, volume X of *SFI Studies in the Sciences of Complexity*, pages 815–840, Redwood City, CA, 1992. Addison-Wesley.

17. D. B. Fogel. *Evolutionary computation: toward a new philosophy of machine intelligence.* IEEE Press, Piscataway, NJ, 1995.

18. P. Galley and E. Sanchez. A hardware implementation of a Tierra processor. Unpublished internal report (in French), Logic Systems Laboratory, Swiss Federal Institute of Technology, Lausanne, 1996.

19. M. S. Gazzaniga. Organization of the human brain. *Science*, 245:947–952, 1989.

20. F. Gers and H. de Garis. CAM-Brain: A new model for ATR's cellular automata based artificial brain project. In *Proceedings of The First International Conference on Evolvable Systems: from Biology to Hardware (ICES96)*, Lecture Notes in Computer Science. Springer-Verlag, Heidelberg, 1996.

21. M. Goeke, M. Sipper, D. Mange, A. Stauffer, E. Sanchez, and M. Tomassini. Online autonomous evolware. In *Proceedings of The First International Conference on Evolvable Systems: from Biology to Hardware (ICES96)*, Lecture Notes in Computer Science. Springer-Verlag, Heidelberg, 1996.

22. D. E. Goldberg. *Genetic Algorithms in Search, Optimization and Machine Learning.* Addison-Wesley, 1989.

23. H. P. Graf and L. D. Jackel. Analog electronic neural network circuits. *IEEE Circuits and Devices magazine*, pages 44–49, July 1989.

24. B. Happel and J. M. Murre. Design and evolution of modular neural network architectures. *Neural Networks*, 7(6/7):985–1004, 1994.

25. H. Hemmi, J. Mizoguchi, and K. Shimohara. Development and evolution of hardware behaviors. In E. Sanchez and M. Tomassini, editors, *Towards Evolvable Hardware*, volume 1062 of *Lecture Notes in Computer Science*, pages 250–265. Springer-Verlag, Berlin, 1996.

26. T. Higuchi, M. Iwata, I. Kajitani, H. Iba, T. Hirao, T. Furuya, and B. Manderick. Evolvable hardware and its application to pattern recognition and fault-tolerant systems. In E. Sanchez and M. Tomassini, editors, *Towards Evolvable Hardware*, volume 1062 of *Lecture Notes in Computer Science*, pages 118–135. Springer-Verlag, Berlin, 1996.

27. T. Hikage, H. Hemmi, and K. Shimohara. Hardware evolution system: Introducing dominant and recessive heredity. In *Proceedings of The First International Conference on Evolvable Systems: from Biology to Hardware (ICES96)*, Lecture Notes in Computer Science. Springer-Verlag, Heidelberg, 1996.

28. J. H. Holland. *Adaptation in Natural and Artificial Systems.* The University of Michigan Press, Ann Arbor, Michigan, 1975.

29. J. E. Hopcroft and J. D. Ullman. *Introduction to Automata Theory Languages and Computation.* Addison-Wesley, Redwood City, CA, 1979.

30. J. O. Kephart. A biologically inspired immune system for computers. In R. A. Brooks and P. Maes, editors, *Artificial Life IV*, pages 130–139, Cambridge, Massachusetts, 1994. The MIT Press.

31. H. Kitano. Morphogenesis for evolvable systems. In E. Sanchez and M. Tomassini, editors, *Towards Evolvable Hardware*, volume 1062 of *Lecture Notes in Computer Science*, pages 99–117. Springer-Verlag, Berlin, 1996.

32. J. R. Koza. *Genetic Programming.* The MIT Press, Cambridge, Massachusetts, 1992.

33. J. R. Koza, F. H Bennett III, D. Andre, and M. A. Keane. Automated WYWIWYG design of both the topology and component values of electrical circuits using genetic programming. In J. R. Koza, D. E. Goldberg, D. B. Fogel, and R. L. Riolo, editors, *Genetic Programming 1996: Proceedings of the First Annual Conference*, pages 123–131, Cambridge, MA, 1996. The MIT Press.

34. C. G. Langton. Self-reproduction in cellular automata. *Physica D*, 10:135–144, 1984.

35. Y. Liu and X. Yao. Evolutionary design of artificial neural networks with different nodes. In *Proceedings of IEEE Third International Conference on Evolutionary Computation (ICEC'96)*, pages 670–675, 1996.

36. D. Mange, M. Goeke, D. Madon, A. Stauffer, G. Tempesti, and S. Durand. Embryonics: A new family of coarse-grained field-programmable gate arrays with self-repair and self-reproducing properties. In E. Sanchez and M. Tomassini, editors, *Towards Evolvable Hardware*, volume 1062 of *Lecture Notes in Computer Science*, pages 197–220. Springer-Verlag, Berlin, 1996. Also available as: Technical Report 95/154, Department of Computer Science, Swiss Federal Institute of Technology, Lausanne, Switzerland, November, 1995.

37. D. Mange, D. Madon, A. Stauffer, and G. Tempesti. Von Neumann revisited: A Turing machine with self-repair and self-reproduction properties. Technical Report

96/180, Department of Computer Science, Swiss Federal Institute of Technology, Lausanne, Switzerland, March 1996. (submitted for publication).

38. D. Mange, E. Sanchez, A. Stauffer, G. Tempesti, S. Durand, P. Marchal, and C. Piguet. Embryonics: A new methodology for designing field-programmable gate arrays with self-repair and self-reproducing properties. Technical Report 95/152, Department of Computer Science, Swiss Federal Institute of Technology, Lausanne, Switzerland, October 1995.

39. D. Mange and A. Stauffer. Introduction to embryonics: Towards new self-repairing and self-reproducing hardware based on biological-like properties. In N. M. Thalmann and D. Thalmann, editors, *Artificial Life and Virtual Reality*, pages 61–72, Chichester, England, 1994. John Wiley.

40. P. Marchal, C. Piguet, D. Mange, A. Stauffer, and S. Durand. Embryological development on silicon. In R. A. Brooks and P. Maes, editors, *Artificial Life IV*, pages 365–370, Cambridge, Massachusetts, 1994. The MIT Press.

41. Z. Michalewicz. *Genetic algorithms + data structures = evolution programs*. Springer, Berlin, third edition, 1996.

42. M. Mitchell. *An Introduction to Genetic Algorithms*. MIT Press, Cambridge, MA, 1996.

43. J. Monod. *Chance And Necessity: An Essay On The Natural Philosophy Of Modern Biology*. Vintage, New York, 1971.

44. J. M. Moreno. *VLSI Architectures for Evolutive Neural Models*. PhD thesis, Universitat Politecnica de Catalunya, Barcelona, 1994.

45. K. Morita and K. Imai. Logical universality and self-reproduction in reversible cellular automata. In *Proceedings of The First International Conference on Evolvable Systems: from Biology to Hardware (ICES96)*, Lecture Notes in Computer Science. Springer-Verlag, Heidelberg, 1996.

46. S. Nolfi, D. Parisi, and J. L. Elman. Learning and evolution in neural networks. *Adaptive Behavior*, 3(1):5–28, 1994.

47. A. Perez and E. Sanchez. FPGA implementation of an adaptable-size neural network. In C. von der Malsburg, W. von Seelen, J. C. Vorbrüggen, and B. Sendhoff, editors, *Proceedings of the International Conference on Artificial Neural Networks (ICANN96)*, volume 1112 of *Lecture Notes in Computer Science*, pages 383–388. Springer-Verlag, Heidelberg, 1996.

48. A. Perez and E. Sanchez. Neural networks structure optimization through on-line hardware evolution. In *Proceedings of the World Congress on Neural Networks (WCNN96)*. INNS (International Neural Networks Society) Press, 1996. (to appear).

49. J. -Y. Perrier, M. Sipper, and J. Zahnd. Toward a viable, self-reproducing universal computer. *Physica D*, 97:335–352, 1996.

50. S. R. Quartz and T. J. Sejnowski. The neural basis of cognitive development: A constructivism manifesto. *Behavioral and Brain Sciences*, 1996. (to appear).

51. T. S. Ray. An approach to the synthesis of life. In C. G. Langton, C. Taylor, J. D. Farmer, and S. Rasmussen, editors, *Artificial Life II*, volume X of *SFI Studies in the Sciences of Complexity*, pages 371–408, Redwood City, CA, 1992. Addison-Wesley.

52. J. A. Reggia, S. L. Armentrout, H.-H. Chou, and Y. Peng. Simple systems that exhibit self-directed replication. *Science*, 259:1282–1287, February 1993.

53. A. Roy, S. Govil, and R. Mirand. A neural network learning theory and a polynomial time RBF algorithm. *IEEE Transactions on Neural Networks*, 1996. (to appear).

54. E. Sanchez. Field programmable gate array (FPGA) circuits. In E. Sanchez and M. Tomassini, editors, *Towards Evolvable Hardware*, volume 1062 of *Lecture Notes in Computer Science*, pages 1–18. Springer-Verlag, Berlin, 1996.

55. E. Sanchez and M. Tomassini, editors. *Towards Evolvable Hardware*, volume 1062 of *Lecture Notes in Computer Science*. Springer-Verlag, Berlin, 1996.

56. H. -P. Schwefel. *Evolution and Optimum Seeking*. John Wiley & Sons, New York, 1995.

57. M. Sipper and M. Tomassini. Co-evolving parallel random number generators. In H. -M. Voigt, W. Ebeling, I. Rechenberg, and H. -P. Schwefel, editors, *Parallel Problem Solving from Nature - PPSN IV*, volume 1141 of *Lecture Notes in Computer Science*, pages 950–959. Springer-Verlag, Heidelberg, 1996.

58. M. Sipper and M. Tomassini. Generating parallel random number generators by cellular programming. *International Journal of Modern Physics C*, 7(2):181–190, 1996.

59. G. Tempesti. A new self-reproducing cellular automaton capable of construction and computation. In F. Morán, A. Moreno, J. J. Merelo, and P. Chacón, editors, *ECAL'95: Third European Conference on Artificial Life*, volume 929 of *Lecture Notes in Computer Science*, pages 555–563, Berlin, 1995. Springer-Verlag.

60. A. Thompson. Silicon evolution. In J. R. Koza, D. E. Goldberg, D. B. Fogel, and R. L. Riolo, editors, *Genetic Programming 1996: Proceedings of the First Annual Conference*, pages 444–452, Cambridge, MA, 1996. The MIT Press.

61. A. Thompson, I. Harvey, and P. Husbands. Unconstrained evolution and hard consequences. In E. Sanchez and M. Tomassini, editors, *Towards Evolvable Hardware*, volume 1062 of *Lecture Notes in Computer Science*, pages 136–165. Springer-Verlag, Berlin, 1996.

62. M. Tomassini. A survey of genetic algorithms. In D. Stauffer, editor, *Annual Reviews of Computational Physics*, volume III, pages 87–118. World Scientific, 1995. Also available as: Technical Report 95/137, Department of Computer Science, Swiss Federal Institute of Technology, Lausanne, Switzerland, July, 1995.

63. M. Tomassini. Evolutionary algorithms. In E. Sanchez and M. Tomassini, editors, *Towards Evolvable Hardware*, volume 1062 of *Lecture Notes in Computer Science*, pages 19–47. Springer-Verlag, Berlin, 1996.

64. J. von Neumann. *Theory of Self-Reproducing Automata*. University of Illinois Press, Illinois, 1966. Edited and completed by A.W. Burks.

65. L. Wolpert. *The Triumph of the Embryo*. Oxford University Press, New York, 1991.

66. S. Xanthakis, R. Pajot, and A. Rozz. Immune system and fault-tolerant computing. In *Evolution artificielle 94*. Cepadues, cop., 1995.

67. R. R. Yager and L. A. Zadeh. *Fuzzy Sets, Neural Networks, and Soft Computing*. Van Nostrand Reinhold, New York, 1994.

68. X. Yao. Evolutionary artificial neural networks. *International Journal of Neural Systems*, 4(3):203–222, 1993.

Promises and Challenges of Evolvable Hardware

Xin Yao[1] and Tetsuya Higuchi[2]

[1] Computational Intelligence Group, School of Computer Science
University College, The University of New South Wales
Australian Defence Force Academy, Canberra, ACT 2600, Australia
[2] Computation Models Section, Electrotechnical Laboratory
1-1-4 Umezono, Tsukuba, Ibaraki, 305 Japan

Abstract. Evolvable hardware (EHW) has attracted increasing attentions since early 1990's with the advent of easily reconfigurable hardware such as field programmable logic array (FPGA). It promises to provide an entirely new approach to complex electronic circuit design and new adaptive hardware. EHW has been demonstrated to be able to perform a wide range of tasks from pattern recognition to adaptive control. However, there are still many fundamental issues in EHW remain open. This paper reviews the current status of EHW, discusses the promises and possible advantages of EHW, and indicates the challenges we must meet in order to develop practical and large-scale EHW.

1 Introduction

Evolvable hardware (EHW) refers to hardware that can change its architecture and behaviour dynamically and autonomously by interacting with its environment. At present, almost all EHW uses an evolutionary algorithm (EA) as their main adaptive mechanism. One of the key motivations behind EHW is to learn from Nature since she has done so well in evolving wonders such as ourselves (i.e., human beings) without external forces. However, *learning* from Nature is quite different from *copying* it. There are many new challenges before us if we want to harness the power of evolution in EHW. This paper discusses the promises and challenges of EHW in more detail in later sections.

There are different views on what EHW is, depending on the purpose of EHW. One view regards EHW as "applications of evolutionary techniques to circuit synthesis" [1](abstract). Another view regards EHW as hardware which is capable of on-line adaptation through reconfiguring its architecture dynamically and autonomously [2]. Although these views are closely related and quite similar to each other, they emphasise different aspects of EHW. The former one uses simulated evolution as an alternative to conventional specification-based electronic circuit design, while the later uses it as an adaptive mechanism. However, the line between the two is grey.

EHW is fundamentally different from the hardware implementation of EAs, where the hardware architecture does not change and is used to implement EA functions such as selection, recombination and mutation [3, 4, 5]. The main motivation for hardware implementation of EAs is to speed up the execution of

EA functions. Such speed-up, however, does not necessarily imply a faster EA application because it does not speed up fitness evaluation which is often the most time-consuming part of an EA application. Discussion of EA's hardware implementation is beyond the scope of this paper.

EHW involves two major aspects — simulated evolution and electronic hardware. According to different EAs, e.g., genetic algorithms (GAs), genetic programming (GP), evolutionary programming (EP), and evolution strategies (ESs), and different electronic circuits, e.g., digital, analogue and hybrid circuits, used, we could classify EHW into different categories along these two dimensions. There are, however, at least two other important dimensions we should consider in investigating EHW, i.e., how the simulated evolution is realised and what the simulated evolution is used for, because they have a direct impact on the future research and development of EHW.

EHW has to be implemented on programmable logic devices (PLDs) such as field programmable logic arrays (FPGAs). The architecture of a PLD and thus its function are determined by a set of architecture bits which can be changed (i.e., reconfigured). In EHW, the simulated evolution is used to evolve a good set of architecture bits in order to solve a particular problem. According to de Garis [6], EHW can be classified into two categories, i.e., extrinsic and intrinsic EHW. Extrinsic EHW simulates evolution by software and only downloads the best configuration to hardware in each generation, i.e., the hardware is only reconfigured once. Intrinsic EHW simulates evolution directly in its hardware, i.e., every chromosome will be used to reconfigure the hardware. The EHW will be reconfigured the same number of times as the population size in each generation. Hirst [1] wrote a good survey paper along this line.

In this paper, we take a much broader view on EHW and address some important issues not covered in Hirst's survey. We argue that there is a difference between EHW used as an alternative to circuit design and EHW as on-line adaptive hardware. Although the techniques used to develop them may be very similar, the criteria used to measure them are different. For EHW which is used as an alternative to conventional circuit design, there are two distinct phases. One is the evolutionary design phase, and the other the execution phase which usually does not require on-line adaptation (although it is possible). To achieve on-line adaptation, EHW must adapt its architecture *while executing in the real environment.* Many new issues arise when on-line adaptation is required.

The rest of this paper is organised as follows: Section 2 reviews the work of evolving hardware as an alternative to designing it from specifications as it is done in conventional electronic circuit design. Section 3 discusses the attempt of developing EHW for on-line adaptation. Sections 4 and 5 present some views towards EHW and other non-evolutionary approachs to EHW. Finally, Section 6 concludes this paper with a summary of the paper and some remarks.

2 EHW as an Alternative to Circuit Design

Although EHW is a relatively new term, evolutionary design of electronic circuits have been attempted for more than a decade [7, 8, 9]. These early attempts did not design the architecture or function of a circuit. They were only used to optimise certain aspects of electronic circuit boards, e.g., cell placement [8, 9] and compaction of symbolic layout [7]. In essence, such work is better described as combinatorial optimisation by EAs.

Most current EHW work concentrates on evolutionary design of electronic circuits although the ultimate goal is to develop on-line adaptive hardware. So far, few studies have been reported on EHW which adapts its architecture and function *while executing in a real physical environment.*

According to the level of chromosome representation, the design approach to EHW can be classified into the direct and indirect ones. The direct approach to EHW encodes circuit's architecture bits as chromosomes, which specify the connectivity and functions of different hardware components (often at the gate level) of the circuit. In contrast, the indirect approach does not evolve architecture bits directly. It uses a higher level representation, such as trees or grammars, as chromosomes. These trees or grammars are then used to generate circuits.

2.1 The Indirect Approach to Evolutionary Circuit Design

Evolving Digital Circuits A typical example of the indirect approach is the evolution of a binary adder using HDL (Hardware Description Language) programs [10]. In this case, programs written in Structured Function Description Language (SFL) were encoded as chromosomes and subject to evolution. The chromosome representation is a derivation tree generated from a context-free grammar, which is called a rewriting system [10]. Each tree can generate one SFL program deterministically if the tree is "well-structured" [10](pp.372). Programs generated by different derivation trees can cover all possible programs in the SFL language which is defined by the grammar.

The root of a derivation tree is the start symbol of the grammar. The internal nodes are non-terminal symbols of the grammar, and the leaves terminal symbols. The crossover and mutation operators applied to derivation trees are similar to those used in GP [11, 12], but with some constraints. A branch (i.e., subtree) in a derivation tree can only be replaced (through either crossover or mutation) by another one with the same non-terminal node as the root (of the subtree). This is equivalent to replace a production (i.e., rewriting) rule in the grammar with another one having the same left-hand side. Such restriction ensures that all offspring produced are legal programs of the language. In addition to crossover and mutation, gene duplication and deletion were also used to modify derivation trees.

The idea of evolving the grammar itself was mentioned [10]. It was hoped that as the tasks to be performed by the hardware got more and more complex, the grammar itself would evolve to cope with the increasing complexity. A grammar was represented by a production diagram, which is a directed graph. The main genetic operator proposed was a kind of node splitting operation. It was unclear from the paper [10] whether they such an idea was tested by experiments.

Software simulation of evolving a binary adder using SFL programs was carried out [10]. The task can be described as follows.

> The target is two input and one output circuit; inputing two sequences of binary numbers from lowest figure, the circuit produces the sum of the binary numbers from the lowest figure in the output terminal. The correct circuit must consider the carry from the lower bit, so it belongs in a class of sequential circuits.

The fitness of each individual (i.e., a derivation tree/program) was calculated by evaluating its correctness in adding two 1536-bit numbers, which include all possible combinations of two 4-bit numbers [10](pp.376). A complete binary adder circuit was found in the experiment.

Related Work on Evolving Derivation Trees The chromosome representation scheme used in evolving the binary adder was also studied by Whigham [13, 14, 15] independently in a very different context. Whigham [13, 14] used a grammar to incorporate biases into GP in order to learn difficult and complex concepts. The major concern was to introduce declarative biases into GP under the general framework of inductive learning. There was no direct connection with EHW.

Whigham [13, 14] used similar crossover and mutation operators to those used by Hemmi et al.'s [10]. A schema theorem under the derivation tree representation and the crossover and mutation was given [16]. The idea of evolving the grammar itself was mentioned but not tested.

Whigham's work provides a different view towards the evolution of grammars. Such evolution can be regarded as the evolution of biases, i.e., certain knowledge or heuristics about what kind of circuits should be evolved. Such biases would be accumulated and obtained through the evolution of SFL programs and used to guide the evolution of grammars at a higher level. The importance of biases in inductive learning has long been recognised in the machine learning field and will not be repeated here.

Evolving Analogue Circuits In comparison with digital circuits, analogue circuits are more difficult to design. Recent work on evolving analogue circuits using GAs [17] and GP [18, 19, 20] shows an alternative to analogue circuit design using the evolutionary approach. One of the key issues in such evolutionary design is to find a suitable chromosome representation of analogue circuits. This problem is quite similar to that in evolutionary artificial neural networks (EANNs) [21] where a good chromosome representation of EANNs is also very important.

In the GP approach to analogue circuit design, trees are used to construct circuits. These circuit-constructing trees are evolved by GP [18, 19, 20]. Each tree can contain connection-modifying functions, component creating functions and automatically defined functions. A number of circuits, such as a lowpass "brick wall" filter, an asymmetric bandpass filter and an amplifier using transistors, have been evolved successfully [18].

The work on evolving analogue circuits described here does not belong to EHW in a strict sense because the evolution was all implemented and simulated by software. The simulation was carried out on a parallel computer system consisting of 64 Power PC 601 processors (80MHz) arranged in a toroidal mesh [18].

2.2 The Direct Approach to Evolutionary Circuit Design

Instead of evolving indirectly HDL programs or trees which specify circuit architecture and function, direct evolution of architecture bits of programmable logic devices (PLDs), such as programmable logic gate arrays, has also been proposed [22, 23, 2]. The architecture bits of a programmable logic gate array refer to those bits which specify its logic function and interconnections. The architecture bits uniquely determine the architecture and function of a programmable gate array. By evolving these bits (i.e., chromosomes), hardware can be evolved.

It is worth pointing out here the distinction made earlier in Section 1 of this paper between EHW used as an alternative to circuit design and EHW as on-line adaptive hardware. Higuchi *et al.* [22, 23, 2] have explicitly emphasised the later although some of the techniques they proposed can also be used in evolutionary circuit design. The hardware evolution described in this subsection is at the gate level since all the hardware functional units are simple logic gates, such as AND and OR gates [24].

A software simulation of evolving a GAL16Z8 chip for the 6-multiplexor problem has been carried out to show the potential of the gate level evolution [22]. The chromosome used in the simulation had 108 bits, of which 12 bits were used to specify the function of the logic cell (an Output Logic Macro Cell, OLMC) and 96 bits used to specify a fuse array that determines interconnections between inputs and the logic cell. A generational GA with uniform crossover, bit-flipping mutation and roulette wheel selection was used to evolve a population of 100 such chromosomes. The fitness of each chromosome was calculated by evaluating the gate array on *all* 64 possible inputs. A correct 6-multiplexor circuit was evolved after about 2000 generations in one run. Experiments on evolving other circuits, such as the exclusive-OR circuit, a 3-bit counter, and a 4-state finite state machine, have also been reported [23, 2].

Thompson *et al.* [25, 26] emphasised the importance of unconstrained evolution of electronic circuits (including both spatial and temporal constraints), and provided both theoretical arguments and experimental evidences to support their points. They viewed electronic circuits more as dynamic systems than as static ones. Such a view enabled them to explore a wide range of potentials of EHW. It also revealed some fundamental issues faced by EHW in general.

One of the experiments carried out by Thompson was the evolution of a slow electronic oscillator using high-speed logic gates [25]. The aim of such an experiment was to find out whether "the high-speed components can somehow be assembled to give rise to slower dynamics, without explicitly providing large time-constant resources or slow-speed clocks." [25] The experiment was quite different from others in that both spatial and temporal constraints were removed. The evolving circuit operated entirely in an asynchronous mode without any clocks. The delay at each gate was assigned a real-value "selected uniformly randomly from the range 1.0 to 5.0 nanoseconds." The delay of connections was ignored, i.e., set to 0.

Thompson [25] fixed the number of logic gates (also called nodes) at 100 in his experiment and defined the fitness of an individual as follows.

The objective was for node number 100 to produce a square wave oscillation of 1kHz, which means alternately spending 0.5×10^{-3} seconds at logic 1 and at logic 0. If k logic transitions were observed on the output of node 100 during the simulation, with the nth transition occurring at time t_n seconds, then the average error in the time spent at each level was calculated as:

$$average\ error = \frac{1}{k-1} \sum_{n=2}^{k} |(t_n - t_{n-1}) - 0.5 \times 10^{-3}| \qquad (1)$$

For the purpose of this equation, transitions were also assumed to occur at the very beginning and end of the trial, which lasted for 10ms of simulated time. The fitness was simply the reciprocal of the average error.

Each node (i.e., logic gate) required a genotype segment of 24 bits in the chromosome representation, which encoded the node function and the sources of its inputs. Each chromosome had a total of 101 segments, i.e., 2424 bits. The GA used was a "generational one with elitism and linear rank-based selection." The population size was 30 [25].

In the 40th generation of one run, the GA was able to evolve an oscillator with approximately 4kHz, while the best individual in the random initial population was one with approximately 18MHz. The experiment did not continue after 40 generations due to "excessive processor time needed to simulate this kind of network" although "fitness was still rising." [25] A total of 68 gates were used in the 4Hz oscillator evolved by the GA.

2.3   Function Level Evolution

As can be seen from the experiments on the gate level evolution, the size of chromosomes grows rapidly as the size of EHW increases. According to Higuchi et al.'s estimation [22], "FPGAs require from 2000 to 30000 architecture bits to configure their circuits." Evolution of chromosomes of such sizes is inefficient even in hardware.

To address the issue, Higuchi et al. [24, 27] proposed the function level evolution for EHW. In the function level evolution, high-level hardware functions, such as addition, subtraction, sine, etc., rather than simple logic functions are used as primitive functions in the evolution. Much more powerful circuits can be evolved using the function level evolution [24, 27]. Since the function level evolution aims at on-line adaptation by EHW, it will be discussed in more details in Section 3.

2.4 Advantages of Evolutionary Design

EHW has been used as an alternative to conventional circuit design although the ultimate goal might be to develop EHW which adapts in a real physical environment. Such an evolutionary design approach offers a number of advantages over the conventional one used by human designers although there are some important issues that remain open.

First, the evolutionary design approach can explore a much wider range of design alternatives than those could be considered by human beings. This has been shown by many experiments in other design tasks, such as evolutionary design of neural networks [28, 21, 29, 30, 31, 32, 33, 34], of building architectures [35], and of arts [36]. These experiments demonstrated how evolutionary techniques could be applied to evolving novel designs which were difficult to discover by human beings. However, all these experiments were carried out by software simulation although some of the techniques used in these software simulations will also be useful for EHW.

Second, the evolutionary design approach does not assume *a priori* knowledge of any particular design domain. It can be applied by users without resorting to domain experts. It can be used in domains where little *a priori* knowledge is available or where such knowledge is very costly to obtain. As the complexity of circuits increases, it becomes extremely difficult to fully understand interactions among various components of the circuits and their dynamics. The conventional design approach tends to break down in such cases, while the evolutionary approach would excel. In essence, the conventional design approach specifies *how* to design and implement a circuit, while the evolutionary approach only specifies *what* the circuit should implement, i.e., what required function or behaviour the circuit should have without worrying how to achieve it.

Third, the evolutionary design approach can work with varying degrees of constraints and special requirements, if necessary, by incorporating them in chromosome representation and fitness function. As mentioned above, the evolutionary approach can work with little *a priori* domain knowledge. However, if some domain knowledge is available, it can be used to improve the efficiency of the evolutionary design. Using domain knowledge to improve EAs has been shown to be achievable and effective [37, 38].

2.5 Some Issues in Evolutionary Design of Electronic Circuits

Scalability of EHW The importance of scalability has been recognised by several researchers [22, 39]. It is a tough problem faced not only by EHW researchers, but also by other researchers in the fields of evolutionary computation, artificial neural networks, and artificial intelligence in general. To our best knowledge, all EHW experiments conducted so far have been on a small scale. That is, the EHW is small with much fewer components in comparison with the circuits designed by the conventional method. Even for these small EHW, researchers have already experienced the high computational cost of evolving circuits [22, 39].

The scalability of EHW could be divided into two related parts. The first part deals with the scalability of the chromosome representation of electronic circuits. At present, the length of chromosomes can be a couple of thousand bits for 100 logic gates [25]. For a circuit with 1000 logic gates, the expected length of chromosomes would be tens of thousands of bits, which is very inefficient to process by the current evolutionary techniques. Roughly speaking, if no constraint is imposed on the connectivity of EHW, i.e., any connectivity is possible, then the length of chromosome would grow in the order of $O(n^2)$ where n is the number of functional components (such as logic gates) in the EHW. If connectivity is constrained to certain local neighbourhood around a functional component, $O(n)$ would be achievable. However, this comes with the cost of *constraining the EHW*, something which we tried to avoid at the beginning when we embarked on EHW.

The second part of EHW's scalability concerns with the computational complexity of an EA. This is a much more important issue, which still remains open, than the scalability of chromosome representation. Neither the worst nor average case time complexity has been established for any EA used to solve any particular problem. At present, it is not unusual to carry out an EHW experiment which runs for days. Yet the EHW used in these experiments contained only 100 functional components or so. The question is: How long will it take to evolve an EHW with 10000 functional components using the current techniques?

The Danger of Relying Too Much on Hardware Speed Using hardware to increase the speed of evolution seems to be an answer to combat the high computational cost. While hardware does offer limited temporary relief on the high computational cost, it does not solve the problem. The sheer speed of dedicated hardware is not the answer to a time complexity issue. The importance of the time complexity and the irrelevance of hardware speed can be seen clearly from the following artificial example. Assume the average time complexity of an EA applied to an EHW is $O(2^n)$ where n is the number of functional components in the EHW. If the EHW with 10 functional components requires $1024 = 2^{10}$ nanoseconds ($\approx 10^{-6}$ seconds) to evolve (in hardware of course), a similar EHW with only 100 functional components would need $2^{100} \approx 10^{30}$ nanoseconds ($> 10^{13}$ years) to evolve. That is certainly not the time we would like to spend on EHW.

The above example shows the danger of relying too much on hardware speed while overlooking the fundamental issue. Fortunately, the time complexity of $O(2^n)$ assumed in the artificial example is not based on any theoretical or empirical evidences. Unfortunately, there is still no result on the time complexity of EHW. The possibility of an $O(2^n)$ time complexity, albeit small, does exist.

Fitness Evaluation and Circuit Verification/Testing An issue arises in the evolutionary design of electronic circuits is how to verify that the evolved circuit, i.e., EHW is correct. This would not be an issue if a fitness function could be defined such that a maximum fitness corresponds to a perfectly correct circuit. For example, in the 6-multiplexor experiment [22], all 64 possible input combinations were used in fitness evaluation. The maximum fitness implies a correct EHW. However, the method will not work for circuits with a large number of inputs since the number of possible input combinations increases exponentially as the number of inputs increases.

Sometimes a fitness function which guarantees the circuit correctness is very difficult to find without incurring heavy computational cost in fitness evaluation. For example, in the sequential binary adder experiment [10], the correctness of evolved circuits had to be confirmed by human beings through "reading the description of the program." [10](pp.376) The maximum fitness value did not guarantee the correctness of a circuit. The fitness of a circuit in the sequential binary adder experiment was defined by considering "all possible combinations of two 4 bit numbers." [10](pp.376) However, this does not imply that a sequential binary adder that operates correctly on all possible combinations of two 4 bit numbers will be correct on all possible combinations of two 5 or more bit numbers. This seems to be a complex problem related to the fitness evaluation and stopping criteria used in EHW. It is difficult for EHW to know when a correct circuit, not just the one with the maximum fitness value, has been evolved because the simulated evolution only manipulates the *syntax* not *semantics* of encoded circuits.

Another example is the fitness definition used in evolving a slow oscillator [25]. The fitness value of a circuit depends on the value of k in Eq.1. A maximum fitness for a particular k does not imply the circuit will operate correctly for larger k's. If a large k is used in the fitness evaluation, the computational cost will no doubt increases. The correctness issue addressed here is related to the generalisation ability of EHW if we viewed EHW as a learning device not an alternative to circuit design.

Unconstrained hardware evolution can cause additional problems in terms of circuit correctness since it exploits every characteristics of electronic circuits and the environment in which it is evaluated, regardless of whether a characteristic is relevant or not. As Thompson pointed out [25], the behaviour of EHW may depend on such factors as fluctuations in temperature and power supply. Exploitation of hardware must be traded against EHW's sensitivity to variations. It was suggested that EHW could be evaluated under various situations to "make sure" that it is not sensitive to small variations [25]. However, it is no simple task to achieve this. First, all characteristics exploited by EHW must be varied. Second, the number of variations for each characteristic must be sufficiently large. A very high computational cost has to be paid for all these. In addition, it is difficult to find out what characteristics would be exploited by EHW in the first place before we could vary them.

Termination of Evolution The difficulty in defining a good fitness function, as mentioned in the previous subsection, also leads to the difficulty in defining a stop criterion for the simulated evolution. EHW does not know when it has found a correct solution and thus should stop since a maximum fitness value does not necessarily guarantee a correct circuit. In the existing EHW experiments [22, 23, 2, 10, 25], the correctness of evolved circuits was established by human beings. Another thing which is unclear from all these experiments is whether the correct circuit is the result of only one run or multiple runs. If every single run can guarantee to produce a correct circuit, then there would be no problem. If not, how many runs on average we have to perform in order to get a correct circuit? When should we stop? Although there are quite a few papers analysing the behaviours of an evolved circuit and showing they are correct, it is unclear whether a circuit with similar behaviours could be evolved from another separate run.

3 EHW for Learning and On-line Adaptation

The real attractiveness and power of EHW comes from its potential as an adaptive hardware which can change its behaviour and improve its performance while executing in a real physical environment (as opposed to simulation). Such on-line adaptation is very difficult to achieve. The difficulty is not caused by EHW, but by the *on-line* requirement. In other words, on-line adaptation would still be very difficult even if a different approach from EHW is adopted.

At present, EHW has mostly been studied in terms of off-line adaptation. That is, EHW is not used in an execution mode while evolving. For example, it is not used to control a real robot in a real physical environment while evolving. This can be regarded as the off-line learning phase of EHW. One of the reasons why off-line learning is used is because of the trial-and-error nature of EAs. It is possible to produce very poor individuals by random mutation or crossover in EAs. These poor individuals could cause severe damages or disasters to EHW or the physical environment in which it is being evaluated, if there is no additional technique to prevent them from happening. For example, an EHW evolved to control a real robot could produce such a poor controller that the robot would hit an obstacle badly. This is certainly not the way to get a fitness value of the controller in a real physical environment.

3.1 EHW Controllers

EHW controllers refer to those EHW which are used primarily as controllers for robots or any other devices (such as ATM switches or multiplexors). Mizoguchi *et al.* [40] used EHW to control an artificial ant to follow the John Muir Trail. The trail was placed on a grid. The controller of the artificial ant, which was implemented by EHW through simulation, had one input and two outputs. The input contained information about whether or not the trail exists in the cell in front of it. The outputs controlled the actions of going straight, turning left and turning right.

The technique used to evolve the EHW controller is the same as that used by Hemmi *et al.* [10], which is described briefly in Section 2.1. Each configuration of the controller was specified by an SFL program which was produced by a derivation tree (i.e., a rewriting tree). Derivations trees generated from the SFL grammar were represented as chromosomes and evolved by Production Genetic Algorithms (PGAs) proposed by Mizoguchi *et al.* [40]. PGAs employed six operators; selection, crossover, mutation, duplication, insertion and deletion. The operators guarantee that all offspring will be legal trees defined by the grammar. The fitness of an artificial ant was determined by the number of cells on the trail which were traversed within a time limit and the number of time steps used. Traversing more cells on the trail with less time steps within a time limit gave higher fitness. All cells of the trail, which were fixed, were used in fitness evaluation. No testing on the generalisation ability of EHW was performed. As pointed out by Mizoguchi *et al.* themselves [40], their system "represents one approach to designing hardware." Adaptivity and generalisation would not be the major concern.

Another experiment on robot control was carried out by Thompson [41], where a real hardware robot controller was evolved for wall-avoidance behaviour. The controller's input came from two sonar heads pointing left and right respectively. Its output went to the motors for controlling two wheels. For the hardware evolution, architecture bits (also called "configuration memory") of the EHW controller, which was implemented in FPGAs, were used as genotypes. They determined functions of the functional blocks in the FPGA and their interconnections. In other words, they determined the whole function and thus behaviour of the EHW controller.

In Thompson's experiment [41], a genotype *directly* encoded all the details of the EHW controller, including the clock information. For the simple wall-avoidance behaviour, the length of genotypes was 32 bits. A genetic algorithm was used to evolve a population of 30 such genotypes. Each genotype was evaluated by evaluating how well the EHW controller performed for four trials of 30 seconds each. The worst performance out of four was used to determined the fitness. "For the final few generations, the evaluations were extended to 90 seconds, to find controllers that were not only good at moving away from walls, but also *staying* away from them." [41]

Although Thompson [41] evolved real hardware controller to control a real physical robot, simulation was still used in fitness evaluation. However, his reason for using simulation appears to be different from our concern about potential risks of evaluating poor controller in a physical environment.

> For convenience, ... The real evolving hardware controlled the real motors, but the wheels were just spinning in the air. The wheels' angular velocities were measured, and used by a real time simulation of the motor characteristics and motor dynamics to calculate how the robot would move. The sonar echo signals were then artificially synthesised and sup-

plied in real time to the hardware DSM. Realistic levels of noise were included in the sensor and motor models, both of which were constructed by fitting curves to experimental measurements, including a probabilistic model for specular sonar reflections. [41]

Such an experiment belongs to the category of evolving real hardware in a simulated environment [25](Section 13). How close the simulated environment (or models) to the real physical one will have a major impact on the performance of evolved hardware in the real physical environment. The good result achieved by Thompson [41] on the transfer from the simulated to the real environment shows that a simulated environment might be a solution to avoiding the potential risks of evaluating poor controller in a physical environment.

3.2 EHW Recognisers and Classifiers

EHW recognisers and classifiers refer to those EHW which are used primarily for pattern recognition and classification. Higuchi *et al.* [22, 23, 2, 27, 24, 42] have carried out a number of experiments using EHW to perform various recognition and classification tasks. They used both the gate and function level evolution.

For the gate level evolution, an EHW pattern recognition system was developed to recognise noisy binary input patterns [2, 24, 42]. The input pattern consisted of 8×8 pixels, which were represented by 64 bits. There were three output classes which were represented by 3 bits. During the learning phase, the EHW recogniser was presented with the training patterns. The chromosome representation scheme used was different from that previously adopted by Higuchi *et al.* [22, 23]. A variable-length chromosome representation scheme was used, which only encoded non-empty (non-nil) entries in the connectivity matrix of EHW (FPGAs) [42, 43]. Such a representation generated substantially shorter chromosomes for sparsely connected FPGAs. The GA used was similar to messy GAs with cut and splice operators [44]. The only difference was that duplicated genes would be removed after the splice operation.

The fitness of each individual (i.e., EHW recogniser) was determined by both the error and complexity of the EHW according to the MDL principle [45]. EHW with lower error and lower complexity had higher fitness. Unlike most of the experiments described previously, the EHW recogniser was tested on a separate testing set which was not used in training. "The test data set consists of 30 patterns which are made by adding some noises into the training patterns. Pixels from 1 to 5 are selected randomly and values of the selected pixels are inverted." [42]. Fairly good results, which were average over 10 runs, were obtained from the experiment [42].

The gate level evolution was also adopted to develop an EHW comparator used in a V-shape ditch tracer of an industrial welding robot [2]. The EHW was used as a backup system for the conventional logic comparator. It would take over control from the conventional logic comparator only when the conventional logic comparator failed due to circuit faults [2].

For the function level evolution, experiments were carried out with four well-known problems, i.e., the two interwined spirals, the Iris data set, 2-D image rotation, and synthesis of a 4-state automaton [27, 24]. For all these experiments, an FPGA model consisting of 100 programmable floating processing units (PFUs) was used, which were arranged on a 5×20 grid in a feed-forward fashion. That is, the output from one column of PFUs would only be fed into the next column. The two inputs to the FPGA could, however, be fed into any PFUs. A chromosome encoded the information about the function selected by each PFU and the interconnections between PFUs. The variable-length chromosome representation scheme proposed by Kajitani *et al.* [43], which was mentioned above, was used in the experiments. However, the GA used did not have any crossover operators. Only three types of mutations were adopted: operand mutation, function mutation and insertion [27].

The fitness of each EHW in the function level evolution only considered the error information [27]. The MDL principle was not used. All these four problems were investigated from the point of view of EHW's generalisation ability. Testing results were given along with the training results. Such experiments were quite different from those aiming at EHW as a design alternative. It was pointed out clearly that the final goal was to achieve on-line adaptation although the current work was only concerned was off-line adaptation [24, 27].

The driving force behind the function level evolution was to partially address the problem of scalability suffered by the gate level evolution, especially for EHW which would be used in industrial applications. Higuchi *et al.*'s work [24, 27] narrowed the gap between EHW research and its applications.

3.3 Some Issues and Related Work in Adaptive EHW

Although adaptive EHW might be accused of being a "seductive" phrase, it is used here to distinguish it from evolutionary design of hardware and to refer to the EHW which requires generalisation ability and on-line adaptation. There are some fundamental and interesting issues in adaptive EHW which are worth probing further. A comparison with some related work would also help to foster cross-fertilisation between EHW and other research areas and identify the potential niches of EHW where its advantages could be fully exploited.

On-line Adaptation In spite of the high hope of EHW, no work has been reported on on-line adaptation by EHW. Only off-line adaptation by EHW has been achieved, where adaptation happens during the learning phase of EHW. It should be noted that on-line adaptation means adaptation of EHW while it is executing in a real physical environment. In a sense, on-line adaptation can also be viewed as real-time adaptation. The meaning of "on-line" here is different from that used in other contexts, such as on-line update of connection weights for a backpropagation neural network.

On-line adaptation requires EHW's learning to be incremental and responsive. Such requirements do not seem to be met by population-based evolutionary learning, which is used by all EHW at present. The current population-based evolutionary learning is not incremental because if a new situation occurs as a result of an environmental change, it would have to re-learn the new *as well as* old situations in order to deal with both.

Evolutionary learning at the population level is slow in responding to environmental changes without local learning at the individual level. The population-based learning is global because learning can only be achieved through interactions among different individuals, although it is possible to restrict such interactions to a neighbourhood. Real-time adaptation would be extremely difficult to achieve.

It appears that "pure" population-based evolutionary learning would not be sufficient to cope with the requirements of on-line adaptation. Local learning at the individual level could be introduced to supplement it. Local learning can respond much faster to environmental changes since such response can be made at the local individual level. It has been shown in the area of evolutionary artificial neural networks (EANNs) that combining evolutionary learning at the population level with local learning at the individual level is feasible and beneficial [46, 47, 28, 21].

Generalisation Generalisation is a key issue for any learning or adaptive systems, including EHW. However, studies on this topic are relatively few in the area of EHW. Some experiments on EHW did not address the issue since the same training and testing data set was used, e.g., the experiments with the artificial ant [40] and the 4-state automaton [27]. It is unclear how well the EHW could generalise to different situations in these cases. In essence, such experiments demonstrated the effectiveness of EHW as an alternative to circuit design, but not necessarily as an adaptive or learning system.

Most work on testing the generalisation ability of EHW was done by Higuchi *et al.* [2, 42, 27, 24]. For the EHW pattern recogniser described in Section 3.2, its performance was tested on a noisy test data set different from the training set [42]. For the Iris data set, different training and testing sets were also used [27].

An issue that arises here is whether the maximum fitness value corresponds to the best generalisation. The issue is somewhat similar to that raised in Section 2.5. For example, a solution learned by the EHW pattern recogniser for identifying three patterns was $O_0 = I_{50}$, $O_1 = \overline{I_{13}}$, and $O_2 = I_{37}$, where O_i's ($0 \leq i \leq 2$) were output and I_j's ($0 \leq j \leq 63$) were input [42]. This was apparently not a good generalisation because the output class was determined by a single pixel value. It meant that a single-bit noise at that particular position would change the output of classification. The patterns used in learning EHW recogniser were digits and letters. Recognising a digit or letter based on the presence or absence of a particular pixel value does not seem to be correct. The fact that the learned EHW had high fitness but not best generalisation implies that the EHW recogniser did not learn what we wanted it to learn. It is possible that the training set used in training the EHW recogniser did not contain data of sufficient variety. A better training set should improve EHW's generalisation.

Evaluating EHW's generalisation can be a difficult task due to different implementations. This difficulty is closely related to that of evaluating the generalisation ability of evolutionary learning systems in general. It is not uncommon to read papers which only report a good system evolved at certain number of generations. It is unclear, however, whether such a good system is the result of one particular run or the average of multiple runs. Statistical analysis of the experimental results seems to be missing. In addition, it is unclear how to decide when to stop in order to get the good system. A more disciplined approach to experimental studies of generalisation in evolutionary learning will greatly help EHW's research.

Adaptive EHW and Evolutionary Artificial Neural Networks Evolutionary artificial neural networks (EANNs) refer to a class of ANNs in which evolution is another form of adaptation in addition to learning [46, 47, 28, 21]. In particular, EANNs which adapt their architectures through simulated evolution and their weights through learning (training) have been shown to be successful in dealing with a number of benchmark problems [29, 30, 31, 32, 33, 34].

Adaptive EHW is closely related to EANNs. For example, both the function level EHW (FEHW) [24, 27] and EPNet [29, 30, 31, 32, 33, 34] evolve feedforward architectures. Both can have different node functions in an architecture [27, 48]. However, node functions in FEHW usually have more variety. There is currently no local learning in FEHW. No weights are associated with connections in FEHW. FEHW's adaptation relies heavily on different compositions of its node functions. In contrast, EPNet uses weights and local learning, but less variety of node functions. It is unknown at present whether FEHW with more node functions without weights would be better than that with less node functions with weights in terms of adaptation and hardware implementation. (It should be pointed out that EPNet is a software package, and is not targeted at hardware implementation.)

Although EPNet is implemented in software, there are some techniques which might be useful for EHW. For example, EPNet uses validation sets and the order of mutations to improve the generalisation ability of learned systems. It grows an ANN by splitting an existing node rather than add a random one. The process is similar to cell splitting in biology. It deletes or adds a connection by evaluating the importance of the connection first. It also maintains a close behavioural link between parents and offspring, which is important for on-line adaptation where we do not want large fluctuations.

Local learning through adjusting weights could be introduced into FEHW since each PFU (i.e., node) in FEHW (implemented by Function level FPGAs) has four constant generators which can be used to produce weights for up to four inputs. Such local learning at the individual level can be realised easily. The approach adopted in EPNet could be borrowed or tailored to FHEW. The potential problem might be the speed of such learning. The various types of node function used in EHW will have a major impact on the speed.

Adaptive EHW and Genetic Programming While FEHW and EPNet manipulate acyclic, weighted and directed graphs by simulated evolution, while genetic programming (GP) [11] mainly manipulates trees. They are closely related to each other because a tree can be regarded as an acyclic directed graph and an acyclic directed graph can be transformed into a tree by repeating nodes and branches.

FEHW and GP share the similarity that both adapt function compositions and/or combinations without weights and local learning. But their representations are different. FEHW and EPNet manipulate acyclic, weighted and directed graphs by simulated evolution, while genetic programming (GP) [11] mainly manipulates trees. Although a tree can be regarded as an acyclic directed graph and an acyclic directed graph can be transformed into a tree by repeating nodes and branches, FEHW is more flexible and general as it can deal with cyclic graphs without much added complexity.

Just as the case in GP, FEHW also requires predefining a set of primitive functions which can be used by each node. One question, which was mentioned briefly in the above subsection, is why we need more than one node function and what the benefits would be. GP systems which use only one type of node function but with weights, such as the STRONGANOFF system [49], seem to work quite well.

Disaster Prevention in Real-Time Evolution It was mentioned in the beginning of Section 3 that evaluating an EHW in a real physical environment could cause severe damages or disasters to EHW or the physical environment. This potential risk restricts possible applications of EHW in domains where evaluating EHW in a real physical environment is impractical and an accurate simulation model of the physical environment is difficult to obtain. In most EHW applications, fitness evaluation is the most time-consuming part of the whole evolutionary process. The distinction between intrinsic and extrinsic EHW does not seem to capture this characteristic of EHW. It is only concerned with whether an EHW is reconfigured once or multiple time for each generation [6].

The aforementioned risk stems from the trial-and-error nature of EAs and the black-box approach used by EAs. An EA only evolves chromosomes *syntactically* not *semantically*. It does not understand evolved systems and the environment as no explicit models are used. A possible way to get around this problem is to develop a knowledge-based adaptive EHW, where constraints and knowledge about the environment in which EHW will be evaluated are incorporated into fitness evaluation as its front end, such that any poor individuals that may cause damages to themselves or the environment could be detected and prevented from being passed to the real physical environment.

4 A Behavioural View of EHW

As pointed out in previous sections, evolving electronic circuits faces many challenges and open issues. Most of them are caused by the confusion between evolving circuits and evolving circuit behaviours. This confusion must be cleared before any further progress can be made in EHW research. On the surface, it does not seem to make much difference when circuits are evolved or their behaviours are evolved. However, conceptually it is inappropriate to evolve circuits. It is circuit behaviours that should be and can be evolved. It is inherently hard to evolve circuits. Why?

Simulated evolution uses a fitness function to evaluate an EHW individual. What does it actually evaluate? Is it really the EHW circuit (connectivity, functional cells, etc.)? The answer is *no*. It is the circuit's behaviours that are evaluated. The fitness function knows nothing directly about the circuit, and it is not supposed to know it. Since it is the circuit behaviours that are evaluated, the fitness value must depend on the environment in which the EHW circuit is evaluated. Hence the fitness value is only a measurement of how good the circuit is in *that* environment. It says nothing about the circuit's behaviours in a different environment. This is where the generalisation and circuit verification issues start coming in and bothering the EHW research, as discussed in Sections 3.3 and 2.5.

In short, EHW should be regarded as an evolutionary approach to behaviour design rather than hardware design. Such a behavioural view of EHW requires a different thinking on EHW circuit design. It is no longer appropriate to talk about what architecture or function a circuit should have. One should start thinking of what behaviours are required from a circuit in certain environments. Then EHW would become a powerful means to evolve and implement such behaviours.

5   EHW That Does Not Change Its Configuration by an EA

This paper is primarily concerned with EHW that uses simulated evolution (notably EAs) to evolve hardware. There are other types of hardware which also use some biological ideas other than evolution. For example, Mange *et al.* [50, 51, 52, 53] have been working on self-reproducing and self-repairing hardware based some ideas from molecular biology (genetics and embryology). The approach used was built on von Neumann's pioneering work on self-reproducing automata [54].

de Garis [55, 56, 57, 58, 59, 60] also used cellular automata in his CAM-BRAIN project exclusively. The aim of the project is to grow and evolve cellular automata based neural networks (i.e., artificial brains). However, it is unclear what the neural networks are used for since little information has been disclosed about experimental studies on CAM-BRAIN. It is also unclear how the artificial brain is going to learn or evolve after it has "grown up."

6 Conclusion

This paper reviews the current research on EHW. Emphasis is given to EHW which employs simulated evolution to evolve hardware. A number of issues are raised and discussed. In particular, EHW research needs to address issues such as scalability, on-line adaptation, generalisation, circuit correctness, and potential risk of evolving hardware in a real physical environment. It is argued that research on a theoretical foundation of EHW needs to be carried out urgently before rushing to large-scale EHW implementation.

This paper also points out some related work, where some of the techniques could be applied to EHW. Two such related areas are EANNs and GP. EHW is a new research area in the intersection between evolutionary computation and electronics. New progresses in these two fields will continue to provide EHW with new opportunities. For example, recent work on fast hardware for computing exponential and trigonometric functions [61] and work on BIST (built-in self-test) test pattern generators [62] will no doubt widen the range of possible EHW applications.

References

[1] A. J. Hirst.
Notes on the evolution of adaptive hardware.
In I. Parmee, editor, *Proc. of the 2nd Int. Conf. on Adaptive Computing in Engineering & Design (ACEDC96)*, 1996.

[2] T. Higuchi, M. Iwata, I. Kajitani, H. Iba, T. Furuya, and B. Manderick.
Evolvable hardware and its applications to pattern recognition and fault-tolerant systems.
In E. Sanchez and M. Tomassini, editors, *Towards Evolvable Hardware: The Evolutionary Engineering Approach, Lecture Notes in Computer Science, Vol. 1062*, pages 118–135. Springer-Verlag, 1996.

[3] P. Graham and B. Nelson.
A hardware genetic algorithm for the traveling salesman problem on splash 2.
In *Proc. of the 5th Int. Workshop on Field Programmable Logic and Applications, Oxford, England*, pages 352–361, August 1995.

[4] S. D. Scott, A. Samal, and S. Seth.
HGA: A hardware based genetic algorithm.
In *Proc. of ACM/SIGDA 3rd Int. Symp. on FPGA's*, pages 53–59, 1995.

[5] M. Salami and G. Cain.
Implementation of genetic algorithms on reprogrammable architectures.
In X. Yao, editor, *Proc. of the Eighth Australian Joint Conference on Artificial Intelligence (AI'95)*, page 581. World Scientific Publ. Co., Singapore, 1995.

[6] H. de Garis.
LSL evolvable hardware workshop report.
Technical report, ATR, Japan, October 1995.

[7] M. P. Fourman.
Compaction of symbolic layout using genetic algorithms.
In J. J. Grefenstette, editor, *Proc. of the First Int'l Conf. on Genetic Algorithms and Their Applications*, pages 141–153. Carnegie-Mellon University, 1985.

[8] J. Cohoon and W. Paris.
Genetic placement.
In *Proc. of Int'l Conf. on CAD*, pages 422–425. IEEE Press, New York, NY 10017-2394, 1986.

[9] R. M. Kling and P. Banerjee.
ESP: Placement by simulated evolution.
IEEE Trans. on CAD, CAD-8:245–256, 1989.

[10] H. Hemmi, J. Mizoguchi, and K. Shimohara.
Development and evolution of hardware behaviours.
In R. Brooks and P. Maes, editors, *Artificial Life IV: Proc. of the 4th Int. Workshop on the Synthesis and Simulation of Living Systems*, pages 371–376. MIT Press, 1994.

[11] J. R. Koza.
Genetic Programming.
The MIT Press, Cambridge, Mass., 1992.

[12] J. R. Koza.
Genetic Programming II.
The MIT Press, Cambridge, Mass., 1994.

[13] P.A. Whigham.
Grammatically-based genetic programming.
In J.Rosca, editor, *Proc. of the Workshop on Genetic Programming: From Theory to Real-World Applications*, pages 33–41. Morgan Kaufmann Publ., July 1995.

[14] P.A. Whigham.
Inductive bias and genetic programming.
In *Proc. of the 1st Int. Conf. on Genetic Algorithms in Engineering Systems: Innovations and Applications*, pages 461–466. IEE, UK, Sept. 1995.

[15] P.A. Whigham and R. McKay.
Genetic approaches to learning recursive relations.
In X. Yao, editor, *Progress in Evolutionary Computation, Lecture Notes in Artificial Intelligence, Vol. 956*, pages 17–27. Springer-Verlag, Heidelberg, 1995.

[16] P.A. Whigham.
A schema theorem for context-free grammars.
In *Proc. of the 1995 IEEE Int. Conf. on Evolutionary Computation (ICEC'95)*, volume 1, pages 178–182. IEEE Press, Piscatawa, NJ, USA, December 1995.

[17] W. Kruiskamp and D. Leenaerts.
DARWIN: CMOS opamp synthesis by means of a genetic algorithm.
In *Proc. of the 32nd Design Automation Conference*, pages 433–438. ACM Press, New York, NY, USA, 1995.

[18] J. R. Koza, F. H Bennett III, D. Andre, and M. A. Keane.
Four problems for which a computer program evolved by genetic programming is competitive with human performance.
In *Proc. of the 1996 IEEE Int. Conf. on Evolutionary Computation (ICEC'96)*, pages 1–10. IEEE Press, Piscatawa, NJ, USA, May 1996.

[19] J. R. Koza, D. Andre, F. H Bennett III, and M. A. Keane.
Use of automatically defined functions and architecture-altering operations in automated circuit synthesis using genetic programming.
In *Proc. of the Genetic Programming 1996 Conference (GP'96)*. The MIT Press, 1996.

[20] J. R. Koza, F. H Bennett III, D. Andre, and M. A. Keane.
Automated WYWIWYG design of both the topology and component values of electrical circuits using genetic programming.
In *Proc. of the Genetic Programming 1996 Conference (GP'96)*. The MIT Press, 1996.

[21] X. Yao.
Evolutionary artificial neural networks.
In A. Kent and J. G. Williams, editors, *Encyclopedia of Computer Science and Technology*, volume 33, pages 137–170. Marcel Dekker Inc., New York, NY 10016, 1995.

[22] T. Higuchi, T. Niwa, T. Tanaka, H. Iba, H. de Garis, and T. Furuya.
Evolving hardware with genetic learning: A first step towards building a Darwin machine.
In *Proc. of 2nd Int. Conf. on the Simulation of Adaptive Behaviour (SAB92)*, pages 417–424. MIT Press, 1992.

[23] T. Higuchi, H. Iba, and B. Manderick.
Evolvable hardware.
In H. Kitano and J. Hendler, editors, *Massively Parallel Artificial Intelligence*, pages 398–421. MIT Press, 1994.

[24] T. Higuchi, M. Iwata, I. Kaijitani, M. Murakawa, S. Yoshizawa, and T. Furuya.
Hardware evolution at gate and function level.
In *Proc. of the Int. Conf. on Biologically Inspired Autonomous Systems: Computation, Cognition and Action, Durham, NC, USA*, 4-5 March 1996.

[25] A. Thompson, I. Harvey, and P. Husbands.
Unconstrained evolution and hard consequences.
In E. Sanchez and M. Tomassini, editors, *Towards Evolvable Hardware: The Evolutionary Engineering Approach, Lecture Notes in Computer Science, Vol. 1062*, pages 136–165. Springer-Verlag, 1996.

[26] A. Thompson.
Silicon evolution.
In *Proc. of the Int. Conf. on Genetic Programming (GP'96)*, page To appear, 1996.

[27] M. Murakawa, S. Yoshizawa, I. Kajitani, T. Furuya, M. Iwata, and T. Higuchi.
Hardware evolution at function level.
In *Proc. of the International Conference on Parallel Problem Solving from Nature (PPSN'96)*, page Accepted, 1996.

[28] X. Yao.
Evolutionary artificial neural networks.
International Journal of Neural Systems, 4(3):203–222, 1993.

[29] X. Yao and Y. Liu.
Evolving artificial neural networks for medical applications.
In *Proc. of 1995 Australia-Korea Joint Workshop on Evolutionary Computation*, pages 1–16. KAIST, Taejon, Korea, September 1995.

[30] X. Yao and Y. Liu.
A new evolutionary system for evolving artificial neural networks.
IEEE Trans. on Neural Networks, 1995.
Submitted.

[31] X. Yao and Y. Liu.
Towards designing artificial neural networks by evolution.
In *Proc. of Int. Symp. on Artificial Life and Robotics (AROB)*, Beppu, Oita, Japan, pages 265–268, 18-20 February 1996.

[32] X. Yao and Y. Liu.
Evolving artificial neural networks through evolutionary programming.
In P. Angeline and T. Bäck, editors, *Evolutionary Programming V: Proc. of the Fifth Annual Conference on Evolutionary Programming*, page To appear. MIT Press, 1996.

[33] X. Yao and Y. Liu.
Evolutionary artificial neural networks that learn and generalise well.
In *1996 IEEE International Conference on Neural Networks, Washington, DC, USA*. IEEE Press, New York, NY, 3-6 June 1996.
Accepted.

[34] Y. Liu and X. Yao.
A population-based learning algorithm which learns both architectures and weights of neural networks.
Chinese Journal of Advanced Software Research (Allerton Press, Inc., New York, NY 10011), 3(1):54–65, 1996.

[35] M. A. Rosenman.
An evolutionary model for non-routine design.
In X. Yao, editor, *Proc. of the Eighth Australian Joint Conference on Artificial Intelligence (AI'95)*, pages 363–370. World Scientific Publ. Co., Singapore, 1995.

[36] K. Sims.
Artificial evolution for computer graphics.
Computer Graphics, 25(4):319–328, 1991.

[37] J. J. Grefenstette.
Incorporating problem specific knowledge into genetic algorithms.

In L. Davis, editor, *Genetic Algorithms and Simulated Annealing*, chapter 4, pages 42–60. Morgan Kaufmann, San Mateo, CA, 1987.

[38] L. Davis.
Handbook of Genetic Algorithms.
Van Nostrand Reinhold, New York, NY 10003, 1991.

[39] H. Hemmi, J. Mizoguchi, and K. Shimohara.
Development and evolution of hardware behaviors.
In E. Sanchez and M. Tomassini, editors, *Towards Evolvable Hardware: The Evolutionary Engineering Approach, Lecture Notes in Computer Science, Vol. 1062*, pages 250–265. Springer-Verlag, 1996.

[40] J. Mizoguchi, H. Hemmi, and K. Shimohara.
Production genetic algorithms for automated hardware design through an evolutionary process.
In Z. Michalewicz *et al.*, editor, *Proc. of the First IEEE Conf. on Evolutionary Computation (ICEC'94), Vol. I*, pages 661–664. IEEE Press, New York, NY 10017-2394, 1994.

[41] A. Thompson.
Evolving electronic robot controllers that exploit hardware resources.
In *Proc. of the 3rd European Conf. on Artificial Life (ECAL95)*, pages 640–656. Springer-Verlag, 1995.

[42] M. Iwata, I. Kajitani, H. Yamada, H. Iba, and T. Higuchi.
A pattern recognition system using evolvable hardware.
In *Proc. of the International Conference on Parallel Problem Solving from Nature (PPSN'96)*, page Accepted, 1996.

[43] I. Kajitani, T. Hoshino, M. Iwata, and T. Higuchi.
Variable length chromosome GA for evolvable hardware.
In *Proc. of the 1996 IEEE Int. Conf. on Evolutionary Computation (ICEC'96)*, pages 443–447. IEEE Press, Piscatawa, NJ, USA, May 1996.

[44] D. E. Goldberg, K. Deb, H. Kargupta, and G. Harik.
Rapid, accurate optimization of difficult problems using fast messy genetic algorithms.
In S. Forrest, editor, *Proc. of the 5th Int. Conf. on Genetic Algorithms (ICGA'93)*, pages 56–64. Morgan Kaufmann, 1993.

[45] J. Rissanen.
Stochastic complexity in statistical inquiry.
World Scientific, Singapore, 1989.

[46] X. Yao.
Evolution of connectionist networks.
In T. Dartnall, editor, *Preprints of the Int'l Symp. on AI, Reasoning & Creativity*, pages 49–52, Queensland, Australia, 1991. Griffith University.

[47] X. Yao.
A review of evolutionary artificial neural networks.
International Journal of Intelligent Systems, 8(4):539–567, 1993.

[48] Y. Liu and X. Yao.
Evolutionary design of artificial neural networks with different nodes.

In *Proc. of the 1996 IEEE Int'l Conf. on Evolutionary Computation (ICEC'96), Nagoya, Japan*, pages 670–675. IEEE Press, New York, NY 10017-2394, 1996.

[49] H. Iba, H. de Garis, and T. Sato.
A numerical approach to genetic programming for system identification.
Evolutionary Computation, 3(4):417–452, 1996.

[50] D. Mange and A. Stauffer.
Introduction to embryonics: Towards new self-repairing and self-reproducing hardware based on biological-like properties.
In N. Magnenat Thalmann and D. Thalmann, editors, *Artificial Life and Virtual Reality*, pages 61–72. John Wiley, Chichester, England, 1994.

[51] P. Marchal, C. Piguet, D. Mange, A. Stauffer, and S. Durand.
Achieving von Neumann's dream: Artificial life on silicon.
In *Proc. of the IEEE International Conference on Neural Networks (ICNN'94), Volume IV*, pages 2321–2326, 1994.

[52] S. Durand, A. Stauffer, and D. Mange.
Biodule: An introduction to digital biology.
Technical Report Preliminary Report, Logic Systems Laboratory, The Swiss Federal Institute of Technology, Lausanne, September 1994.

[53] D. Mange, S. Durand, E. Sanchez, A. Stauffer, G. Tempesti, P. Marchal, and C. Piguet.
A new self-reproducing automaton based on a multi-cellular organization.
Technical Report 95/114, Logic Systems Laboratory, The Swiss Federal Institute of Technology, Lausanne, April 1995.

[54] J. von Neumann.
The Theory of Self-Reproducing Automata.
University of Illinois Press, Urbana, 1966.

[55] H. de Garis.
Circuits of production rule GenNets: the genetic programming of artificial nervous systems.
In R. F. Albrecht, C. R. Reeves, and N. C. Steele, editors, *Proc. of the Int. Conf. on Artificial Neural Nets and Genetic Algorithms, Innsbruck, Austria*, pages 699–705. Springer-Verlag/Wien, 1993.

[56] H. de Garis.
Evolvable hardware: Genetic programming of a darwin machine.
In R. F. Albrecht, C. R. Reeves, and N. C. Steele, editors, *Proc. of the Int. Conf. on Artificial Neural Nets and Genetic Algorithms, Innsbruck, Austria*, pages 441–449. Springer-Verlag/Wien, 1993.

[57] H. de Garis.
An artificial brain.
New Generation Computing, 12:215–221, 1994.

[58] H. de Garis.
CAM-BRAIN — the evolutionary engineering of a billion neuron artificial brain by 2001 which grows/evolves at electronic speeds inside a cellular automata machine (CAM).

In D. W. Pearson, N. C. Steele, and R. F. Albrecht, editors, *Proc. of the Int. Conf. on Artificial Neural Nets and Genetic Algorithms, Alés, France,* pages 84–87. Springer-Verlag/Wien, 1995.

[59] H. de Garis.
CAM-BRAIN: the evolutionary engineering of a billion neuron artificial brain by 2000 which grows/evolves at electronic speeds inside a cellular automata machine (cam).
In E. Sanchez and M. Tomassini, editors, *Towards Evolvable Hardware: The Evolutionary Engineering Approach, Lecture Notes in Computer Science, Vol. 1062,* pages 76–98. Springer-Verlag, 1996.

[60] H. de Garis.
CAM-BRAIN issues: implementation and performance-scaling issues concerning the genetic programming of a cellular automata based artificial brain.
In *Proc. of the IEEE International Conference on Neural Networks (ICNN'94), Volume III,* pages 1714–1720, 1994.

[61] V. Kantabutra.
On hardware for computing exponential and trigonometric functions.
IEEE Trans. on Computers, 45(3):328–339, 1996.

[62] C.-A. Chen and S. K. Gupta.
BIST test pattern generators for two-pattern testing — theory and design algorithms.
IEEE Trans. on Computers, 45(3):257–269, 1996.

Evolware

Designing Evolware by Cellular Programming

Moshe Sipper

Logic Systems Laboratory, Swiss Federal Institute of Technology, IN-Ecublens, CH-1015 Lausanne, Switzerland. E-mail: Moshe.Sipper@di.epfl.ch

Abstract. A major impediment preventing ubiquitous computing with cellular automata (CA) stems from the difficulty of utilizing their complex behavior to perform useful computations. In this paper *non-uniform* CAs are studied, presenting the *cellular programming* algorithm for co-evolving such systems to perform computations. The algorithm is applied to five computational tasks: density, synchronization, ordering, boundary computation, and thinning; our results show that non-uniform CAs can attain high computational performance, and furthermore, that such systems can be evolved rather than designed. We believe that cellular programming holds potential for attaining 'evolving ware', *evolware*, which can be implemented in software, hardware, or other possible forms, such as bioware. We have recently implemented an evolving, online, autonomous hardware system based on the approach described herein.

1 Introduction

Cellular automata (CA) are dynamical systems in which space and time are discrete. A cellular automaton consists of an array of cells, each of which can be in one of a finite number of possible states, updated synchronously in discrete time steps according to a local, identical interaction rule. The state of a cell is determined by the previous states of a surrounding neighborhood of cells [34, 30].

CAs exhibit three notable features: massive parallelism, locality of cellular interactions, and simplicity of basic components (cells). They perform computations in a distributed fashion on a spatially extended grid. As such they differ from the standard approach to parallel computation in which a problem is split into independent sub-problems, each solved by a different processor, later to be combined in order to yield the final solution. CAs suggest a new approach in which complex behavior arises in a bottom-up manner from non-linear, spatially extended, local interactions [14].

A major impediment preventing ubiquitous computing with CAs stems from the difficulty of utilizing their complex behavior to perform useful computations. Designing CAs to have a specific behavior or perform a particular task is highly complicated, thus severely limiting their applications; automating the design (programming) process would greatly enhance the viability of CAs [14]. A prime motivation for studying CAs stems from the observation that they are naturally suited for hardware implementation, with the potential of exhibiting extremely fast and reliable computation that is robust to noisy input data and component failure [7].

The model investigated in this paper is an extension of the CA model, termed *non-uniform cellular automata* [20]. Such automata function in the same way as uniform ones, the only difference being in the cellular rules that need not be identical for all cells. Our main focus is on the *evolution* of non-uniform CAs to perform computational tasks, employing a local, co-evolutionary algorithm, an approach referred to as *cellular programming*. We believe that cellular programming holds potential for attaining 'evolving ware', *evolware*, which can be implemented in software, hardware, or other possible forms, such as bioware. Of particular interest is the issue of evolving hardware, which has recently made its appearance on the artificial evolution scene [19]. We have recently implemented an evolving, online, autonomous hardware system based on the approach described herein [8].

Our aim in this paper is to introduce the cellular programming approach, toward which end we shall delineate the basic methodology and present examples of co-evolved, non-uniform CAs. In Section 2 we present previous work on non-uniform CAs and evolving CAs. The cellular programming algorithm is delineated in Section 3, and applied to five computational tasks in Section 4: density, synchronization, ordering, boundary computation, and thinning. We demonstrate that high performance is attained on these tasks using one-dimensional grids as well as previously unstudied two-dimensional ones. Our findings are discussed in Section 5.

2 Previous work

The application of genetic algorithms to the evolution of *uniform* cellular automata was initially studied by [16] and recently undertaken by the EVCA (evolving CA) group [15, 14, 13, 6, 5]. They carried out experiments involving uniform, one-dimensional CAs with $k = 2$ and $r = 3$, where k denotes the number of possible states per cell and r denotes the radius of a cell, i.e., the number of neighbors on either side (thus, each cell has $2r + 1$ neighbors, including itself). Spatially periodic boundary conditions are used, resulting in a circular grid. A common method of examining the behavior of one-dimensional CAs is to display a two-dimensional space-time diagram, where the horizontal axis depicts the configuration at a certain time t and the vertical axis depicts successive time steps (e.g., Figure 2). The term 'configuration' refers to an assignment of 1 states to several cells, and 0s otherwise.

The EVCA group employed a standard genetic algorithm to evolve uniform CAs to perform two computational tasks, namely density and synchronization (see Section 4). The algorithm uses a randomly generated initial population of CAs with $k = 2$, $r = 3$. Each CA is represented by a bit string, delineating its rule table, containing the output bits for all possible neighborhood configurations (i.e., the bit at position 0 is the state to which neighborhood configuration 0000000 is mapped to and so on until bit 127 corresponding to neighborhood configuration 1111111). The bit string, known as the "genome", is of size $2^{2r+1} = 128$, resulting in a huge search space of size 2^{128}. Each CA in the population was run for a maximum number of M time steps, after which its fitness was evaluated, defined as the fraction of cell states correct at the last time step. Using the genetic algorithm highly successful CA rules were found for both tasks [15, 14, 5].

The model investigated in this paper is that of non-uniform CAs, where cellular rules need not be identical for all cells. As noted in Section 1, CAs lend themselves naturally to hardware implementation, which is one of the primary motivations for their study. Note that from a hardware point of view the resources required by non-uniform CAs are identical to those of uniform ones since a cell in both cases contains a rule (albeit not necessarily the same one in our case). A prime motivation for studying non-uniform CAs stems from the observation that the uniform model is essentially "programmed" at an extremely low-level [18]. A *single* rule is sought that must be applied universally to all cells in the grid, a task which may be arduous even for evolutionary approaches. For non-uniform CAs search space sizes are vastly larger than with uniform CAs, a fact that initially seems as an impediment; however, we have found that this model presents novel dynamics, offering new and interesting paths in the evolution of complex systems.

We have previously applied the non-uniform CA model to the investigation of artificial life issues, presenting multi-cellular "organisms" that display several interesting behaviors, including reproduction, growth and mobility. We also studied evolution in the context of various environments, observing genotypic as well as phenotypic effects [20, 23, 21]. In [22, 25] we examined the issue of universal computation in two-dimensional CAs. We demonstrated that universality can be attained in non-uniform, 2-state, 5-neighbor cellular space (i.e., with a minimal number of states as well as a minimal neighborhood), which is not universal in the uniform case [2]. The universal systems we presented are simpler than previous ones and are *quasi*-uniform, meaning that the number of distinct rules is extremely small with respect to rule space size; furthermore, the rules are distributed such that a subset of dominant rules occupies most of the grid [22, 25, 24].

The co-evolution of non-uniform, one-dimensional CAs to perform computations was undertaken in [24, 25]. In [24] we presented results pertaining to the density task, showing that high performance, non-uniform CAs can be co-evolved not only with radius $r = 3$, as studied by [15, 14], but also for smaller radiuses, most notably $r = 1$ which is minimal. In [25] we showed that high performance can be attained for the synchronization task as well, with $r = 1$, using the cellular programming algorithm. It was also found that evolved systems exhibiting high performance are quasi-uniform.

The one-dimensional density and synchronization tasks are elaborated upon in Section 4, along with recent results pertaining to two-dimensional grids and novel tasks. Our main findings from our previous work, as well as that reported in this paper are:

1. Universal computation can be attained in simple, non-uniform cellular spaces that are not universal in the uniform case. This is accomplished by utilizing a small number of different rules (quasi-uniformity).

2. Non-uniform CAs can attain high performance on non-trivial computational tasks.
3. Non-uniform CAs can be co-evolved to perform computations, with high performance systems exhibiting quasi-uniformity.
4. Non-uniformity may reduce connectivity requirements, i.e., the use of smaller cellular neighborhoods is made possible.

3 The cellular programming algorithm

We study 2-state, non-uniform CAs, in which each cell may contain a different rule. A cell's rule table is encoded as a bit string, known as the "genome", containing the output bits for all possible neighborhood configurations (as in Section 2). Rather than employ a *population* of evolving, uniform CAs, as with genetic algorithm approaches, our algorithm involves a *single*, non-uniform CA of size N, with cell rules initialized at random. Initial configurations are then generated at random, in accordance with the task at hand. For each initial configuration the CA is run for M time steps. Each cell's *fitness* is accumulated over $C = 300$ initial configurations, where a single run's score is 1 if the cell is in the correct state after M iterations, and 0 otherwise. After every C configurations evolution of rules occurs by applying crossover and mutation. This evolutionary process is performed in a completely *local* manner, where genetic operators are applied only between directly connected cells. It is driven by $nf_i(c)$, the number of fitter neighbors of cell i after c configurations. The pseudo-code of our algorithm is delineated in Figure 1.

Crossover between two rules is performed by selecting at random (with uniform probability) a single crossover point and creating a new rule by combining the first rule's bit string before the crossover point with the second rule's bit string from this point onward. Mutation is applied to the bit string of a rule with probability 0.001 per bit.

There are two main differences between our algorithm and the standard genetic algorithm: (a) A standard genetic algorithm involves a population of evolving, uniform CAs; all CAs are *ranked* according to fitness, with crossover occurring between *any* two individuals in the population. Thus, while the CA runs in accordance with a local rule, evolution proceeds in a *global* manner. In contrast, our algorithm proceeds *locally* in the sense that each cell has access only to its locale, not only during the run but also during the evolutionary phase, and no global fitness ranking is performed. (b) The standard genetic algorithm involves a population of *independent* problem solutions; the CAs in the population are assigned fitness values independent of one another, and interact only through the genetic operators in order to produce the next generation. In contrast, our CA *co-evolves* since each cell's fitness depends upon its evolving neighbors.

This latter point comprises a prime difference between our algorithm and parallel genetic algorithms, which have attracted attention over the past few years. These aim to exploit the inherent parallelism of evolutionary algorithms,

```
for each cell i in CA do in parallel
    initialize rule table of cell i
    f_i = 0 { fitness value }
end parallel for
c = 0 { initial configurations counter }
while not done do
    generate a random initial configuration
    run CA on initial configuration for M time steps
    for each cell i do in parallel
        if cell i is in the correct final state then
            f_i = f_i + 1
        end if
    end parallel for
    c = c + 1
    if c mod C = 0 then { evolve every C configurations}
        for each cell i do in parallel
            compute nf_i(c) { number of fitter neighbors }
            if nf_i(c) = 0 then rule i is left unchanged
            else if nf_i(c) = 1 then replace rule i with the fitter neighboring rule,
                                  followed by mutation
            else if nf_i(c) = 2 then replace rule i with the crossover of the two fitter
                                  neighboring rules, followed by mutation
            else if nf_i(c) > 2 then replace rule i with the crossover of two randomly
                                  chosen fitter neighboring rules, followed by mutation
            end if
            f_i = 0
        end parallel for
    end if
end while
```

Fig. 1. Pseudo-code of the cellular programming algorithm.

thereby decreasing computation time and enhancing performance [32]. A number of models have been suggested, among them coarse-grained, island models [28, 3, 29], and fine-grained, grid models [31, 12]. The latter resemble our system in that they are massively parallel and local; however, the co-evolutionary aspect is missing. As we wish to attain a system displaying global computation, the individual cells do not evolve independently as with genetic algorithms (be they parallel or serial), i.e., in a "loosely-coupled" manner, but rather co-evolve, thereby comprising a "tightly-coupled" system.

4 Results

In this section we study five computational tasks using one-dimensional grids as well as previously unstudied two-dimensional ones: density (Section 4.1), synchronization (Section 4.2), ordering (Section 4.3), rectangle-boundary (Section 4.4), and thinning (Section 4.5). Minimal cellular spaces are used: 2-state, $r = 1$ for the one-dimensional case and 2-state, 5-neighbor for the two-dimensional one. The total number of initial configurations per evolutionary run was in the range $[10^5, 10^6]$. Performance values reported hereafter represent the average fitness of all grid cells after C configurations, normalized to the range $[0, 1]$; these are obtained during execution of the cellular programming algorithm.

4.1 The density task

The one-dimensional density task is to decide whether or not the initial configuration contains more than 50% 1s, relaxing to a fixed-point pattern of all 1s if the initial density of 1s exceeds 0.5, and all 0s otherwise. As noted by [14], the density task comprises a non-trivial computation for a small radius CA ($r \ll N$, where N is the grid size); the density is a global property of a configuration whereas a small-radius CA relies solely on local interactions. Since the 1s can be distributed throughout the grid, propagation of information must occur over large distances (i.e., $O(N)$). The minimum amount of memory required for the task is $O(\log N)$ using a serial scan algorithm, thus the computation involved corresponds to recognition of a non-regular language.

We have studied this task in [24, 25] using non-uniform, one-dimensional, minimal radius $r = 1$ CAs of size $N = 149$. The search space involved is extremely large; since each cell contains one of 2^8 possible rules this space is of size $(2^8)^{149} = 2^{1192}$. In contrast, the size of *uniform*, $r = 1$ CA rule space is small, consisting of only $2^8 = 256$ rules. This enabled us to test each and every one of these rules, a feat not possible for larger values of r, finding that the maximal performance of uniform, $r = 1$ CAs on the density task is 0.83.

For the cellular programming algorithm we used randomly generated initial configurations, uniformly distributed over densities in the range $[0, 1]$, with the CA being run for $M = 150$ time steps (thus, computation time is linear with grid size). We found that non-uniform CAs had co-evolved that exhibit peak performance values as high as 0.93; furthermore, these consist of a grid in which one rule dominates, a situation referred to as quasi-uniformity (Section 2). Figure 2 demonstrates the operation of one such co-evolved CA along with a rules map, depicting the distribution of rules by assigning a unique color to each distinct rule.

The results obtained by us using a minimal radius of $r = 1$ are comparable to those of [15], who used uniform CAs of radius $r = 3$; furthermore, we attain notably higher performance than any *possible* uniform, $r = 1$ CA. This suggests that non-uniformity reduces connectivity requirements, i.e., the use of smaller cellular neighborhoods is made possible. A detailed investigation of the one-dimensional density task can be found in [24, 25].

The density task can be extended in a straightforward manner to two-dimensional grids. Applying our algorithm to such grids yielded notably higher performance than the one-dimensional case, with peak values of 0.99. Figure 3 demonstrates the operation of one such co-evolved CA. Qualitatively, we observe the CA's

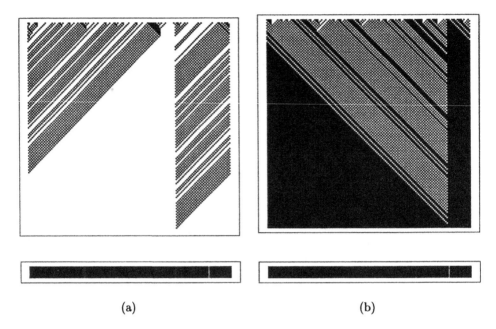

(a)   (b)

Fig. 2. One-dimensional density task: Operation of a co-evolved, non-uniform, connectivity radius $r = 1$ CA. Grid size is $N = 149$. White squares represent cells in state 0, black squares represent cells in state 1. The pattern of configurations is shown through time (which increases down the page). Top figures depict space-time diagrams, bottom figures depict rule maps. Initial configurations were randomly generated. (a) Initial density of 1s is 0.40, final density is 0. (b) Initial density of 1s is 0.60, final density is 1. The CA relaxes in both cases to a fixed pattern of all 0s or all 1s, correctly classifying the initial configuration.

"strategy" of successively classifying local densities, with the locality range increasing over time; "competing" regions of density 0 and density 1 are manifest, ultimately relaxing to the correct fixed point.

4.2   The synchronization task

The one-dimensional synchronization task was introduced by [5] and studied by us in [25] using non-uniform CAs. In this task the CA, given any initial configuration, must reach a final configuration, within M time steps, that oscillates between all 0s and all 1s on successive time steps. This task comprises a non-trivial computation for a small radius CA; it belongs to a class of problems studied in other domains, such as distributed computing, known as firing squad problems [11].

In [25] we studied non-uniform, one-dimensional, minimal radius $r = 1$ CAs of size $N = 149$. As for the density task, we first tested all possible uniform, $r = 1$ CAs on the synchronization task, finding that the maximal performance is 0.84. For the cellular programming algorithm we used randomly generated initial

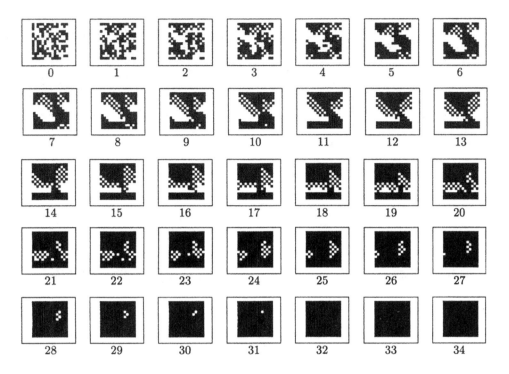

Fig. 3. Two-dimensional density task: Operation of a co-evolved, non-uniform, 2-state, 5-neighbor CA. Grid size is $N = 225$ (15×15). Initial density of 1s is 0.51, final density is 1. Numbers at bottom of images denote time steps.

configurations, uniformly distributed over densities in the range $[0, 1]$, with the CA being run for $M = 150$ time steps. We found that quasi-uniform CAs had co-evolved that exhibit perfect performance, thereby surpassing any possible uniform CA. Figure 4 depicts the operation of two such co-evolved CAs, along with rule maps. A detailed investigation of the one-dimensional synchronization task can be found in [25]. This task can also be extended in a straightforward manner to two-dimensional grids, an investigation of which we have carried out; our results show that perfect performance can be co-evolved for such CAs as well.

4.3 The ordering task

In this task, the one-dimensional CA, given any initial configuration, must reach a final configuration in which all 0s are placed on the left side of the grid and all 1s on the right side. The ordering task may be viewed as a variant of the density task and is clearly non-trivial using similar arguments to those of Section 4.1. It is interesting in that the output is not a uniform configuration of all 0s or all 1s as with the density and synchronization tasks.

Testing all uniform, $r = 1$ CAs on the ordering task we found that the

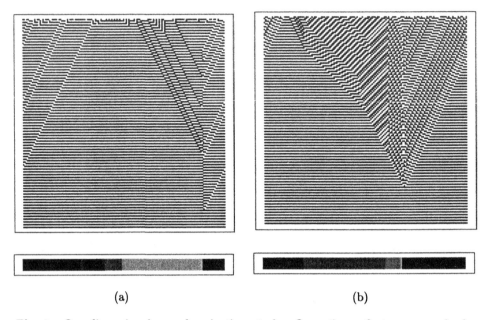

(a)   (b)

Fig. 4. One-dimensional synchronization task: Operation of two co-evolved, non-uniform, $r = 1$ CAs. Grid size is $N = 149$. Top figures depict space-time diagrams, bottom figures depict rule maps.

maximal performance is 0.71. Our algorithm yielded quasi-uniform CAs with fitness values as high as 0.93, one of which is depicted in Figure 5. As with the previous two tasks we find that non-uniform CAs can be co-evolved to attain high performance, exceeding that of the best uniform CA.

4.4 The rectangle-boundary task

The possibility of applying CAs to perform image processing tasks arises as a natural consequence of their architecture; in a two-dimensional CA, a cell (or a group of cells) can correspond to an image pixel, with the CA's dynamics designed so as to perform a desired image processing task. Earlier work in this area, carried out mostly in the 1960s and the 1970s, was treated in [17], with more recent applications presented in [1, 10].

The final two tasks we study involve image processing operations. In this section we discuss a two-dimensional boundary computation: given an initial configuration consisting of a non-filled rectangle, the CA must reach a final configuration in which the rectangular region is filled, i.e., all cells within the confines of the rectangle are in state 1, and all other cells are in state 0. Initial configurations consist of random-sized rectangles placed randomly on the grid (in our simulations, cells within the rectangle in the initial configuration were set to state 1 with probability 0.3; cells outside the rectangle were set to 0). Note that boundary cells can also be absent in the initial configuration. This operation can

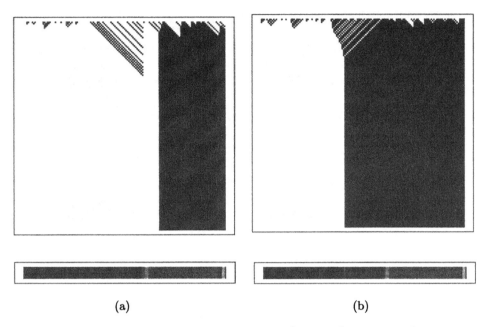

Fig. 5. One-dimensional ordering task: Operation of a co-evolved, non-uniform, $r = 1$ CA. Top figures depict space-time diagrams, bottom figures depict rule maps. (a) Initial density of 1s is 0.315, final density is 0.328. (b) Initial density of 1s is 0.60, final density is 0.59.

be considered a form of image enhancement, used, e.g., for treating corrupted images. Using cellular programming, non-uniform CAs were evolved with peak performance values of 0.99, one of which is depicted in Figure 6.

Upon studying the (two-dimensional) rules map of the co-evolved, non-uniform CA, we found that the grid is quasi-uniform, with one dominant rule present in most cells. This rule maps the cell's state to zero if the number of neighboring cells in state 1 (including the cell itself) is less than two, otherwise mapping the cell's state to one[1]. Thus, growing regions of 1s are more likely to occur within the rectangle confines than without.

4.5 The thinning task

Thinning (also known as skeletonization) is a fundamental preprocessing step in many image processing and pattern recognition algorithms. When the image consists of strokes or curves of varying thickness it is usually desirable to reduce them to thin representations located along the approximate middle of the original figure. Such "thinned" representations are typically easier to process in later stages, entailing savings in both time and storage space [9].

---

[1] This is referred to as a totalistic rule, in which the state of a cell depends only on the sum of the states of its neighbors at the previous time step, and not on their individual states [33].

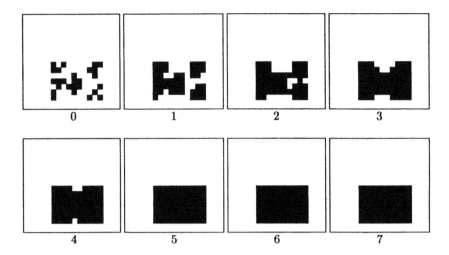

Fig. 6. Two-dimensional rectangle-boundary task: Operation of a co-evolved, non-uniform, 2-state, 5-neighbor CA. Grid size is $N = 225$ (15×15). Numbers at bottom of images denote time steps.

While the first thinning algorithms were designed for serial implementation, current interest lies in parallel systems, early examples of which were presented in [17]. The difficulty of designing a good thinning algorithm using a small, local cellular neighborhood, coupled with the task's importance has motivated us to explore the possibility of applying the cellular programming algorithm.

In [9] four sets of binary images were considered, two of which consist of rectangular patterns oriented at different angles. The algorithms presented therein employ a two-dimensional grid with a 9-cell neighborhood; each parallel step consists of two sub-iterations in which distinct operations take place. The set of images considered by us consists of rectangular patterns oriented either horizontally or vertically; while more restrictive than that of [9], it is noted that we employ a smaller neighborhood (5-cell) and do not apply any sub-iterations.

Figure 7 demonstrates the operation of a co-evolved CA performing the thinning task. Although the evolved grid does not compute perfect solutions, we observe, nonetheless, good thinning "behavior" upon presentation of rectangular patterns as defined above (Figure 7a); furthermore, partial success is demonstrated when presented with more difficult images involving intersecting lines (Figure 7b).

5 Discussion

A major impediment preventing ubiquitous computing with CAs stems from the difficulty of utilizing their complex behavior to perform useful computations. We presented the cellular programming algorithm for co-evolving computation in

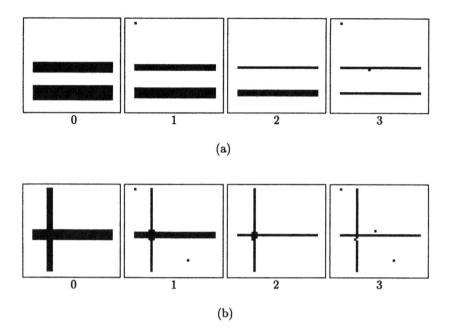

Fig. 7. Two-dimensional thinning task: Operation of a co-evolved, non-uniform, 2-state, 5-neighbor CA. Grid size is $N = 1600$ (40×40). Numbers at bottom of images denote time steps. (a) Two separate lines. (b) Two intersecting lines.

non-uniform CAs, demonstrating that high performance systems can be evolved for a number of non-trivial computational tasks. Our results suggest that non-uniformity reduces connectivity requirements, i.e., the use of smaller cellular neighborhoods is made possible.

An important issue when considering systems such as ours is that of scaling, where two separate matters are of concern: the evolutionary algorithm and the evolved solutions. As for the former, namely how does the evolutionary algorithm scale with grid size, we note that as our algorithm is local, it scales better in terms of hardware resources than the standard (global) genetic algorithm; adding grid cells requires only local connections in our case whereas the standard genetic algorithm includes global operators such as fitness ranking and crossover. The second issue is how can larger grids be obtained from smaller (evolved) ones, i.e., how can evolved solutions be scaled? This has been purported as an advantage of uniform CAs, since one can directly use the evolved rule in a grid of any desired size. However, this form of *simple* scaling does not bring about *task* scaling; as demonstrated, e.g., by [4] for the density task, performance decreases as grid size increases. For non-uniform CAs, quasi-uniformity may facilitate scaling since only a small number of rules must ultimately be considered. To date we have attained successful systems for some tasks using a simple scaling scheme involving the duplication of the rules grid; we are currently exploring a more sophisticated scaling approach, with preliminary encouraging results.

Our work to date suggests that the use of non-uniform CAs coupled with a local, co-evolutionary algorithm offers a number of advantages, including: (1) increased rule variability, thereby entailing easier "adaptation" to a possible change in the "environment", i.e., task, (2) easier implementation as evolware, (3) fault tolerance arising from the insensitivity to minor differences between cellular rules, and (4) better scalability (as noted above).

We found that markedly higher performance is attained for the density task with two-dimensional grids along with shorter computation times, as compared with one-dimensional grids. It is readily observed that a two-dimensional, locally connected grid can be embedded in a one-dimensional grid with local and distant connections. Since the density task is global, it is likely that the observed superior performance of two-dimensional grids arises from the existence of distant connections that enhance information propagation across the grid. This result has motivated the study of a modified model, involving the concomitant evolution of cellular rules and cellular connections. We have found that performance can be markedly increased for global computational tasks by co-evolving connectivity architectures [26, 27].

The nature of computation in CAs is a question of primary importance that has been gaining attention in recent years. We wish to enhance our understanding of the ways CAs perform computations, attempting to gain insight into the laws and mechanisms by which they operate. It is important to learn how CAs may be evolved, rather than designed, to perform computational tasks and what kinds of classes of tasks are most suited for such a computational paradigm. We seek to understand how evolution creates complex, global behavior in locally interconnected systems of simple parts. These goals are significant both from a scientific standpoint as well as from an applicative one.

Evolving, non-uniform CAs hold potential for studying phenomena of interest in areas such as complex systems, artificial life and parallel computation. This work has shed light on the possibility of computing with such CAs, and demonstrated the feasibility of their programming by means of co-evolution. We believe that cellular programming holds potential for attaining evolware which can be implemented in software, hardware, or other possible forms, such as bioware.

Acknowledgments

I am grateful to Daniel Mange, Eytan Ruppin, Eduardo Sanchez, and Marco Tomassini for helpful discussions.

References

1. A. Broggi, V. D'Andrea, and G. Destri. Cellular automata as a computational model for low-level vision. *International Journal of Modern Physics C*, 4(1):5–16, 1993.
2. E. F. Codd. *Cellular Automata*. Academic Press, New York, 1968.

3. J. P. Cohoon, S. U. Hedge, W. N. Martin, and D. Richards. Punctuated equilibria: A parallel genetic algorithm. In J. J. Grefenstette, editor, *Proceedings of the Second International Conference on Genetic Algorithms*, page 148. Lawrence Erlbaum Associates, 1987.

4. J. P. Crutchfield and M. Mitchell. The evolution of emergent computation. *Proceedings of the National Academy of Sciences USA*, 92(23):10742–10746, 1995.

5. R. Das, J. P. Crutchfield, M. Mitchell, and J. E. Hanson. Evolving globally synchronized cellular automata. In L. J. Eshelman, editor, *Proceedings of the Sixth International Conference on Genetic Algorithms*, pages 336–343, San Francisco, CA, 1995. Morgan Kaufmann.

6. R. Das, M. Mitchell, and J. P. Crutchfield. A genetic algorithm discovers particle-based computation in cellular automata. In Y. Davidor, H. -P. Schwefel, and R. Männer, editors, *Parallel Problem Solving from Nature- PPSN III*, volume 866 of *Lecture Notes in Computer Science*, pages 344–353, Berlin, 1994. Springer-Verlag.

7. P. Gacs. Nonergodic one-dimensional media and reliable computation. *Contemporary Mathematics*, 41:125, 1985.

8. M. Goeke, M. Sipper, D. Mange, A. Stauffer, E. Sanchez, and M. Tomassini. On-line autonomous evolware. In *Proceedings of The First International Conference on Evolvable Systems: from Biology to Hardware (ICES96)*, Lecture Notes in Computer Science. Springer-Verlag, Heidelberg, 1996.

9. Z. Guo and R. W. Hall. Parallel thinning with two-subiteration algorithms. *Communications of the ACM*, 32(3):359–373, March 1989.

10. G. Hernandez and H. J. Herrmann. Cellular-automata for elementary image-enhancement. *CVGIP: Graphical Models and Image Processing*, 58(1):82–89, January 1996.

11. L. Lamport and N. Lynch. Distributed computing: Models and methods. In J. Van Leeuwen, editor, *Handbook of Theoretical Computer Science, Volume B: Formal Models and Semantics*, pages 1159–1199. Elsevier, 1990.

12. B. Manderick and P. Spiessens. Fine-grained parallel genetic algorithms. In J. D. Schaffer, editor, *Proceedings of the Third International Conference on Genetic Algorithms*, page 428. Morgan Kaufmann, 1989.

13. M. Mitchell, J. P. Crutchfield, and P. T. Hraber. Dynamics, computation, and the "edge of chaos": A re-examination. In G. Cowan, D. Pines, and D. Melzner, editors, *Complexity: Metaphors, Models and Reality*, pages 491–513. Addison-Wesley, Reading, MA, 1994.

14. M. Mitchell, J. P. Crutchfield, and P. T. Hraber. Evolving cellular automata to perform computations: Mechanisms and impediments. *Physica D*, 75:361–391, 1994.

15. M. Mitchell, P. T. Hraber, and J. P. Crutchfield. Revisiting the edge of chaos: Evolving cellular automata to perform computations. *Complex Systems*, 7:89–130, 1993.

16. N. H. Packard. Adaptation toward the edge of chaos. In J. A. S. Kelso, A. J. Mandell, and M. F. Shlesinger, editors, *Dynamic Patterns in Complex Systems*, pages 293–301. World Scientific, Singapore, 1988.

17. K. Preston, Jr. and M. J. B. Duff. *Modern Cellular Automata: Theory and Applications*. Plenum Press, New York, 1984.

18. S. Rasmussen, C. Knudsen, and R. Feldberg. Dynamics of programmable matter. In C. G. Langton, C. Taylor, J. D. Farmer, and S. Rasmussen, editors, *Artificial Life II*, volume X of *SFI Studies in the Sciences of Complexity*, pages 211–254, Redwood City, CA, 1992. Addison-Wesley.

19. E. Sanchez and M. Tomassini, editors. *Towards Evolvable Hardware*, volume 1062 of *Lecture Notes in Computer Science*. Springer-Verlag, Berlin, 1996.

20. M. Sipper. Non-uniform cellular automata: Evolution in rule space and formation of complex structures. In R. A. Brooks and P. Maes, editors, *Artificial Life IV*, pages 394–399, Cambridge, Massachusetts, 1994. The MIT Press.

21. M. Sipper. An introduction to artificial life. *Explorations in Artificial Life (special issue of AI Expert)*, pages 4–8, September 1995. Miller Freeman, San Francisco, CA.

22. M. Sipper. Quasi-uniform computation-universal cellular automata. In F. Morán, A. Moreno, J. J. Merelo, and P. Chacón, editors, *ECAL'95: Third European Conference on Artificial Life*, volume 929 of *Lecture Notes in Computer Science*, pages 544–554, Berlin, 1995. Springer-Verlag.

23. M. Sipper. Studying artificial life using a simple, general cellular model. *Artificial Life Journal*, 2(1):1–35, 1995. The MIT Press, Cambridge, MA.

24. M. Sipper. Co-evolving non-uniform cellular automata to perform computations. *Physica D*, 92:193–208, 1996.

25. M. Sipper. Complex computation in non-uniform cellular automata, 1996. (submitted).

26. M. Sipper and E. Ruppin. Co-evolving architectures for cellular machines. *Physica D*, 1996. (to appear).

27. M. Sipper and E. Ruppin. Co-evolving cellular architectures by cellular programming. In *Proceedings of IEEE Third International Conference on Evolutionary Computation (ICEC'96)*, pages 306–311, 1996.

28. T. Starkweather, D. Whitley, and K. Mathias. Optimization using distributed genetic algorithms. In H. -P. Schwefel and R. Männer, editors, *Parallel Problem Solving from Nature*, volume 496 of *Lecture Notes in Computer Science*, page 176, Berlin, 1991. Springer-Verlag.

29. R. Tanese. Parallel genetic algorithms for a hypercube. In J. J. Grefenstette, editor, *Proceedings of the Second International Conference on Genetic Algorithms*, page 177. Lawrence Erlbaum Associates, 1987.

30. T. Toffoli and N. Margolus. *Cellular Automata Machines*. The MIT Press, Cambridge, Massachusetts, 1987.

31. M. Tomassini. The parallel genetic cellular automata: Application to global function optimization. In R. F. Albrecht, C. R. Reeves, and N. C. Steele, editors, *Proceedings of the International Conference on Artificial Neural Networks and Genetic Algorithms*, pages 385–391. Springer-Verlag, 1993.

32. M. Tomassini. A survey of genetic algorithms. In D. Stauffer, editor, *Annual Reviews of Computational Physics*, volume III, pages 87–118. World Scientific, 1995. Also available as: Technical Report 95/137, Department of Computer Science, Swiss Federal Institute of Technology, Lausanne, Switzerland, July, 1995.

33. S. Wolfram. Statistical mechanics of cellular automata. *Reviews of Modern Physics*, 55(3):601–644, July 1983.

34. S. Wolfram. Universality and complexity in cellular automata. *Physica D*, 10:1–35, 1984.

Online Autonomous Evolware

Maxime Goeke, Moshe Sipper, Daniel Mange, Andre Stauffer,
Eduardo Sanchez, and Marco Tomassini

Logic Systems Laboratory, Swiss Federal Institute of Technology, IN-Ecublens,
CH-1015 Lausanne, Switzerland. E-mail: {Name.Surname}@di.epfl.ch

Abstract. We present the *cellular programming* approach, in which parallel cellular machines evolve to solve computational tasks, specifically demonstrating that high performance can be attained for the *synchronization* problem. We then described an FPGA-based implementation, demonstrating that 'evolving ware', *evolware*, can be attained; the implementation is facilitated by the cellular programming algorithm's local dynamics. The machine's only link to the outside world is an external power supply, thereby exhibiting online autonomous evolution.

1 Introduction

The idea of applying the biological principle of natural evolution to artificial systems, introduced more than three decades ago, has seen an impressive growth in the past few years; usually grouped under the term *evolutionary algorithms* or *evolutionary computation*, we find the domains of genetic algorithms, evolution strategies, evolutionary programming, and genetic programming [1, 7, 8]. Research in these areas has traditionally centered on proving theoretical aspects, such as convergence properties, effects of different algorithmic parameters, and so on, or on making headway in new application domains, such as constraint optimization problems, image processing, neural network evolution, and more. The implementation of an evolutionary algorithm, an issue which usually remains in the background, is quite costly in many cases, since populations of solutions are involved coupled with computationally-intensive fitness evaluations. One possible solution is to parallelize the process, an idea which has been explored to some extent in recent years (see reviews by [3, 25]); while posing no major problems in principle, this may require judicious modifications of existing algorithms or the introduction of new ones in order to meet the constraints of a given parallel machine.

In this paper we consider the general issue of evolving machines; while this idea finds its origins in the cybernetics movement of the 1940s and the 1950s, it has recently resurged in the form of the nascent field of bio-inspired systems and evolvable hardware [14]. In what follows we present the *cellular programming* approach, in which parallel cellular machines evolve to solve computational tasks [17, 18, 19, 20, 21, 22]. We describe the algorithm and its hardware implementation, demonstrating that 'evolving ware', *evolware*, can be attained; while current evolware is hardware-based, future ware may include other forms, such as *bioware*. Our primary goal in this paper is to demonstrate that online autonomous evolware can be attained, which operates without any reference to an external device or computer; toward this end we shall concentrate on a specific, well-defined synchronization problem.

The machine model we employ is based on the cellular automata model. Cellular automata (CA) are dynamical systems in which space and time are discrete. They consist of an array of cells, each of which can be in one of a finite number of possible states, updated synchronously in discrete time steps according to a local, *identical* interaction rule. The *state* of a cell at the next time step is determined by the current states of a surrounding neighborhood of cells; this transition is usually specified in the form of a *rule table*, delineating the cell's next state for each possible neighborhood configuration [24, 26]. The cellular array (grid) is n-dimensional, where $n = 1, 2, 3$ is used in practice; in this work we shall concentrate on $n = 1$, i.e., one-dimensional grids.

CAs exhibit three notable features, namely massive parallelism, locality of cellular interactions, and simplicity of basic components (cells); thus, they present an excellent point of departure for our forays into the evolution of parallel cellular machines. The machine model we employ is an extension of the original CA model, termed *non-uniform cellular automata* [15]. Such automata function in the same way as uniform ones, the only difference being in the cellular rules that need not be identical for all cells.

The evolware implementation is based on FPGA (Field-Programmable Gate Array) technology. An FPGA circuit is an array of logic cells, laid out as an interconnected grid, with each cell capable of realizing a logic function [13]. The cells, as well as the interconnections, are programmable "on the fly", thus offering an attractive technological platform for realizing, among others, evolware.

In Section 2 we present previous work on evolving cellular machines; in particular, we present the synchronization problem, a non-trivial, global computational task. Section 3 delineates the cellular programming algorithm used to evolve non-uniform CAs; as opposed to the standard genetic algorithm, where a population of *independent* problem solutions *globally* evolves [8], our approach involves a grid of rules that *co-evolves locally*. In Section 4 we describe the FPGA-based evolware; evolution takes place within the machine itself, with no reference to or aid from any external device (e.g., a computer that carries out genetic operators) apart from a power supply, thus attaining *online autonomous evolware*. Finally, our conclusions are presented in Section 5.

2 Evolving parallel cellular machines

The application of genetic algorithms to the evolution of *uniform* cellular automata was initially studied by [12] and recently undertaken by the EVCA (evolving CA) group [4, 5, 6, 9, 10, 11]. They carried out experiments involving one-dimensional CAs with $k = 2$ and $r = 3$, where k denotes the number of possible states per cell and r denotes the radius of a cell, i.e., the number of neighbors on either side (thus, each cell has $2r + 1$ neighbors, including itself).

Spatially periodic boundary conditions are used, resulting in a circular grid. A common method of examining the behavior of one-dimensional CAs is to display a two-dimensional space-time diagram, where the horizontal axis depicts the configuration at a certain time t and the vertical axis depicts successive time steps (e.g., Figure 1). The term 'configuration' refers to an assignment of 1 states to several cells, and 0s otherwise.

The EVCA group employed a genetic algorithm to evolve uniform CAs to perform two computational tasks, density and synchronization, the latter of which we shall consider in this paper. In the synchronization task the CA, given any initial configuration, must reach a final configuration, within M time steps, that oscillates between all 0s and all 1s on successive time steps. As noted by [5], this is perhaps the simplest, non-trivial synchronization task. Oscillation is a global property of a configuration, whereas a small radius CA employs only local interactions; thus, while local regions of synchrony can be directly attained, it is more difficult to design CAs in which spatially distant regions are in phase. Since out-of-phase regions can be distributed throughout the lattice, transfer of information must occur over large distances (i.e., $O(N)$, where N is the grid size) to remove these phase defects and produce a globally synchronous configuration. [5] reported that in 20% of the evolutionary runs the genetic algorithm discovered CAs that successfully solve the task.

It is interesting to point out that the phenomenon of synchronous oscillations occurs in nature, a striking example of which is exhibited by fireflies; thousands such creatures may flash on and off in unison, having started from totally uncoordinated flickerings [2]. Each insect has its own rhythm, which changes only through local interactions with its neighbors' lights. Another interesting case involves pendulum clocks; when several of these are placed near each other, they soon become synchronized by tiny coupling forces transmitted through the air or by vibrations in the wall to which they are attached (for a review on synchronous oscillations in nature see [23]).

The model investigated in this paper is that of non-uniform CAs, where cellular rules need not be identical for all cells. Thus, rather than seek a *single* rule that must be applied universally to all cells in the grid, we allow each cell to "choose" its own rule through evolution. As we shall see, the removal of the uniformity constraint from the original CA model lends itself to a novel algorithm which is more amenable to implementation as evolware, in comparison to standard evolutionary algorithms [17, 18, 19, 20, 21, 22].

Using the cellular programming algorithm, delineated in the next section, we have shown that non-uniform CAs can be evolved to perfectly[1] solve the synchronization task. This is achieved with minimal radius, $r = 1$ CAs (i.e., each cell is connected to its two immediate left and right neighbors), as opposed to the aforementioned uniform CAs, where $r = 3$. We have shown that the performance

---

[1] The term 'perfect' is used here in a stochastic sense since we cannot exhaustively test all 2^{149} possible initial configurations nor are we in possession to date of a formal proof; nonetheless, we have tested our best-performance CAs on numerous configurations, for all of which synchronization was attained.

level attained by evolved, non-uniform, $r = 1$ CAs is better than *any* possible *uniform*, $r = 1$ CA, none of which can solve the synchronization problem. The evolved systems were observed to be *quasi*-uniform, meaning that the number of distinct rules is extremely small with respect to rule space size; furthermore, the rules are distributed such that a subset of dominant rules occupies most of the grid [16, 17]. Figure 1 demonstrates the operation of two co-evolved CAs along with the corresponding rule maps; these maps depict the distribution of rules by assigning a unique color to each distinct rule.

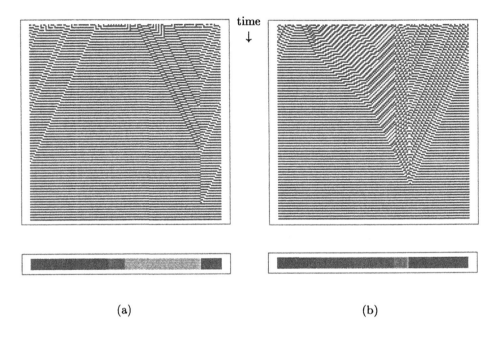

(a)   (b)

Fig. 1. The one-dimensional synchronization task: Operation of two co-evolved, non-uniform, 2-state CAs, with connectivity radius $r = 1$. Grid size is $N = 149$. White squares represent cells in state 0, black squares represent cells in state 1. The pattern of configurations is shown through time (which increases down the page). Top figures depict space-time diagrams, bottom figures depict rule maps.

3 The cellular programming algorithm

We study 2-state, non-uniform CAs, in which each cell may contain a different rule. A cell's rule table is encoded as a bit string, known as the "genome", containing the next-state (output) bits for all possible neighborhood configurations, listed in lexicographic order; e.g., for CAs with $r = 1$, the genome consists

```
for each cell i in CA do in parallel
    initialize rule table of cell i
    f_i = 0 { fitness value }
end parallel for
c = 0 { initial configurations counter }
while not done do
    generate a random initial configuration
    run CA on initial configuration for M time steps
    for each cell i do in parallel
        if cell i is in the correct final state then
            f_i = f_i + 1
        end if
    end parallel for
    c = c + 1
    if c mod C = 0 then { evolve every C configurations}
        for each cell i do in parallel
            compute nf_i(c) { number of fitter neighbors }
            if nf_i(c) = 0 then rule i is left unchanged
            else if nf_i(c) = 1 then replace rule i with the fitter neighboring rule,
                                 followed by mutation
            else if nf_i(c) = 2 then replace rule i with the crossover of the two fitter
                                 neighboring rules, followed by mutation
            else if nf_i(c) > 2 then replace rule i with the crossover of two randomly
                                 chosen fitter neighboring rules, followed by mutation
                                 (this case can occur if the cellular neighborhood includes
                                 more than two cells)
            end if
            f_i = 0
        end parallel for
    end if
end while
```

Fig. 2. Pseudo-code of the cellular programming algorithm.

of 8 bits, where the bit at position 0 is the state to which neighborhood config-uration 000 is mapped to and so on until bit 7 corresponding to neighborhood configuration 111. Rather than employ a *population* of evolving, uniform CAs, as with genetic algorithm approaches, our algorithm involves a *single*, non-uniform CA of size N, with cell rules initialized at random. Initial configurations are then generated at random, and for each one the CA is run for M time steps (in our simulations we used $M \approx N$ so that computation time is linear with grid size). Each cell's *fitness* is accumulated over $C = 300$ initial configurations, where a single run's score is 1 if the cell is in the correct state after $M + 4$ iterations, and 0 otherwise. The (local) fitness score for the synchronization task is assigned to each cell by considering the last four time steps (i.e., $[M + 1..M + 4]$); if the se-quence of states over these steps is precisely $0 \rightarrow 1 \rightarrow 0 \rightarrow 1$ (i.e., an alternation of 0s and 1s, starting from 0), the cell's fitness score is 1, otherwise this score is 0. After every C configurations evolution of rules occurs by applying crossover and mutation. This evolutionary process is performed in a completely *local* manner, where genetic operators are applied only between directly connected cells. It is driven by $nf_i(c)$, the number of fitter neighbors of cell i after c configurations. The pseudo-code of our algorithm is delineated in Figure 2.

Crossover between two rules is performed by selecting at random (with uniform probability) a single crossover point and creating a new rule by combining the first rule's bit string before the crossover point with the second rule's bit string from this point onward. Mutation is applied to the bit string of a rule with probability 0.001 per bit.

As opposed to the standard genetic algorithm, where a population of *independent* problem solutions *globally* evolves [8], our approach involves a grid of rules that *co-evolves locally* [17]. As noted in Section 1, the CA performs computations in a completely local manner, each cell having access only to its immediate neighbors' states; in addition, the *evolutionary process* in our case is local since application of genetic operators as well as fitness assignment takes place locally. This renders our approach more amenable to implementation as evolware, in comparison to other approaches, e.g., the standard genetic algorithm.

4 Implementing evolware

The cellular programming algorithm presented in Section 3 was studied extensively through software simulation; in this section we present its online, autonomous hardware implementation, resulting in evolving ware, evolware. To facilitate implementation, the algorithm is slightly modified (with no loss in performance); the two genetic operators, one-point crossover and mutation, are replaced by a single operator, *uniform crossover*. Under this operation, a new rule, i.e., an "offspring" genome, is created from two "parent" genomes (bit strings) by choosing each offspring bit from one or the other parent, with a 50% probability for each parent [8, 25]. The changes to the algorithm are therefore as follows (refer to Figure 2):

> **else if** $nf_i(c) = 1$ **then** replace rule i with the fitter neighboring rule,
>   *without mutation*
> **else if** $nf_i(c) = 2$ **then** replace rule i with the *uniform* crossover of the
>   two fitter neighboring rules, *without mutation*

The evolutionary process ends following an arbitrary decision by an outside observer (the '**while** not done' loop of Figure 2).

The cellular programming evolware is implemented on a physical board whose only link to the "outside world" is an external power supply. The features distinguishing this implementation from previous ones [14] are: (1) an ensemble of individuals (cells) is at work rather than a single one; (2) genetic operators are all carried out on-board, rather than on a remote, offline computer; (3) the evolutionary phase does not necessitate halting the machine's operation, but is rather intertwined with normal execution mode. These features entail an *online autonomous* evolutionary process.

The active components of the evolware board comprise exclusively FPGA (Field-Programmable Gate Array) circuits, with no other commercial processor whatsoever. An LCD screen enables the display of information pertaining to the evolutionary process, including the current rule and fitness value of each cell. The parameters M (number of time steps a configuration is run) and C (number of configurations between evolutionary phases, see Section 3) are tunable through on-board knob selectors; in addition, their current values are displayed. The implemented grid size is $N = 56$ cells, each of which includes, apart from the logic component, a LED indicating its current state (on=1, off=0), and a switch by which its state can be manually set[2]. We have also implemented an on-board global synchronization detector circuit, for the sole purpose of facilitating the external observer's task; this circuit is *not* used by the CA in any of its operational phases. A schematic diagram of the board is depicted in Figure 3.

The architecture of a single cell is shown in Figure 4. The binary state is stored in a D-type flip-flop whose next state is determined either randomly, enabling the presentation of random initial configurations, or by the cell's rule table, in accordance with the current neighborhood of states. Each bit of the rule's bit string is stored in a D-type flip-flop whose inputs are channeled through a set of multiplexors according to the current operational phase of the system:

1. During the initialization phase of the evolutionary algorithm, the (eight) rule bits are loaded with random values; this is carried out once per evolutionary run.
2. During the execution phase of the CA, the rule bits remain unchanged. This phase lasts a total of $C * M$ time steps (C configurations, each one run for M time steps).
3. During the evolutionary phase, and depending on the number of fitter neighbors, $nf_i(c)$ (Section 3), the rule is either left unchanged ($nf_i(c) = 0$), replaced by the fitter left or right neighboring rule ($nf_i(c) = 1$), or replaced by the uniform crossover of the two fitter rules ($nf_i(c) = 2$).

To determine the cell's fitness score for a single initial configuration, i.e., after the CA has been run for $M + 4$ time steps (Section 3), a four-bit shift register is used (Figure 5); this register continuously stores the states of the cell over the last four time steps ($[t + 1..t + 4]$). An AND gate tests for occurrence of the "good" final sequence (i.e., $0 \rightarrow 1 \rightarrow 0 \rightarrow 1$), producing the HIT signal, signifying whether the fitness score is 1 (HIT) or 0 (no HIT).

Each cell includes a fitness counter and two comparators for comparing the cell's fitness value with that of its two neighbors. Note that the cellular connections are entirely local, a characteristic enabled by the local operation of the cellular programming algorithm. In the interest of cost reduction, a number of resources have been implemented within a central control unit, including the

---

[2] This is used to test the evolved system after termination of the evolutionary process, by manually loading initial configurations.

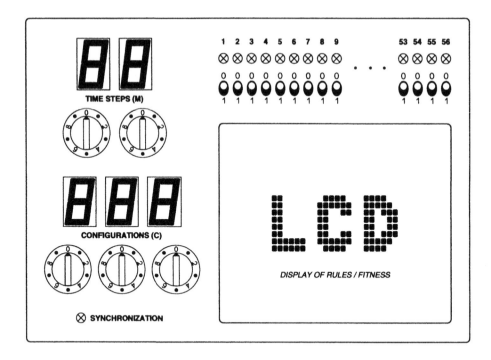

Fig. 3. Schematic diagram of evolware board: (1) LED indicators of cell states (upper right); (2) switches for manually setting the initial configuration (upper right, below LEDs); (3) display and knobs for controlling the M parameter (time steps) of the cellular programming algorithm (upper left); (4) display and knobs for controlling the C parameter (number of initial configurations between evolutionary phases) of the cellular programming algorithm (middle left); (5) synchronization indicator (lower left); (6) LCD display of evolved rules and fitness values (lower right).

random number generator and the M and C counters. Note that these are implemented *on-board* and do not comprise a breach in the machine's autonomous mode of operation.

The random number generator is implemented with a linear feedback shift register (LFSR), producing a random bit stream that cycles through $2^{32} - 1$ different values (the value 0 is excluded since it comprises an undesirable attractor). As a cell uses at most eight different random values at any given moment, it includes an 8-bit shift register through which the random bit stream propagates. The shift registers of all grid cells are concatenated to form one large stream of random bit values propagating through the entire CA. Cyclic behavior is eschewed due to the odd number of possible values produced by the random number generator ($2^{32} - 1$) and to the even number of random bits per cell.

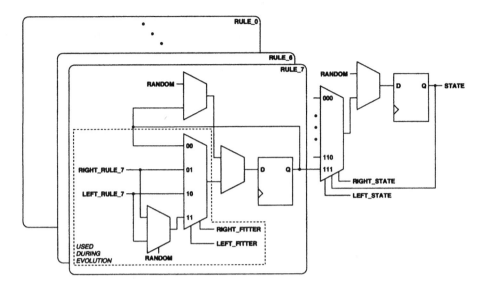

Fig. 4. Circuit design of a cell.

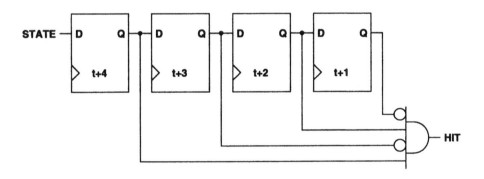

Fig. 5. Circuit used (in each cell) after execution of an initial configuration to detect whether a cell receives a fitness score of 1 (HIT) or 0 (no HIT).

5 Conclusions

In this paper we considered the general issue of evolving machines. We presented the cellular programming approach, in which parallel cellular machines evolve to solve computational tasks, specifically demonstrating that high performance can be attained for the synchronization problem. We then described an FPGA-based implementation, demonstrating that 'evolving ware', *evolware*, can be attained;

the implementation was facilitated by the cellular programming algorithm's local dynamics. The machine's only link to the outside world is an external power supply, thereby exhibiting online autonomous evolution.

Evolving, cellular machines hold potential both scientifically, as vehicles for studying phenomena of interest in areas such as complex adaptive systems and artificial life, as well as practically, showing a range of potential future applications ensuing the construction of adaptive systems. Our primary goal in this paper was to demonstrate that online autonomous evolware can be attained, toward which end we concentrated on a specific, well-defined problem. The success of our system raises the possibility of constructing more complex evolware, able to tackle real-world problems, that call for adaptive behavior.

References

1. T. Bäck. *Evolutionary algorithms in theory and practice: evolution strategies, evolutionary programming, genetic algorithms.* Oxford University Press, New York, 1996.

2. J. Buck. Synchronous rhythmic flashing of fireflies II. *The Quarterly Review of Biology*, 63(3):265–289, September 1988.

3. E. Cantú-Paz. A summary of research on parallel genetic algorithms. Technical Report 95007, Illinois Genetic Algorithms Laboratory, University of Illinois at Urbana-Champaign, Urbana, IL, July 1995.

4. J. P. Crutchfield and M. Mitchell. The evolution of emergent computation. *Proceedings of the National Academy of Sciences USA*, 92(23):10742–10746, 1995.

5. R. Das, J. P. Crutchfield, M. Mitchell, and J. E. Hanson. Evolving globally synchronized cellular automata. In L. J. Eshelman, editor, *Proceedings of the Sixth International Conference on Genetic Algorithms*, pages 336–343, San Francisco, CA, 1995. Morgan Kaufmann.

6. R. Das, M. Mitchell, and J. P. Crutchfield. A genetic algorithm discovers particle-based computation in cellular automata. In Y. Davidor, H. -P. Schwefel, and R. Männer, editors, *Parallel Problem Solving from Nature- PPSN III*, volume 866 of *Lecture Notes in Computer Science*, pages 344–353, Berlin, 1994. Springer-Verlag.

7. Z. Michalewicz. *Genetic algorithms + data structures = evolution programs.* Springer, Berlin, third edition, 1996.

8. M. Mitchell. *An Introduction to Genetic Algorithms.* MIT Press, Cambridge, MA, 1996.

9. M. Mitchell, J. P. Crutchfield, and P. T. Hraber. Dynamics, computation, and the "edge of chaos": A re-examination. In G. Cowan, D. Pines, and D. Melzner, editors, *Complexity: Metaphors, Models and Reality*, pages 491–513. Addison-Wesley, Reading, MA, 1994.

10. M. Mitchell, J. P. Crutchfield, and P. T. Hraber. Evolving cellular automata to perform computations: Mechanisms and impediments. *Physica D*, 75:361–391, 1994.

11. M. Mitchell, P. T. Hraber, and J. P. Crutchfield. Revisiting the edge of chaos: Evolving cellular automata to perform computations. *Complex Systems*, 7:89–130, 1993.

12. N. H. Packard. Adaptation toward the edge of chaos. In J. A. S. Kelso, A. J. Mandell, and M. F. Shlesinger, editors, *Dynamic Patterns in Complex Systems*, pages 293–301. World Scientific, Singapore, 1988.

13. E. Sanchez. Field programmable gate array (FPGA) circuits. In E. Sanchez and M. Tomassini, editors, *Towards Evolvable Hardware*, volume 1062 of *Lecture Notes in Computer Science*, pages 1–18. Springer-Verlag, Berlin, 1996.

14. E. Sanchez and M. Tomassini, editors. *Towards Evolvable Hardware*, volume 1062 of *Lecture Notes in Computer Science*. Springer-Verlag, Berlin, 1996.

15. M. Sipper. Non-uniform cellular automata: Evolution in rule space and formation of complex structures. In R. A. Brooks and P. Maes, editors, *Artificial Life IV*, pages 394–399, Cambridge, Massachusetts, 1994. The MIT Press.

16. M. Sipper. Quasi-uniform computation-universal cellular automata. In F. Morán, A. Moreno, J. J. Merelo, and P. Chacón, editors, *ECAL'95: Third European Conference on Artificial Life*, volume 929 of *Lecture Notes in Computer Science*, pages 544–554, Berlin, 1995. Springer-Verlag.

17. M. Sipper. Co-evolving non-uniform cellular automata to perform computations. *Physica D*, 92:193–208, 1996.

18. M. Sipper. Designing evolware by cellular programming. In *Proceedings of The First International Conference on Evolvable Systems: from Biology to Hardware (ICES96)*, Lecture Notes in Computer Science. Springer-Verlag, Heidelberg, 1996.

19. M. Sipper and E. Ruppin. Co-evolving architectures for cellular machines. *Physica D*, 1996. (to appear).

20. M. Sipper and E. Ruppin. Co-evolving cellular architectures by cellular programming. In *Proceedings of IEEE Third International Conference on Evolutionary Computation (ICEC'96)*, pages 306–311, 1996.

21. M. Sipper and M. Tomassini. Co-evolving parallel random number generators. In H. -M. Voigt, W. Ebeling, I. Rechenberg, and H. -P. Schwefel, editors, *Parallel Problem Solving from Nature - PPSN IV*, volume 1141 of *Lecture Notes in Computer Science*, pages 950–959. Springer-Verlag, Heidelberg, 1996.

22. M. Sipper and M. Tomassini. Generating parallel random number generators by cellular programming. *International Journal of Modern Physics C*, 7(2):181–190, 1996.

23. S. H. Strogatz and I. Stewart. Coupled oscillators and biological synchronization. *Scientific American*, pages 102–109, December 1993.

24. T. Toffoli and N. Margolus. *Cellular Automata Machines*. The MIT Press, Cambridge, Massachusetts, 1987.

25. M. Tomassini. Evolutionary algorithms. In E. Sanchez and M. Tomassini, editors, *Towards Evolvable Hardware*, volume 1062 of *Lecture Notes in Computer Science*, pages 19–47. Springer-Verlag, Berlin, 1996.

26. S. Wolfram. Universality and complexity in cellular automata. *Physica D*, 10:1–35, 1984.

Speeding-up Digital Ecologies Evolution Using a Hardware Emulator: Preliminary Results

Pierre Marchal[1], Pascal Nussbaum[1], Christian Piguet[1]
Moshe Sipper[2]

[1] CSEM Centre Suisse d'Electronique et de Microtechnique SA, Neuchâtel
[2] Laboratoire de Systèmes Logiques, Swiss Federal Institute of Technology, Lausanne

"The creatures cruise silently, skimming the surface of their world with the elegance of ice skaters. They move at varying speeds, some with the variegated cadence of vacillation, others with what surely must be firm purpose."

Steven Levy - Artificial Life - The Quest for a New Creation

Abstract. For a more than a decade, the idea of applying the biological principle of natural evolution to artificial systems in order to create or to improve digital ecologies has emerged from different laboratories. During the past couple of years, a new trend consists in applying these investigations to hardware design. This concept is called "Evolvable Hardware". For this quest, hardware emulation offers an alternative approach to the development of a generic evolvable system including fitness evaluation. Compared to a software solution, emulation can be on the order of a million times faster which is of higher interest when billion steps of evolution are necessary. A further advantage of emulation is to provide the description of the VLSI to be implemented as well as a validation of its behavior.

In this paper, we describe the way followed to implement the system (cellular automata and the surrounding evolutionary control logic) as a hardware description in an emulator. For different examples presented in this paper, reasonable with respect to simulation, processing time of hardware emulation versus software simulation are compared. The time saved by hardware emulation has given the opportunity to increase the complexity of the "evolving organism" by including the selection of intervening neighbors in the parameter selected by evolution.

1 Introduction

Recent advances in the field of Computer Engineering (circuit synthesis, programmable devices and artificial ecologies) as well as in Molecular Biology (embryology, genetics and immune systems) combined with a better understanding of dynamical systems have paved the way of re-breathing life into the old dream of constructing biological-like machines. This theme, first raised almost fifty years ago by one of the founding father of cybernetics, John von Neumann, is based on the concepts of self-reproduction and self-repair (von Neumann 1966). Unfortunately, the technologies available at the time as well as the "molecular level" he had addressed (Marchal 1994) was far removed from that necessary to implement his idea. The remarkable increase in computional power and more recently, the appearance of a new generation of programmable logic devices, i.e., Field Programmable Gate Arrays (Brown 1992, Moore 1991, Moore 1994), have made it possible to couple genetic encoding and artificial evolution. We have hence reached and crossed a technological barrier, beyond which we no longer need to content ourselves with traditional approaches to engineering design, we rather can now evolve machines to attain the desired behavior (Sanchez 1996).

Our main focus, in this paper, is to increase the speed of "digital ecologies" evolution. As a consequence, the saved time has been used to increase the complexity of the "evolving organism", so the selection of intervening neighbors has been added as a new possible parameter in the evolution process. Hence, the evolution process acts on the transition table and on the neighborhood as well. Preliminary results shown in this paper address the evolution of non-uniform cellular automata (CA) to perform computational algorithms, an approach referred to as cellular programming (Sipper 1994). Section 2 reviews the investigations performed on digital ecologies evolution. Section 3 briefly describes the advantages offered by the hardware emulation versus software simulation. Section 4 presents the necessary steps to implement hardware evolution on the emulator. Section 5 describes preliminary results before concluding remarks of section 6.

2 Evolving Digital Ecologies

Is it possible to actually evolve a real creature? To start with some inert lump of information and, compressing billions of years of activity into something a bit more manageable (a night, a week, even a year) to wind up with life? Can one indeed follow the path apparently taken on earth, so that something as simple as a bacterium could make its way up the evolutionary ladder into something as complex as a multicellular organism? These are some of the questions that lead research investigations on evolving digital ecologies.

For more than a decade, the idea of applying the biological principle of natural evolution to artificial systems in order to create or to improve digital ecologies has emerged from different laboratories. Three schools of thought may be distinguished. The very first one, related to the evolution of one individual, is led by Stuart Kauffman (Kauffman 1993, Kauffman 1995). This research was first oriented towards the emergence of life on earth. Kauffman's approach is driven by the occurrence of self-organization, i.e., the spontaneous emergence of order: a type of energy-sink in which an ergodic dynamical system will fall after a certain transient period. For instance, if oil-droplets in water manage to be spherical, or if snowflakes assume their evanescent sixfold symmetry: physiochemical reasons must be invoked, none of these effects have anything to do with natural selection. This research led him to consider that complex systems, poised on the boundary between order and chaos, are the ones best fit to adapt by mutation and selection. Such systems appear not only to be able to coordinate complex and flexible behavior but also to respond to changes in their environment. Kauffman pointed out that promising indications that linked coevolving complex systems are led by selection to form ecosystems whose members mutually attain the edge of chaos.

The second school of thought, related to the evolution of the species, is led by Thomas Ray (Ray 1992, Thearling 1994). Ray models his system on a later stage in life's development, the explosion of biological diversity that signaled the onset of the Cambrian Era, roughly six hundred million years ago. From a relative paucity of phyla, the earth teemed with unprecedented new life forms. Ray has developed an ecological system, called Tierra, in which computer programs (digital ecologies) compete for survival in a "physical" environment consisting of the energy resource (CPU time) and the memory space. The implicit fitness function favors the evolution of creatures which are able to replicate with less CPU time. However, must of the evolution consists of creatures discovering ways to exploit one another.

The third school of thought lies somewhere in between the previous ones; it comes after the appearance of life on earth, but long before the Cambrian Era. It nearly corresponds to the emergence of multicellular organisms by gathering single cells into a colony, something starting with the symbiosis phenomenon. The activists in this field include M. Mitchell, (Mitchell 1993, 1994, 1996), J.P. Crutchfield (Crutchfield 1995) working on uniform one-dimensional CAs, and M. Sipper (Sipper 1994, 1995, 1996a) working on non-uniform (heterogeneous) CAs. In order to realize computational tasks of the same complexity, an homogeneous environment implies a larger neighborhood. So heterogeneity enables to decrease the size of the neighborhood and hence to decrease the amount of both computation time and memory space.

2.1 Evolvable Hardware

During the past years, a new trend consists in applying these investigations to hardware design. This concept is called "Evolvable Hardware" (Hemmi 1994, Higuchi 1995). Evolution may be realized on-line or off-line. In the on-line hardware evolution, each individual is an autonomous physical entity, ideally capable of modifying itself; this occurs as a result of directly sensing feedback signals communicated by a suitable physical environment and possibly by other members of a population of similar entities. In the off-line case, evolution design is carried out as a software simulation, with the resulting satisfactory solution (design) used to configure the programmable hardware. To date, on-line evolution presents practical difficulties and the genetic operations (selection, mutation, re-combination) as well as fitness evaluation are usually performed off-line in software. This paper, together with its companion paper (Sipper 1996b) present two different ways of implementing truly on-line evolution.

2.2 ASIC design phase

During the design phase of an Application Specific Integrated Circuits (ASIC), after the schematics have been created and captured with an appropriate CAD-tool, some functional verifications are necessary to ensure that the circuit correctly implements the given specs. Most often, this verification phase is purely virtual since it takes the form of simulation. Behavioral models of gates, flip-flops and all the other primitive components of the design, along with the netlist describing their interconnections, are analyzed and checked by the circuit simulator program. Special files describing the input stimuli as well as awaited responses are introduced so the simulator is able to check circuit dynamics and report (graphically) behavioral discrepancies as a function of time.

2.3 Improving simulation by using emulation

Hardware emulation offers an alternative approach to function verification that can be on the order of a million times faster than simulation, which is of higher interest when billion steps of evolution are necessary. A hardware emulator contains a large pool of programmable devices. General purpose logic functions can be configured and interconnected to exactly match the functional behavior of a given design. Since hardware emulation is by essence a hardware implementation of the design, each and every part runs concurrently, hence leading to a solution much faster than simulation. Although an emulator does not provide a gate-to-gate implementation of the design (it produces a logical equivalent implementation), it provides a more important advantage: the description of the VLSI to be implemented as well as a validation of its behavior.

3 Hardware Emulator

The Meta-Systems Simexpress is a new original digital emulation solution, for which full custom circuits (called Meta) have been designed to optimize the mapping and routing of the netlists to be emulated. The emulator acts like a giant FPGA on which the circuit, to be tested and debugged, can be mapped.

3.1 Emulator description

The emulator is based on a building bloc called BLP (French acronym for Programmable Logic Block), which resembles some FPGA solutions but has been optimized for emulation. It contains a 16-bit LUT which allows 4-input 1-output logic functions to be configured. Each Meta-chip contains 128 BLPs, with the necessary control and interconnect logic. The boards of the emulator (logic cards) contain 24 times the trio composed of one Meta-chip, one 32k 8-bit static RAM and one 1Mbit VRAM. The static RAMs provide possibilities to map memories described in the netlist. The VRAMs sample all the internal nodes for logic analysis of the netlist. A logic card allows the mapping of around 20kgates. The emulator itself is composed of 1 to 6 racks. Each rack contains up to 23 logic cards. Various other cards are provided in addition, like I/O cards (enabling external connections for on-board emulation), memory cards (emulation of huge memories of up to 64Mbytes/card), and prototype cards used when special logic has to be inserted into the emulator. The version used at CSEM is a 2 racks emulator equipped with 31 logic-cards plus one 336-p I/O board. This gives a total amount of nearly 600 kgates.

3.2 Compilation

The emulator must be configured like a huge FPGA. The global configuration of the emulator is created by compilation of an ANF netlist (ANF netlist language allows designers to describe hierarchical designs). The netlist relies on four objects (models, instances of models, signals and connectors). In our experiments we have exclusively made use of the primitives provided with the machine: Meta-lib (Meta-lib is a library of logic gates and logic blocks for which a direct mapping on the Meta-chip is provided) or Meta-memories (Meta-memories enable the user to make use of memories available on each board). The compilation tool, called XMCI, handles the ANF netlist, analyzes it, re-synthesizes and optimizes it in different ways according to the user's need. Optimizations are of 2 types: area and delay. Three levels of optimization are available for each of them. Finally, XMCI computes the maximum emulation frequency of the design. This feature is possible because inter-gate delays are fixed: between 2 gates internal to a chip, between 2 gates belonging to different chips on the same board, between 2 gates belonging to different chips on the same rack and between 2 gates belonging to 2 different racks. So knowing the size of the design and how it is downloaded on the machine, makes it is possible to compute the maximum emulation frequency.

3.3 Emulation

Emulation is performed using the MetaSystem Emulation Language (MEL) tool, which loads the emulator with the configuration file, and allows to run control, to perform logic analysis, to fine tune triggering features, and to create the necessary files needed for patterns verification. MEL can be driven by procedures written in a C-like code, for complex control with repetitive operations. All the signals or vectors (busses) to be probed can be displayed in a waveform. Input control can be done through monitors, where any vector or signal can be displayed and modified. To fill-up these two forms, a navigator gives hierarchical access to any node or instance of the netlist. Each node can be sampled, without recompilation. As a consequence of optimization, some nodes may automatically be removed by XMCI. To avoid this, a Meta device, called Meta Visibility, is added in the netlist. It can be connected in the original schematic to force the node to appear in the final netlist. Debugging the system must be done at a maximum 1 MHz frequency, due to the limitations of the VRAMs handling the probing during digital analysis. Indeed, standard 1 Mbit VRAM serial pipes work at 32 MHz on 4 channels (128 Mbit/s). Hence, the maximum speed to sample 128 BLPs outputs is 1 MHz.

4 First Experiments

In a first investigation phase, we have addressed the co-evolution of a cellular automaton to perform computations and apply it to different computational tasks: density, synchronization and sorting. The goal is to let the global function emerge from local interactions. Evolution consists of modifying the transition function of each automaton according to the local fitness - adequacy of each cell with respect to the awaited response. Simulating this kind of automata can be considered a complex task, since the number of necessary evolution cycles is too huge for a reasonable amount of time. All explanations concerning the simulation of these tasks may be found in the companion paper (Sipper 1996b).

4.1 Dynamics and evolution

The first step in this domain has been to apply genetic algorithms to a uni-dimensional, 256 cells, 4-state cellular automata. The initial state is loaded at the beginning as an input pattern (Cf. Fig. 1).

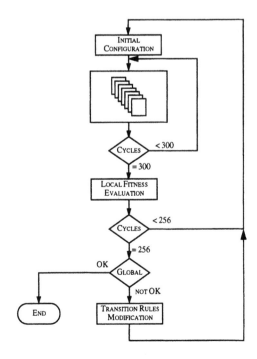

Fig. 1. Block scheme of the dynamic process

We then let the system go through its dynamics for a period at least as long as the network's size. This is the mean time necessary to leave the transients due to the circular bordering conditions we have chosen. So after 300 iterations, the pattern has changed, depending on the initial value of the cell and on the transition rule of each cell (randomly choosen at start). This "final state[1]" is considered as the result of the function performed by the CA network. The final state of each cell is compared with the expected result. If they are equal, the fitness value of the cell is incremented, else the value remains untouched. This experiment is repeated for 256 times (over 8k possible initial configurations), providing each cell with a different fitness value. The evaluation of hundreds of initial configurations, guarantees not to be stuck in some local minimum corresponding to a precise initial seed and given transition rules. After these experiments, we can be confident in the value shown by the fitness evaluator located in each cell. Here takes place the evolution by production of a new transition rule. The genetic algorithm changes the transition rule of each cell according to this definition:

1. If the fitness of the current cell is higher or equal to the fitness of its two neighbours, then the transition rule remains the same.
2. If one and only one cell has a better fitness value, then its transition rule is completely copied into the current cell transition rule.
3. If the two neighbours have a better fitness value, then their rules are copied randomly from one or the other into the cell (this operation is called cross-over).

It is clear that simulation of such kind of system requires huge amounts of computation time: thousands of iteration steps for hundreds of automata running concurrently. This is the major reason why emulation has been chosen instead of simulation to perform the evaluation. After these first computational tasks (leading to fixed-point attractors for which two successive states are sufficient), we have addressed a counting task, for which the first neighborhood was no longer sufficient, and for which we were able to choose the length (greater than 2) of the cyclic attractor.

4.2 System description

Figure 2 shows the main constituent of the system. It consists of:

1. The sequencer, called EVOLVER, responsible of both dynamics and evolution. It includes the INITIALIZATION control block, the DYNAMICS control block, the FITNESS control block, the EVOLUTION control block.
2. The SEEDS memory. It stores a set of possible seeds for the evolution. This strategy enable to repeat the same experimentation more than once.

---

[1] Note that this final state, should be stable for a convenient period of time. Convenient means that, depending on the complexity of the awaited result, it can be 2 iterations for fixed-point attractor or a complete cycle plus one iteration for a cyclic attractor.

3. The awaited response memory, called RESULTS. For each seed stored in the seed memory, this memory provides the corresponding results.

4. The noise memory, called RANDOM. It is used to store random patterns used for the genetic operation of cross-over[2]. This strategy enables to repeat experiment with the same random numbers.

5. The array of cells. It contains 256 instantiations of the cell schematic (Fig. 3).

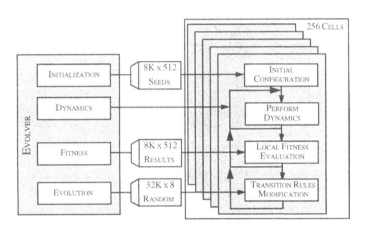

Fig. 2. Top-level schematic of the system

Figure 3 depicts the block scheme of one element of the cell array. Each cell computes its new state (S_{T+1}) by using its current state (S_T) and the one of the nearest neighbors (left and right), to address a look-up table (LUT). This LUT provides the next state of the considered cell. The set of values stored in this LUT, that define the dynamics of the automata, is called TRANSITION RULES. Means are provided to check if the final state and the AWAITED RESULTS are identical. In this case, the FITNESS COUNTER is incremented. The values of the FITNESS COUNTER is then used to modify the TRANSITION RULES.

4.3 Compilation time and speed

In our case, the compilation gave a result file filling between 20 and 24 logic cards (all kinds and levels of optimization scanned). The emulation frequencies were between 1MHz and 1.8Mhz. The only problem, coming from the hundreds of memory used, was the inability to route connections for some versions of the schematic. In fact, the memory address bus had to be propagated through all the 256 cells, consuming a lot of interconnection resources. A wide range of options allows to fine control parameters for compilation. We reduced the filling factor,

---

[2] The mutation has not been implemented in this very first experiments. Hence, it may be possible to get stuck in a local minimum

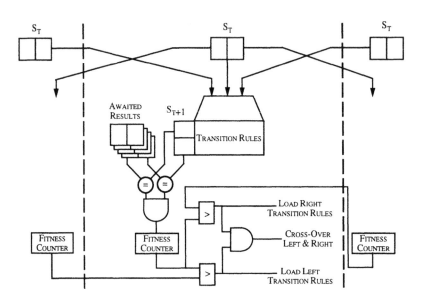

Fig. 3. Cell's schematic

which determines the level of compaction of parts into a Meta-chip. This operation gave improvements. A problem that remains is that a network of iterative cells quickly requires extensive interconnection resources, due to the necessity of having global nets. This necessity of having global nets was a consequence of our choice to be able to repeat the same experiments with same initial values and same random numbers. A solution, using LFSRs (Linear Feedback Shift Registers), could easily overcome this problem. The compilation time oscillated between 30 and 40 minutes.

4.4 Counting/ Macro-automaton

As mentioned previously, all results concerning the tasks of density, sorting and corresponding discussions may be found in the companion paper (Sipper 1996b). We just briefly report the result of the synchronization task as a starting point to our investigations.

Figure 4 demonstrates the operation of a co-evolved, non-uniform, $r = 1$ CA. White squares represent cells in state 0, black squares represent cells in state 1. The pattern of configurations is shown through time (which increases down the page). Note that upon presentation of a random initial configuration the grid converges to an oscillating pattern, alternating between an all-0 configuration and an all-1 one; this period-2 cycle may be considered a 1-bit counter. Building upon this evolved CA, 2- and 3-bit counters can be constructed, as demonstrated in Fig. 5 and 6.

Figure 5 depicts the operation of a one-dimensional synchronization task: a 2-bit counter. The resulting non-uniform CA converges into a period-4 cycle upon

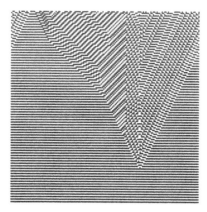

Fig. 4. The one-dimensional synchronization task: 1-bit counter

presentation of a random initial configuration. Due to memory requiremeents problems, the software solution is based on a non-uniform, 2-state CA, with connectivity radius $r = 2$, derived from the co-evolved, $r = 1$ CA, while the hardware implementation directly uses 4-state CA, with connectivity radius $r = 1$. The software implementation is achieved by "interlacing" two $r = 1$ CAs, in the following manner: Each cell in the $r = 1$ CA is transformed into an $r = 2$ cell, two duplicates of which are placed next to each other (the resulting grid's size is thus doubled). This transformation is carried out by "blowing up" the $r = 1$ rule table into an $r = 2$ one, creating from each of the (eight) $r = 1$ table entries four $r = 2$ table entries, resulting in the 32-bit $r = 2$ rule table. For example, entry $110 \rightarrow 1$ specifies a next-state bit of 1 for an $r = 1$ neighborhood of 110 (left cell is in state 1, central cell is in state 1, right cell is in state 0). Transforming it into an $r = 2$ table entry is carried out by "moving" the adjacent, distance-1 cells to a distance of 2, i.e., $110 \rightarrow 1$ becomes $1X1Y0 \rightarrow 1$; filling in the four permutations of (X, Y), i.e., $(X, Y) = (0,0), (0,1), (1,0), (1,1)$, results in the four $r = 2$ table entries. The clock of the odd numbered cells functions twice as fast as that of the even-numbered cells; this means that the latter update their states every second time step with respect to the former.

Figure 6 shows the operation of a one-dimensional synchronization task: a 3-bit counter. The resulting non-uniform CA converges into a period-8 cycle upon presentation of a random initial configuration. The software solution is based on a 2-state CA, with connectivity radius $r = 3$, derived from the co-evolved, $r = 1$ CA. This is achieved by "interlacing" three $r = 1$ CAs (thus, the grid size is multiplied by 3), in a similar manner to that used for obtaining the 2-bit counter. The clock of cells 0, 3, 6, ... functions normally, that of cells 1, 4, 7, ... is divided by two (i.e., these cells change state every second time step with respect to the "fast" cells), and the clock of cells 2, 5, 8, ... is divided by four (i.e., these cells change state every fourth time step with respect to the fast cells).

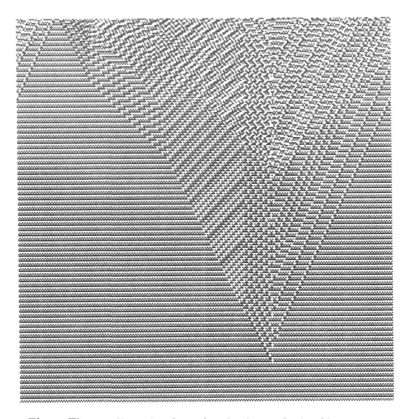

Fig. 5. The one-dimensional synchronization task: A 2-bit counter

5 Results

The original idea of co-evolving cellular-automata to perform complex tasks is due to Moshe Sipper (Sipper 1994, and main of his most recent investigation may found in the companion paper 1996b. He has performed his experiments by simulating uni-dimensional (and bi-dimensional) networks of 150 two-state automata. His simulations are running on UltraSparc 1 workstation. He programmed a 256 cells network and adapted the generation cycle to fit our working conditions. On an UltraSparc 1 workstation, the time consumed by 1 generation is about 60 to 70 seconds. As any part off the process is software simulated, displaying results is not a considerable effort, which is not the case with hardware emulation.

5.1 Handling crude data

Althought the Meta-Box allows ASIC designers and "evolutionist controlers" to share the same hardware, with respect to the data to be handled as well as the way they are handled, their repective points of view completely differ. For an ASIC builder, a glance at some waveform is sufficient to his verifying task (Cf.

Fig. 6. The one-dimensional synchronization task: A 3-bit counter

Fig. 7). Conversely, the "evolutionist" absolutely needs to go throught a complete sequence of "snapshots" targetting the cell states. In order to control the evolution of the CA network, two types of data are necessary to be considered:

1. The history of a pattern along the working cycles of the automata (from 256 iterations up to 512 depending on the size of the cell array (Cf. Fig. 4 – 6).
2. The content of the 256 transition rules memories (in order to compare the evolution all-over the different generations).

The first type of data is generated by launching 256 times the emulator for one step, and reading the states of the 256 cells. A MEL procedure (DumpPattern) has been written to perform these operations. Thousands of values displayed along the waveform do not give any pertinent information (Cf. Fig. 7). So we choose to write a viewing tool able to display pixelmap files, and to create such a file in the MEL procedure.

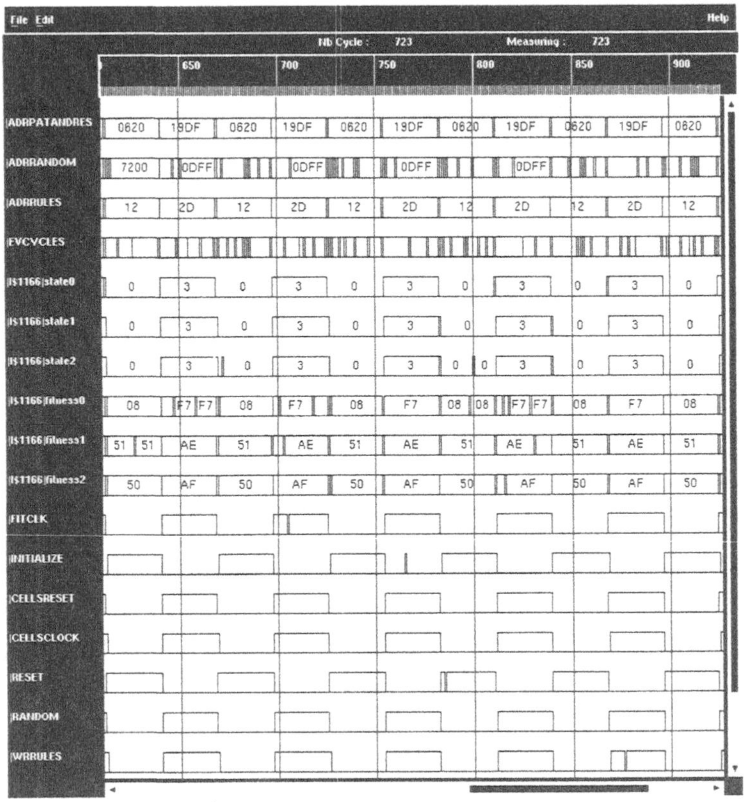

Fig. 7. Waveform showing probed signals and vectors

The second type of data requires to dump all the rules memories into a file. Again, a bidimensionnal representation being needed for clarity, the values are read into the memories and dumped into a pixelmap file by a MEL procedure (DumpRules). When both history and rules patterns are displayed together, they give a good idea of the behavior of each cell and a good representation of the rules transfer among the generations cycles. Figure 8 shows such data extracted during a run. In this figure, the trend for a cell to copy fitter neighbor rules appears clearly in the bottom part. Between regions of same rules, interfaces appear, that show the cross-over behavior (copy randomly the rules of the two neighbors if they are both fitter). In this figure, the function to be fitted is density. The result (top part) after 100 generations is not perfect yet, but the tendency is correct.

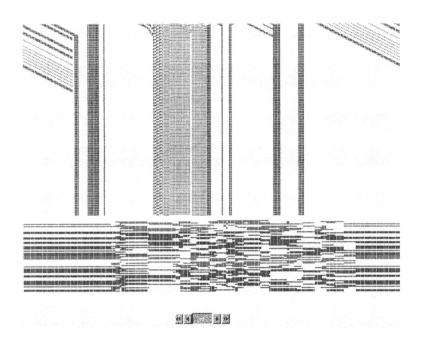

Fig. 8. Pattern execution and rules state after 100 generations

The full cycle (generation) of the system in the emulator is done after 66048 clock steps:

1. 256 working cycles are executed for each one of 256 patterns.
2. Two additional cycles are used at the end of each pattern execution, to increment the fitness of the cells which have the right result.
3. 256 cycles are needed to perform the cross-over of the 64 locations of the rules memories.

The last version of the design which has been compiled gave a limit frequency of 1.362MHz. So a generation takes 50ms. Consequently, in our case, the speed-up ratio between emulation and simulation is at least 1200 (Cf. Table 1). However, the system has been simulated and optimized at high level, and this for a 2-state automaton (ours is 4-state). For standard digital circuit simulations, like those of Quicksim, an additional time factor of 10^2 to 10^3 is expected, that will push up the ratio to near a million. Finally, in emulation, increasing the number of cells or the number of states doesn't reduce the speed, contrary to simulation.

Table 1. Speed-up factor (Emulation versus Simulation)

Criterion	Emulation S500M	Simulation Ultra-1	Ration E/S
Compilation time	30 min	40 s	1/45
One generation cycle	50 ms	60 s	1200

6 Conclusions

In this paper, we have shown that the remarquable increase in computational power and the new generation of FPGAs have made it possible to couple gnetic encoding and artificial evolution. Furthermore, instead of perfoming off-line evolution using software to realize the genetic operations, we have opportunistically taken benefit from the nature of the programming devices to create and implement the evolution tool itself. The system has been designed and tested (debugged) and interesting results have emerge from these preliminary experiments. This evolution tool reveals to be particularly efficient in computation time. A speed-up ratio of about 1200 has been one of our key results. A second adavantage oriented towards the next step of the evolving process is that the results given by the emulator not only concern the evolution itself: a description of the hardware implementation of the whole system is provided at the same time as the validation of its behavior.

All the experiments given in this papers have been realized using the CSEM Meta-Box (a 2-rack 31-board machine). The experiments we have realized can be compared to some symbiosis phenomenon or certain processsus happening in living systems. Ordering and density provide symbiosis-like process. Synchronization can be compared with the dynamic process leading to the emergence of the myocard muscle in which cells start to pulse at a certain frequency. In fact, all cells end by exchanging their genetic information so that nearly all cells in a region are sharing the same genetic information. Between different regions, membrane-like cells (with a typical clustering behavior) represent hard delimiter. They introduce a non-linearity in a sea of totally identicall cells. Emergent behaviors come out of the clusters in which a suffisant large amount of identical elements interconnected leads to global behavior much more complex than those of elemntary units. Membrane-like cells enable to limit interference between different clusters, so any dramatic events occurring inside one cluster will not have any significant impact on the complete system .

Some results about the sensitivity of GAs have already been obtained from these preliminary experiments:

1. Uniform cross-over versus one-point cross-over speeds-up convergence.
2. Evolution is very sensitive (like the learning process as mentioned in the neural network studies) to the input patterns. Using the same input patterns cycle during each generation causes the network to reach a fitness equilibrium that blocks the evolution.

During these experiments, we have had some difficulties with the hardware:

1. The connections. Global connections necessary for the input and output were much ressource consumming
2. The size of the machine. the machine is modular so we can increase the configuration at any time.
3. The time to first experiment and time to instruct a new users. Very efficient for any any ASIC designer.
4. the ease of the compilator and the ease of managing experiments. The C-like control language is particularly efficient and enable a very large gain in the necessary time to desribe and control experiment.

More complex experiments can take place now. The system is now ready to experiment the fitting of complex functions (sorting, decoding, compression-decompression), that would need 10^3 to 10^5 generations. The system described here is more an "academic problem" than an industrial one. In order to tackle our problem of Evolving Hardware, we have used the emulator more as an application specific supercomputer than a real circuit emulator. However, the emulation flow has been tested, and emulation promises a lot of new possibilities.

7 Acknowledgments

This design for emulation was the first at CSEM and in Switzerland. Passing through this new kind of methodology caused the fix-up of several minor bugs or misunderstanding of the tool philosophy. For their strong help, we would like to thank:

At MGC-Meta-Systems: Jean-Marc Brault for his experience and availability Gerard Morisset for software support

At MGC-Switzerland: Dominique Yerly for general support and key people access Nish Parikh for software support

At CSEM: Vincent Rikkink and all his team (CLT), for local tools facilities, very fast library make-up and strong help for bugs fixing.

References

Brown S. D., Francis R. J., Rose J., Vranesic Z. G.: *Field-Programmable Gate Arrays*. Kluwer Academic Publishers, 1992

Codd E. F.: *Cellular Automata*. Academic Press, 1968

Collins R. J. and Jefferson D. R.: "Antfarm: Towards Simulated Evolution" in *Artificial Life II* Santa Fe Institute Series, Studies in the Sciences of Complexity, Volume X, 579–603, Addison-Wesley, 1992

Crutchfield J.P. and Mitchell M.: "The Evolution of Emergent Computation", *Proceedings of the National Academy of Sciences USA*, 92(23), 1995

Gutowitz H.: *Cellular Automata - Theory and Experiment*. Elsevier, 1990

Hemmi H., Mizoguchi J. and Shimohara K.: "Development and evolution of hardware behaviors" in *Artificial IV*, Brooks R. A. and Maes P. (Eds.), MIT Press, Cambridge, MA, 371–376, 1994

Higuchi T. and Hirao Y.: "Evolvable Hardware with Genetic Learning - Toward Fault-tolerant Systems", in Proc. of the Second Workshop on Synthetic World, Paris (F), 1995

Kauffman S.: *The Origins of Order - Self-organization and Slection in Evolution*. Oxford University Press, New York, 1993

Kauffman S.: *At home in the Universe*. Oxford University Press, New York, 1995

Langton C.: *Cellular Automata* Physica 10D, North-Holland, 1984

Langton C.: *Artificial Life* Santa Fe Institute Series, Studies in the Sciences of Complexity, Volume IV Addison-Wesley, 1989

Langton C.: "Artificial Life" in *Artificial Life II* Santa Fe Institute Series, Studies in the Sciences of Complexity, Volume X, Addison-Wesley, 1992

Marchal P., Piguet C., Mange D., Stauffer A., Durand S.: "Embryological Development on Silicon" in *Artificial IV*, Brooks R. A. and Maes P. (Eds.), MIT Press, Cambridge, MA, 365–370, 1994

Mitchell M., Hraber P.T. and Crutchfield J.P.: "Revisiting the Edge of Chaos: Evolving Cellular Automata to Perform Computations", *Complex Systems*, 7:89–130, 1993

Mitchell M., Crutchfield J.P. and Hraber P.T.: "Evolving Cellular Automata to Perform Computations: Mechanisms and Impediments", *Physica 75D*, 361–391, 1994

Mitchell M.: *An Introduction to Genetic Algorithmss*, MIT Press, Cambridge, MA, 1996

Moore W. and Luk W.: *FPGAs*. Abingdon, 1991

Moore W. and Luk W.: *More FPGAs*. Abingdon, 1994

Ray T.S.: "An Approach to the Synthesis of Life", in *Artificial Life II* Santa Fe Institute Series, Studies in the Sciences of Complexity, Volume X, 371–408, Addison-Wesley, 1992

Sanchez E., Tomassini M. (Eds): *Towards Evolvable Hardware - The Evolutionary Engineering Approach*. Springer-Verlag, 1996

Sipper M.: "Non-uniform Cellular Automata: Evolution in Rule Space and Formation of Complex Structures" in *Artificial Life IV*, R.A. Brooks and P. Maes (Eds), MIT Press, 1994

Sipper M.: "Quasi-uniform Computation-Universal Cellular Automata" in *Lecture Notes in Computer Science*, Moreno A., J.J. Merelo and P. Chacón (Eds), Springer-Verlag, 1995

Sipper M.: "Co-evolving non-uniform Celullar Automata to Perform Computations" in *Physica 92 D*, 193–208, North-Holland, 1996a

Sipper M.: "Designing Evolware by Cellular Programming" in *Proceedings of the First International Conference on Evolvable Systems: from Biology to Hardware*, Tsukuba (Japan), 1996b

Taub A. H.: *John von Neumann - Collected Works*. Volume V, 288–328. Macmillan, New York, 1961-1963

Thearling K. and Ray T.S.: "Evolving Multi-Cellular Artificial Life", in *Artificial IV*, Brooks R. A. and Maes P. (Eds.), MIT Press, Cambridge, MA, 283–288, 1994

Ulam S.: "On Some Mathematical Problems Connected with Patterns of Growth of Figures", in *Essays on Cellular Automata*, Burks A. W. (Ed.), Univ. of Illinois Press, 1970

von Neumann J.: *Theory of Self-Reproduction Automata*. Edited and completed by A.W. Burks, Univ. of Illinois Press, 1966

Wolfram S.: *Theory and Applications of Cellular Automata*. World Scientific Publishing Co. Pte. Ltd., 1986

Zeleny M., Klir G. J. and Hofford K. D.: "Precipitation Membranes, Osmotic Growths and Synthetic Biology", in *Artificial Life* Santa Fe Institute Series, Studies in the Sciences of Complexity, Volume IV, 125–139, Addison-Wesley, 1989

Challenges of Evolvable Systems: Analysis and Future Directions

Hiroaki Kitano

Sony Computer Sicence Laboratory
3-14-13 Higashi-Gotanda, Shinagawa
Tokyo, 141 Japan
kitano@csl.sony.co.jp

Abstract. The goal of research in evolutionary systems is to establish technologies for building highly complex functional systems using evolutionary apporachs. Ideally, such a system should exhibit a certain level of 'intelligence.' Evolvable hardware research is an effort to accomplish direct hardware implementation of such a system. In this paper, we analyze fundermental problems in current resaerch and provide perspectives for evolving intelligent systems.

1 Goal of Evolutionary Systems Research

The ultimate goal of the evolutionary approach is to develop intelligent systems. There has been many research projects and applications for using the evolutionary approach, particularly genetic algorithms for engineering optimization problems. However, most research efforts have searched for optimal parameters in a fixed dimension parameter space. In order for the evolutionary approach to impact engineering beyond simple optimization techniques, research must focus on the evolutionary design of systems that exhibit substantially complex behaviors. Before discussing specific technical problems, it is best to detail the types of problems which should be focused on in the evolable systems research.

1.1 Problem Class Hierarchy

Problems which are considered to be a target of intelligent systems can be broadly classified into the following four classes:

Linear decomposition problems (Class-I): The problem can be decomposed into a set of linear problem. Airline scheduling, and some engineering design problems fall into this class. Expert systems and traditional AI are capable of solving this class of problem effectively.

Linear approximation problems (Class-II): Such problems are essentially nonlinear, but can be solved by linear approximation in practice, allowing for certain levels of error. Limited-domain natural language processing, speech recognition, vision, and many engineering problems fall into this class.

Non-linear problems (Class-III): Such problems are highly non-linear. Thus, they cannot be solved by linear approximation. Many difficult real-world problems, such as full-scale natural language processing, speech recognition, auditory scene understanding, and computer vision come under this class.

Non-equilibrium environment problems (Class-IV): Here, the environment in which the system operates changes dynamically, to the extent that the system's original design must be reformulated. Evolution has proved an effective means of coping with this problem. Deep space exploration and other projects, which must cope with unknown and dynamically changing environments, are faced with this type of problem.

Many linear decoposition problems and linear approximation problems can be solved by applying traditional AI techniques and algorithmic methods. Traditional AI techniques, as represented by expert systems, can be characterized by:

− Formal representation
− Rule-driven reasoning
− Strong methods.

The basic assumptions of this approach are; experts knew necessary and sufficient knowledge for the task, and this expert knowledge can be expressed in symbolic form. It also assumes that the knowledge acquired is complete, correct, and consistent. Provided these assumptions hold, traditional AI techniques are a powerful means of problem solving. Class-I problems include features that satisfy these assumptions. The salient features of class-I problems are:

Discreteness: Objects and features in the domain have a high degree of discreteness, allowing the domain to be mapped into a symbolic representations.
Explicitness: Rules governing the domain exist in an explicit form.
Completeness: A complete set of rules can be obtained.

To date, artificial intelligence has been applied to class I, II, and III problems. We already have solutions to classes I and II. Finding a solution to class III is the current focus of attention in the AI community. Work on finding a solution to the class-IV problems has just begun.

1.2 Real World Constraints

In order for animals and artificial systems to survive in a dynamically changing environment, it must recognize and evaluate the environment and react properly. However, it is impossible to pre-program behaviors for all possible situations in the open environment. Thus, the learning capability of the system is critically important. The type of learning capability necessary can be determined by the characteristics of the environment the system operates. The real world environment is often dynamic and nonlinear, which can be characterized as the dissipative system. Such a system exhibits chaotic and unpredictable behavior. The following constraints will be imposed on the system for learning:

Unpredictability: No training data is available. Since the behavior of the environment is unpredictable, it is not possible to prepare training data, which consists of inputs and correct outputs. Thus, supervised learning cannot be used.

Undecidability: An evaluation function to control the learning process cannot be prepared beforehand. Since the actions and behaviors which contribute to survival are not known, so that pre-defining an evaluation function is not possible. Even if such a function can be defined for a certain time point, the effectiveness of this function is limited, because optimal behavior may change as the environment changes.

Frequency/Importance Disparity: The importance of the event is not related to the frequency of the event. Even if the frequency of a certain event happening to the system is less than 1%, the event can be vital for the survival of the system. We call this *frequency/importance disparity*. For example, the frequency of being exposed to the danger of a traffic accident is very small, but such an event is extremely important for survival. However, important events should be memorized despite their low frequency of occurance. Traditional training methods for neural networks, such as back propagation are not suitable for this task, since it relates frequency with importance.

These constraints are common to systems that operate in the real world for a long period of time. For details on discussions in this chapter refer to publications elsewhere [Kitano, 1993, Kitano, 1994b].

1.3 The Evolutionary Pathway towards Intelligence

The implicit committment made in evolvable systems research is to create "intelligent systems" through an evolutionary approach. Over the past 30 years, the field of artificial intelligence has been struggling to build "intelligent systems" using the classical AI approach, or GOFAI (good old fashioned AI). This approach is characterized in Fig. 1, which creates a system performing complex tasks in a highly restricted environment and attempts to apply it to a broader domain. However, this approach encounters the scaling-up problem. The real world task is too complicated, noisy, and full of uncertaintity, as discussed already. The alternative approach is to create a system which exhibits a simple behavior, and then scale it up vertically to exhibit a higher level of intelligence [Brooks, 86, Brooks, 91]. However, this approach also encounters the scaling-up problem in building up intelligence. The problem with these approachs was that the system was designed manually. Thus, it lost its capability to adapt.

We speculate that one feasible path is to make a system adaptive and then increase its intelligence (Figure 1). Obviously, following this path is not a trivial task, either. However, any attempt to increase the adaptability of an existing system would be far more difficult.

It is hard to say just how complex and adaptive a system can be built through evolution. For example, any intelligent system should be able to use language,

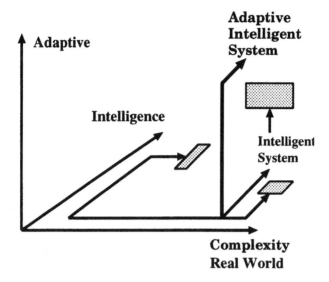

Fig. 1. The path to adaptive intelligent systems

which is a highly complex symbol manipulation system. The issue is one of how circuits acquired by evolution climb up the ladder of intelligent behavior. However, we should make it very clear that this pathway is the implicit committment that the evolvable system community made.

2 Issues on Evolvable Systems

From the evolutionary systems perspectives, there are three fundermental issues:

Genotype-Phenotype Mapping: The first issue is the genotype-phenotype mapping. A simple direct mapping of genotype into phenotype was already recognized to be too inefficient to evolve complex and large-scale systems. The alternative approach was introduced by the author in 1990, which used grammatical encoding to represent structure of neural networks. It encodes graph grammar in a chromosome and a final network structure is generated by recursively applying graph grammars [Kitano, 1990]. After the introduction of this approach, a number of improvments and applications has been proposed [Gruau and Whitley, 1993, Sims, 1994, Hemmi et al, 1994].

Components Functionalities: The second issue is the functionalities of each component, which consists of the final structure. The current model for neural networks and logic-based systems are based on a very simplified assumption of discrete-state-based information processing. For example, almost no neural networks account for the delay of signal transmission. Delay and phase are essential elements in actual central nervous systems for auditory functions.

Co-Evolution of Agents and Environments: The third issue is co-evolution of individuals and environment. If we defined evaluation functions, which are too hard from the begining of the evolution, the genetic search turns into a random search of huge design space. It is almost impossible for the evolutionary approach to generate individuals which exhibit a high level of functionality. The evaluation environment should start from a relatively simple task so that the evolutionary process can find a near-optimal solution, thereby forming solution basis of more complex structures.

3 Development — Embryogenics

In order to implement a basic scheme for complex sturcture formation, the use of sophisticated genotype-phenotype mapping is essential. In the biological system, this is called development, or embryogenesis. We need to create an engineering counterpart — embryogenics. We will take a look at the real biological systems. Aside from chaotic dynamics, expression of a specific genes are critical in forming complex structures. Expression of gene must be controlled spatio-temporally so that the specific gene will be expressed at a specific time and location. Thus, finding fundermental principles for coordinating transcriptions is essential.

Our hypothesis is that symmetry breakdown and constraints are two fundermental principles in biological pattern formation. In this paper, we focus on symmetry breakdown. We assume that diversity of cell types, thus related morphogenesis, are created due to symmetry breakdown at four levels. these four levels are:

DNA: Disparity of mutation level between the leading strand and the ragging strand in DNA replication [Wada et al., 1993],

Genome: Asymmetric distribution of chromatin, or repressors, in DNA replication [Kitano and Imai, ms.],

Cell: Asymmetric distribution of cytoplasmic materials [Rhyu et al., 1994, Knoblich et al., 1994], and

System: Cell-cell interactions [Kitano, 1994a, Kitano, 1995b, Kondo, 1992, Kondo, 1995].

In designing an engineering model of development, there are a few interesting biological models which can be applied immidiately.

3.1 Replication Dependent Transcriptional Control

While these four levels of symmetry breakdown can be introduced in the machinary, DNA replication-dependent transcriptional control (RDTC) plays a key role, particularly in regulating temporal-specific gene expression [Kitano and Imai, ms.].

Figure 2 shows a schematic model of this theory, adapted to the evolvable system field. At each cell division, the chromatin structure changes and areas

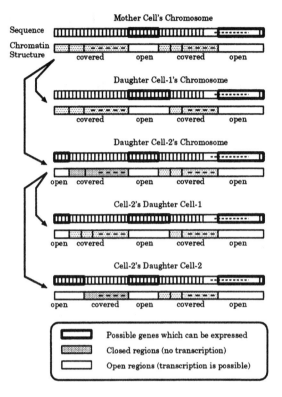

Fig. 2. Asymmetric distribution of chromatin structure

covered by chromatin and areas open for transcription change. This mechanism can be a temporal control mechanism for gene regulation because certain genes will be expressed only after a certain number of cell divisions.

3.2 Cellular-Based Reaction-Diffusion

Another factor which should be taken into account is the reaction-diffusion model of morphogenesis. In recent experiments by Melton, diffusion was shown to be a less likely mechanism for cell fate determination. However, it was shown that certain relay mechanisms may enable communication between distant cells. Some computational models of morphogenesis [Kondo, 1992, Kondo, 1995] fit well with this findings and are potentially useful for an engineering model of development.

3.3 Genetic Control of Neural Growth

The regulatory mechanisms of a gene are important to understand how central nervous systems are created, and to apply these ideas to engineering methodology. In fact, recent findings on fas3 and *gcm* indicate that the connection topology of central nervous systems are highly governed by genetic information, though not fully determined [Chiba et al., 1995, Hosoya et al., 1995]. These genes determine target cells that neurons extend their axons to in a very specific manner. Actually, part of their relationship can be described using Boolian algebra. Thus, it is a mistake to consider that neural systems are generally wired using a chemical gradient, which provides low specificity. However, it is also misleading to believe that all connections are deterministic. Sophisticated modeling of this aspect of neuroscience would greatly aid the engineering of development.

4 Component Functionalities — Asynchronous Pulse-Coded Neural Systems

The current models for neural networks is a crude abstraction of real neural systems, proposed over 30 years ago. Many biological findings have been made since then, not only in the area of detailed neural systems, but also on the systematic understanding of the brain. While it is appearent that the current model of a neural network falls short of creating intelligence, we need to investigate a new model of neural networks along with the evolutionary computing research. After all, what evolvable systems do — at least in current research — is find the strcuture of a network that fits well with certain tasks. This situation is the same for logic circuit design and genetic programming. It is clear that we need to investigate new features and functionalities that exhibits the power of an evolutinary approach.

One approach is to look into biological findings more carefully and model their essential parts. Addition of the following three features may help improve potential functionalities of the components which make up the final system:

- Delay of signal transmission between neurons
- Pulse coding
- Chemical messengers

In the real neural systems, information is carried as pulses propagating from one neuron to the other. Naturally, there is a limitation in the speed of propagation, which creates temporal differences in signal arrival time for neurons connected at different axon lengths (Fig. 3). This feature has not been actively incorporated in current neural network models. However, the transmission delay has been used for echolocation in many animals [Konishi, 1986].

In current neural network models, weight is the single parameter which determines the relationship between two neurons. In the new model, weight and delay are two parameters that define the relationship between two neurons (Fig. 4).

In the new model, an average firing rate of the neuron j (a_j) is defined as:

$$a_j(t) = \sum_{i=0}^{N} w_{ij} a_i(t - d_{ij}) - h \tag{1}$$

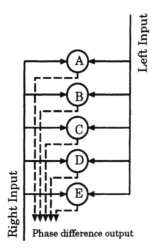

Fig. 3. A circuit to detect the phase difference between two inputs

Fig. 4. A neural network model with delay

where w_{ij} is a connection weight between neuron i and j, and d_{ij} is the transmission delay time between neuron i and j.

Finally, the effects of chemical messengers can be modeled as a modulator of signal transmission strength and weight update strength during learning [Kitano, 1995a].

5   Co-Evolution

The goal of evolutionary systems research is to evolve intelligent systems, one of the major research topics should be the evolution of systems that manipulate language. Since the use of language requires substatial and complex symbol manipulation, memory, recall, and planning capabilities, an evolutionary approach to design such an agent should be given special attention. Language, however, emerged from a simple form of communication, and evolved to a broad and complex set of syntactic structures. For example, one way of classifying the evolution of language is:

No Symbol (Level-0): No effective language exists.
Single Token (Level-1): A very primitive form of language, or communication via chemical substances such as feromones.
Token Combination (Level-2): The power of expression increases by combining a number of primitive signs. Apes appear to use this level of language.

Embedding (Level-3): The power of expression increases drastically as this class of language allows complex structure operations.

Recursion (Level-4): Maximum expressive power can be obtained by recursion. Human languages are of this level.

There is a fundermental need for serious investigation into how evolution forced the development of the complex neural circuits that enable us to use complex language [Hawkins and Gell-Mann, 1992]. The emergence of language and its development (in a linguistic sense) has been investigated in the context of interaction [Steels, 1996].

It is essentail that co-evolution of both agent and environment be attained. Abundant empirical experiments suggest that an evolutionary design on systems using difficult evaluaiton functions from the initial stage of the evolution are doomed to fail. The evaluaiton function must start from easy tasks that gradually become more difficult. It is not feasible to provide gradual changes in the evaluation function in a top-down manner. It is essential that such a function emerge from the interaction between agents in the competitive environment.

After a certain period of evolution, capable systems should exhibit a highly dynamic internal process. It is expected that at this stage, the system embodies the complex dynamics as investigated in the chaotic nature of neural activities research [Tsuda, 1992, Pollack, 1991], self-organization [Bak, et al., 1988], and symbol dynamics [Hao, 1991]. It would be biologically exciting if the results obtained can be linked with biological findings [Eckhorn, et al., 1988, Gray, et al., 1989, Rolls, 1989].

6 Conclusion

The goal of this paper was to analyse the challenges for evolvable systems research, and provide a perspective for future research. Evolvable systems are a new and promising decipline. However, in order for the field to have an impact, it is absolutely necessary that evolvable systems be able to perform highly intelligent tasks, or to exhibit radically new functionality which cannot be obtained using a conventional approach. We believe that we should look into recent biological findings more carefully, and extract their essential principles to adapt them to the engineering domain. In this paper, we pointed out that three essential aspects of evolvable systems; genotype-phenotype mapping, components functionality, and co-evolution of systems and enviornment. These three aspects require special attention for the future of evolvable systems. The author acknowledges that this paper is speculative, rather than reporting concrete technical results. However, it is essential that an overall picture of the field be discussed to put detailed technical results into proper perspective.

References

[Bak, et al., 1988] Bak, P., Tang, C. and Wiesenfeld, K., "Self-Organized Criticality," *Physical Review A*, Vol. 38, No. 1., 364-374, 1988.

[Brooks, 91] Brooks, R., "Intelligence without Reason," *Proc. of IJCAI-91*, 1991.

[Brooks, 86] Brooks, R. (1986). "A Robust Layered Control System For A Mobile Robot," IEEE Journal of Robotics and Automation, Vol. RA-2, no. 1, 1986.

[Chiba et al., 1995] Chiba, A., Snow, P., Keshishian, H., and Hotta, Y., "Fasciclin III as a synaptic target recognition molecular in Drosophila," *Nature,* 374, 166-168, 1995.

[Eckhorn, et al., 1988] Eckhorn, R., Bauer, R., Jordan, W., Brosch, M., Kruse, W., Munk, M. and Reitboeck, H., "Coherent oscillations: A mechanism of feature linking in the visual cortex?", *Biological Cybernetics,* 60, 121-130, 1988.

[Gray, et al., 1989] Gray, C., Koenig, P., Engel, A. and Singer, W., "Oscillatory responses in cat visual cortex exhibit inter-columnar synchronization which reflects global stimulus properties," *Nature,* 338, 334-337, 1989.

[Gruau and Whitley, 1993] Gruau, F., and Whitley, D., "Adding Learning to the Cellular Development of Neural Networks: Evolution and the Baldwin Effect," *Evolutionary Computation,* 1(3): 213-233, 1993.

[Hao, 1991] Hao, B., "Symbolic dynamics and characterization of complexity," *Nonlinear Science,* The MIT Press, 1991.

[Hawkins and Gell-Mann, 1992] Hawkins, J. and Gell-Mann, M., *The Evolution of Human Languages,* Addison-Wesley, 1992.

[Hemmi et al, 1994] Hemmi, H., Mizoguchi, J., Shimohara, K., "Development and evolution of hardware behaviors", *Proc. of Artificial Life IV*, MIT Press, 1994.

[Higuchi and Manderic, 1994] Higuchi, T., Iba, H., Manderick, B., "Evolvable Hardware", in *Massively Parallel Artificial Intelligence,* (ed. H.Kitano), MIT Press, 1994.

[Hillis, 1991] Hillis, D., "Co-evolving parasite improve simulated evolution as an optimization procedure," *Emergent Computation,* The MIT Press, 1991.

[Hosoya et al., 1995] Hosoya, T., Takizawa, K., Nitta, K., and Hotta, Y., "glial cells missing: A Binary Switch between Neuronal and Glial Determination in Drosophila," *Cell,* 82, 1025-1036, 1995.

[Ikegami and Kaneko, 1991] Ikegami, T. and Kaneko, K., "Computer symbiosis — Emergence of symbiotic behavior through evolution," *Emergent Computation,* The MIT Press, 1991.

[Kitano, 1996] Kitano, H., "Morphogenesis for Complex Systems", *Toward Evolvable Hardware,* Springer-Verlag, 1996.

[Kitano, 1995a] Kitano, H (1995). "Hormonal Modulation of Learning," *Proc. of IJCAI-95,* Montreal, 1995.

[Kitano, 1995b] Kitano, H., "A Simple Model of Neurogenesis and Cell Differentiation Based on Evolutionary Large-Scale Chaos", *Artificial Life,* 2: 79-99, 1995.

[Kitano, 1994a] Kitano, H., "Evolution of Metabolism for Morphgenesis", *Proc. of Artificial Life IV,* 1994.

[Kitano, 1994b] Kitano, H., "Toward Adaptive Intelligent Systems", *Proc. of IEEE International Conference on Evolutionary Computation,* Orland, 1994.

[Kitano, 1993] Kitano, H (1993). "Challenges of Massive Parallelism," Proc. of IJCAI-93. (The Computers and Thought Award Lecture)

[Kitano, 1990] Kitano, H., "Designing Neural Networks using Genetic Algorithms with Graph Generation System," *Complex Systems,* Vol. 4, No. 4, 1990.

[Kitano and Imai, ms.] Kitano, H. and Imai, S., "Two distinct instrinstic mechanism regulate the stochastic and catastriphic phases in cellular senescence," manuscript, 1996.

[Knoblich et al., 1994] Knoblich, J., Jan, L., and Jan, Y., "Asymmetric segregation of Numb and Prospero during cell division," *Nature,* Vol. 377, 624-626, 1995.

[Kondo, 1995] Kondo, S., "A reaction-diffusion wave on the skin of the marine angelfish Pomacanthus," *Nature*, Vol. 376, No. 6543, pp. 765-768, 1995.

[Kondo, 1992] Kondo, S., "A mechanistic model for morphogenesis and regeneration of limbs and imaginal discs," *Mechanisms of Development*, 39, 161-170, 1992.

[Konishi, 1986] Konishi, M., *Trends in Neurosci.*, 9, 163-168, 1986.

[Langton, 1989] Langton, C. (1989). "Artificial Life," Artificial Life, Addison Wesley.

[McGaugh, 1989] McGaugh, J., "Involvement of hormonal and neuromodulatory systems in the regulation of memory storage," *Ann. Rev. Neurosci.* 12:255-87, 1989.

[Newell and Simon, 1976] Newell, A. and Simon, H., "Computer science as empirical inquiry: Symbols and search," *Communications of the ACM*, 19(3), 113-126, 1976.

[Pollack, 1991] Pollack, J., "The Induction of Dynamical Recognizers," *Machine Learning*, 7, 227-252, 1991.

[Rolls, 1989] Rolls, E., "The representation and storage of information in neuronal networks in the primate cerebral cortex and hippocampus," Durbin, et al. (Eds.), *The Computing Neuron*, Addison-Wesley, 1989.

[Rhyu et al., 1994] Rhyu, M., Jan, L., and Jan, Y., "Asymmetric Distribution of Numb Protein during Division of the Sensory Organ Precursor Cell Confers Distinct Fates to Daughter Cells," *Cell*, Vol. 76, 477-491, 1994.

[Sims, 1994] Sims, K., "Evolving Virtual Creatures," *Proc. of SIGGRAPH-94*, 1994.

[Steels, 1996] Steels, L., "Self-Organizing Vocabularies," *Proc. of Alife-V*, 1996.

[Thompson, 1995] Thompson, A., "Evolving electronic robot controllers that exploit hardware resources", *Proc. of the 3rd European Conf. on Artificial Life*, 1995.

[Tsuda, 1992] Tsuda, I., "Dynamic Link of Memory — Chaotic Memory Map in Nonequilibrium Neural Networks," *Neural Networks*, Vol. 5, 313-326, 1992.

[Wada et al., 1993] Wada, K., Doi, H., Tanaka, S., Wada, Y., and Furusawa, M., "A neo-Darwinian algorithm: Asymmetrical mutations due to semiconservative DNA-type replication promote evolution," *Proc. Natl. Acad. Sci. USA*, Vol. 90, 11934-11938, 1993.

Cellular Systems

Functional Organisms Growing on Silicon

Pascal Nussbaum, Pierre Marchal, Christian Piguet

CSEM Centre Suisse d'Electronique et de Microtechnique SA,
Jaquet-Droz 1
CH-2007 Neuchâtel
email: <name>@csemne.ch

Abstract. This paper describes a novel architecture inspired from the multicellular organizations found in Nature. This architecture is tailored to let functional organisms (logical functions) grow on silicon. To this aim, the silicon surface is populated with an array of identical programmable cells, which may be configured by a bitstream. By analogy with the biological world, the concatenation of the bitstreams used to program the cells composing a given function is called the "genome" of that function. In addition to conventional BIST (Built-in Self-Test) structures addressing signal line faults, this new version tolerates failures affecting power supply. It also allows the growth of differentiated organisms on the same surface by including a code in the genome to distinguish them. As a testbed, we have developped an integrated circuit prototype, code name Genom*IC*. It contains only a single 4-cell structure, but prefigures which kind of structure can be massively integrated in very large circuits in order to manage complexity (multicellular organization), evolvability (genetic data manipulation) as well as fault tolerance.

1   Introduction

In a will to extend the work of John von Neumann [1], E. F. Codd [2] and later Chris Langton [3] develop self-reproducing automaton. A scale change from "molecular description of reproduction" [1], to "cellular reproduction" [4] as well as its application to programmable devices [5] make it possible to let electronic functions reproduce on silicon [6]. This technique is interesting in different points of view:

1. Compared with self-reproducing automaton of Langton [3], Byl [7], Reggia [8], which does nothing but propagates, the growth of our "organisms" ends with useful logical functions.
2. Compared with conventional FPGAs, cell functions are not tied to a physical address but rely on a logical one, as learned from biology.
3. Furthermore, the positive inheritances from the biological counterpart lead to emerging properties like developing, maintaining and reproducing.

The basics reside in the use of a totally homogeneous array of cells. In a previous work [4], each cell contained the entire description of the organism (genome) and was hence able to achieve the functionality of any deficient neighbouring cell in a one step process. In order to keep development tools to a minimum, we focused on a very fine grained cell: a one-variable DMUX. Such a fine-grained cell is able to efficiently implement the different tiles necessary to describe a Binary Decision Diagram (Cf. §3.2). The functional cell also includes programmable connections with neighbourhood, and a flip-flop for sequential purposes. The choice of such cell allows to easily map any logical function on a 2-dimension array and to derive the genome information straight from any truth table (combinational system) or from any transition table (sequential system).

The way each cell chooses the right part of its patrimony (genome) is to numerate (coordinate computation) from a starting cell in the array (the equivalent of a zygote cell). This simple computation originates from gradient-like processes found in embryological development. The locally computed coordinates serve as a pointer to extract the right gene out of the genome (cell differentiation). The numeration, a continually achieved task, allows to reconfigure the organism once a cell dies (due to any kind of error) as depicted in Fig. 1.a. Autonomy, obviously, requires the use of spare parts (cells). In our implementation, spare parts are universal, and can be shared by different organisms. One major advantage is that a spare cell is not devoted to a given function as in conventional fault-tolerance approaches. So, as long as spare parts are available, any given cell can fail and still be replaced.

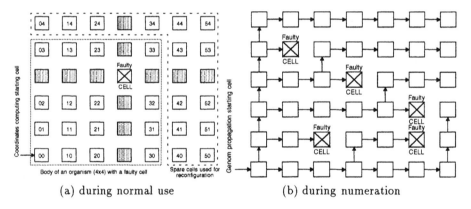

(a) during normal use   (b) during numeration

Fig. 1. Faulty cell cell avoidance

2 New concepts

2.1 Propagated versus stored genome

Previously, the genome was propagated and then stored. As selected by nature, we have chosen to store the complete genome in each cell, in RAMs for reprogrammable cellular spaces or in ROMs for dedicated ones. Unfortunately, storing the entire genome in each cell is not only tremendously expensive in terms of memory but it also bounds the complexity of organisms to the available memory size. In this paper, we describe the continuous propagation of the genome and hence we can restrict the local memory of the cells to only store a single gene. The genome is injected in the array in the left-most bottom operational cell. This cell sends the genome stream to its top and right neighbours, which do the same, etc. The strategy of double propagation of the genome enable turns around dead cells as shown in Fig. 1.b. The stream is responsible both of growth and reconfiguration of organisms.

When a fault occurs in a given cell, the first faultfree cell needs to swap its gene to express the one of the suddenly dead neighbour. To do so, the faulty cell becomes transparent with respect to coordinate computation. As soon as the coordinates of fault-free cells are re-computed, they can grasp the correct genes in the continuously propagating genome.

2.2 Organisms discrimination

As local coordinates are used to grasp local genes, a code is needed in order to discriminate one organism from another as depicted in Fig. 2. In order to distinguish host genes from the ones belonging to foreign bodies, Nature has selected the same principle in the immune system: the Major Histocompatibility Complex (MHC) also called Human Leucocyte Antigenes (HLA) in humans. In our implementation, the organism code, also called MHC, is located in the header of each gene.

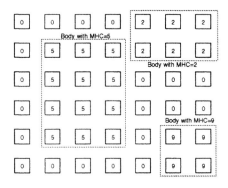

Fig. 2. Organisms coexisting in an array of cells

2.3 Organisms coated with a membrane

A physical boundary is necessary to delimit the development of an organism to a given size. Still taking inspiration from biology, we developed an equivalent of the epithelial membrane on silicon. As in biology, the membrane assures three main tasks:

1. The first task is to isolate one organism from the external world (other organisms). The membrane plays a key role during the development of an organism, it protects fully developed organisms from the colonization by the genes of the organism in development.
2. The second task is to provide autonomy in maintaining the function by internal self-reconfiguration. Depending on the severity of the default affecting a cell, the amount of hardware to be bypassed may be: a single cell, a complete line or column, a complete line and column. The presence of the membrane enable to restrict the bypassed zone to the inside of the organism (Fig. 3).

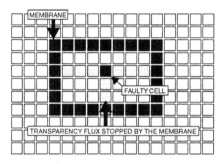

Fig. 3. Reconfiguration restricted to the inside of the organism

3. The third task is to promote exchange of information with other organisms. This solution may introduce a mismatch of communication between organisms after an internal reconfiguration. For instance, when two neighbouring organisms are exchanging data, the reconfiguration of one organism slide away input-output port on the membrane and a failure occurs Fig. 4.

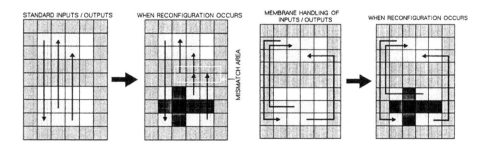

Fig. 4. Membranes handling the communication mapping problem when reconfiguration occurs

To avoid such problems, we rally inputs and outputs at defined sites, whatever happens in the organ itself. This makes the exchanges independent from reconfiguration. Notice that if a faulty cell appears in the membrane or in the communication channels, the organism doesn't recover (it can't reconfigure), and the whole system has to move somewhere else (organism reproduction).

2.4 Programmed death

When a cell dies in the membrane, the organism is no longer protected against foreigner gene intrusion. It must be totally erased first (or killed), and then produce another copy of itself where the membrane integrity can be insured. A programmed death has been developed. It induces an inflammation of the entire membrane as soon as one membrane cell dies. Once the membrane is inflamed, by neighbourhood propagation, the organism erases all the data in cells, releasing its area free to receive another organism. The different phases of this process appear in Fig. 5.

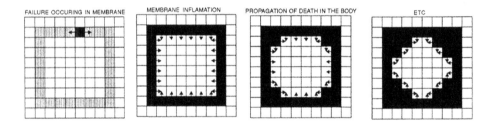

Fig. 5. Organism death when membrane is damaged

2.5 Powerfailure tolerance

A self-detection phase takes place prior to reconfiguration. This phase, based on conventional Built-In-Self-Test (BIST) techniques, enables to detect unreliable signals. These techniques can only be applied to faults affecting signal lines. Faults affecting global signals, such as power lines and clocks, are much more complex to detect (good reference). Although long term developments are investigating self-timed logic and local autonomous power supplies, for present circuits we have developed a fault-tolerant power supply strategy. It is mainly based on redundant supplies and electronic fuses. It's organization, as shown in Fig. 6, is specially tailored for homogenous arrays.

Each cell is connected to two power distribution rails (row and column), through a fuse and a selector. If the cell contains a short, the electronic fuse disconnect the cell from the power network. If a rail fails, the selector switches to the other one. What is true for a cell is true for a rail, so each rail is connected to its side-rail through an electronic fuse. The propagation of the redundancy out

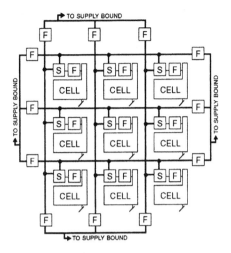

Fig. 6. Fail-safe power supply distribution

of the chip by "two bounds per side-rail" removes the bottleneck which makes the circuit entirely dependent to a single bound or a single solder for an entire circuit.

3   The circuit

In order to check the validity of our concepts in bio-inspiration, we have extended our investigations downto silicon. Using of-the-shelves circuits, a complete demonstrator populated with 20 cells has been realized [10]. Unfortunately, available reticles and technology sizes do not permit yet to integrate thousands of cells at fair prices. So we have integrated a cluster of 4 cells sharing the same genomic nucleus Fig. 10. A bigger cellular space can obtained by simple abutment of several chips.

3.1   The genomic level

The cell nucleus is mainly in charge of copying and translating the genome information. To this purpose, we have embedded the ribo-processors in the nucleus. Under precise circumstances, the ribo-processors may also modify the genome content (inverse transcriptase like).

As mentioned above, hormonal gradients necessary to differentiate cells have been replaced by a coordinate computation. Each nucleus computes its own coordinates by incrementing those obtained by its neighbours. Coordinates are used to select the correct gene in the whole genome of an organism, which in turn configures the functional part of the cell.

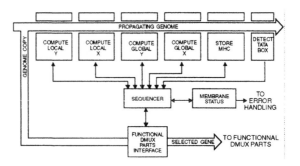

Fig. 7. Nucleus architecture

Architecture The nucleus architecture resembles to a bunch of ribosomes sticked to the propagating genome as depicted in Fig. 7.

Each ribo-processor acts as a shift register, and is responsible for a group of simple tasks applied to a specific part of the genome, such as value storage or comparison, increment, equal zero testing...

The first 4 ribo-processors are devoted to the computation and storage of the global (whole array) and local (organism) coordinates of the cell. The 5th one stores the MHC (Major Histocompatibility Complex), used to distinguish different organisms.

In the first silicon, we have considered 6-bit fields. So a maximal array of 64x64 for organisms measuring up to 64x64 is available. The 6-bit MHC allows up to 64 organisms coexisting on the same cellular space. As the genome continuously flows through the shift registers, extra registers are necessary to store temporary items. A sequencer is required to control the different operations taking place during the embryological development, such as:

1. storage,
2. incrementing,
3. insertion (of the stored value in the genome flow),
4. comparison with the stored value,
5. comparison with 0.

The error handling unit manages the propagation and generation of error and transparency signals as depicted in Fig. 8. This combinational logic block is controlled by membrane and MHC signals. An error signal is used to inform neighbours that the cell is faulty and can no longer fulfill its tasks. A transparency signal is used to bypass a fault-free cell.

Membrane cells propagate error signals but stop transparency signals, while body cells propagate transparency signals and translate an error input signal into transparency output signal.

Main Principles As mentioned above, the genomes are continuously propagated through the network to minimize useless storage of data. So forth, the

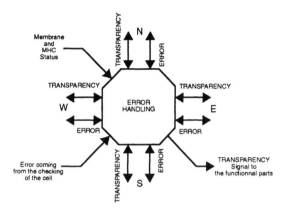

Fig. 8. Error handling unit

nucleus of each cluster must grasp relevant data on the fly. The genome data structure is depicted in Fig. 9.

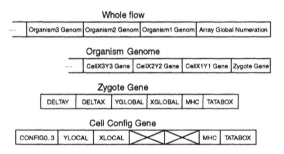

Fig. 9. Structure of the genome flow

The first event caused by the flow is the numeration of the global coordinates contained in the cells nuclei. Once done, cells are able to detect if the zygote gene of an organism should start a body growth at their place. This is the reason why zygote genes contain global coordinates, and XY extensions (DELTAX - DELTAY) of the body. It also contains the MHC, stored in all the cells of the organism, distinguishing between config genes having the same local coordinates, but belonging to different organisms.

The config gene addresses one cell of one organism, so when caught by the ribo-processor, its local coordinates and MHC are checked, and in case of correct fitting, the corresponding configuration section is downloaded in appropriate storage registers.

The configuration of the cell is done by filling the 4 serial registers, each one controling its own DMUX3 cell as shown in Fig. 10.

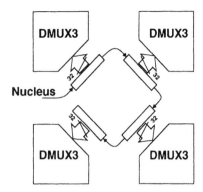

Fig. 10. Registers configuring DMUX3 functional parts

Typical Sequencement One must keep in mind the parallel distributed processing of such a system: the working engine is an array of cells, and not only a single cell or nucleus. The main process initializes the array (global address computation), and download the genomes bitstream, i.e. growth of organisms and continuous reconfiguration of them as soon as cells become faulty. The overview of the different phases is:

1. Whole array cells global coordinates computation.
2. Interception of a zygote gene by the correct local cell.
3. Growth of the organism (and exception processing).
4. Grasping of config genes by the body cells.
5. Next grasping of an organism's zygote gene induces reconfiguration.
6. Serious injuries cause the organism death and cells deletion.

Let us examine each case in details, describing the major steps followed by the nucleus sequencer.

Phase 1: Global numeration The header of the flow, called "Array Global Numeration" could be considered as a global zygote gene. This gene resembles to a standard zygote gene, differing only in its header (TATAbox). Its global and local fields as well as its MHC field are empty (filled with 0). The flow being injected to the first healthy left-most bottom cell, this one takes the X=0, Y=0 coordinates. The gene is then shifted out to the two N and E neighbours. At the same time, two signals are activated: ELOAD and NLOAD. These signals inform the neighbours that the incoming gene must be taken in the numeration mode Fig. 11:

The propagation of the flow is horizontal, since the W neighbour is considered able to deliver it. If not, the S neighbour is then selected. In the nucleus, the global fields are captured, incremented, stored and then re-injected in the flow. The global zygote gene is changing when travelling across the array, always carrying the last computed coordinates.

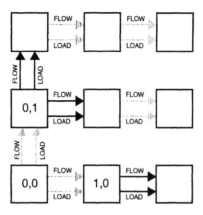

Fig. 11. Global numeration

Phase 2: Interception of a zygote gene

After global numeration, all the cells have fixed global coordinates, starting in the left-most bottom and ending in the right-most top cell. The organism genomes are directly following the global zygote gene. All organisms start with a zygote gene Fig. 9 which contains global coordinates that determine where the embryological development must take place in the array, what are its extents in both X and Y direction, and what is the MHC (number) of the organism to grow. Cells are continuously checking the flow to detect if a zygote try to start growth at their place.

Phase 3: Growth of the organism

Once the cell of some global coordinates has intercepted the corresponding zygote, a local numeration, limited to the growing organism, is performed. The processus stops when local coordinates equals the extents of the organism (DELTAX, DELTAY) carried in the zygote gene. During the growth, the surrounding cells of the organism are set to the membrane status, by testing if their local coordinates are peripheral to the organism.

Phase 4: Grasping a config gene

A cell belonging to an organism has 3 characteristics that determine the config gene to intercept and store. First of all, the MHC code that determine the identity of the organism, and then the X and Y local coordinates that determine its position inside the body of the organism.

Phase 5: Reconfiguration

The structure of an organism is frozen as long as the genome has not been re-propagated. As soons as the genome is downloaded again, the reconfiguration takes place, if necessary (pesence of dead cells). When a cell dies, it sends an active error signal to its neighbours. These ones propagate a transparency signal, as described in the previous section. The cells numeration process skips the transparent cells by making no increment on coordinates. The re-numeration also takes progress in vertical membrane cells. These cells being the interface with the inside of the body, they must undergo reconfigurations.

Phase 6: Organism death

Like living creatures, artificial organisms cannot tolerate any amounts of wounds nor wounds of any size. So, once the organism can't continue to fulfill its tasks. Some wounds are more serious than others, like cell death in the membrane. The membrane is the interface of the organism with outside world, and any problem occurring inside it corrupts exchanges with the outside. So a special process takes place when the membrane is altered. This process is identical to the one that takes place when a collision occurs between organisms during growth. As depicted in Fig. 5, a dead cell in the membrane emits error signals, and its membrane neighbours propagate them in all the membrane (inflammation), and to the body. When the body is coated with error signals, the error state propagates and fills the body. Once a clock period is done, all cells MHC fields are reset: they return to neutral state, and an other organism can use them to grow.

3.2 The functional level

As noticed earlier, the genomic part can be used to configure any homogeneous functional part. In our case, the implementation of binary decision diagram in demultiplexers (DMUXs) is the simplest to work with (no huge software tools for function implementation).

DMUX basements The DMUX grain has been chosen to quickly implement binary decision diagram. All logic function can be implemented using these diagrams. They are close a compressed truth-table description Fig. 12:

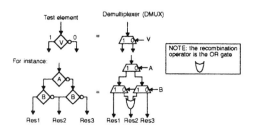

Fig. 12. Binary decision trees implemented with DMUXs

The original cell was based on a DMUX, a neighbourhood configurable connectivity, a flip-flop (for sequenced logic) and a long distance bus crosspoint as shown in Fig. 13:

This cell, repeated a huge amount of time in a big array, is able to implement any logic function, from elementary combinatorial logic upto finite state machines or processors.

Enhanced version 3 The last version of this cell is still based on a DMUX. A second bus crosspoint has been added for long distance transfer. The bus

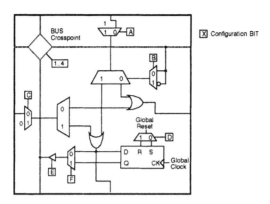

Fig. 13. First version of a DMUX cell

is fully duplexed. The neighbourhood connectivity is bidirectional horizontally, allowing signal recombination through the OR gate in both directions as depicted in Fig. 14.

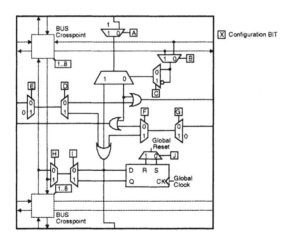

Fig. 14. Last version (DMUX3) of the demultiplexer cell

This new architecture makes it possible to implement more compact binary decision trees. In addition, the improvment of long distance buses help to solve the extraction of signals from a dense region.

4   Conclusions

Balancing between the features provided by biological systems and state of the art electronic, the Genom*IC* gives a new point of view on what could be silicon devices in the future.

Another feature of this approach is the high repetitivity of identical cells: this kind of circuit is the first to get success on new technologies, like the RAM chips. When a technological shrink occurs, only one single cell has to be redesigned.

Problems are remaining when using huge arrays of cells. For instance, the long distance communication being repeated in all cells, the delays become very long from side to side of the array. Universal signals shared by all cells like clocks become difficult to distribute.

Finally, the fail-safe concept looses its signification as early as only one part of the system has not been included in self-test and repair systems. It transforms even simple design in a Graal quest from where only few people came back. This is why biological systems have to be investigated. The intrinsic resistance to failures of such systems is tremendous, but forces us to reconsider our design methodology from the basements.

The present sacrifices in terms of complexity and gates number will bear fruits once circuits containing billions of transistor will be available.

To verify the IC, a demo board will be developed early next year. The cell array size will be around 10x10. This means that the functional parts (DMUX3 parts) will be around 20x20. Each IC has a test unit, powered independently, in order to verify how the cell behaves even power failure occurs.

The verification will consist to initialize the array and grow some organisms filling an interesting task (implantable in 200-300 DMUX3 parts) in group.

References

1. von Neumann J., The Theory of Self-Reproducing Automata, A.W. Burks, ed. Univ. Illinois Press, Urbana, (1966)
2. Codd E. F., Cellular Automata, Academic Press, New-York, (1968).
3. Langton C. G., Self-Reproduction in Cellular Automata, Physica 10D, 135-144, (1984)
4. Marchal P., Nussbaumm P., Piguet C., Durand S., Mange D., Sanchez E., Stauffer A., Tempesti G. Genomic Cellular Automata Transposed in Silicon, in *IPCAT'95*, World Scientific, 1996
5. Marchal P., Stauffer A. Binary Decision Diagram oriented FPGAs, ACM International Workshop on Field Programmable Gate Arrays, 2(1-10) (1994).
6. Marchal P., Piguet C., Mange D., Stauffer A. and Durand S. Embryological Development on Silicon, Proceeding of Artificial Life IV - MIT Press, 365-370 (1994).
7. Byl J., Self-Reproduction in Small Cellular Automata, Physica34D, 295-299, (1989)
8. Reggia J. A., Armentrout S. L., Chou H. H., Peng Y., Science 259, 1282 (1993)
9. Marchal P., Embryonics: the Birth of Synthtic Life, in *Towards Evolvable Hardware*, Springer-Verlag, 1996
10. Mange D., Stauffer A., Marchal P., Embryonics: Designing Programmable Circuits with Biological-like Properties, 12th IASTED International Conference, Annecy, France, 1994

Logical Universality and Self-Reproduction in Reversible Cellular Automata

Kenichi MORITA and Katsunobu IMAI

Faculty of Engineering, Hiroshima University, Higashi-Hiroshima-shi, 739, Japan
morita@ke.sys.hiroshima-u.ac.jp

Abstract. A reversible cellular automaton (RCA) is a "backward deterministic" CA in which every configuration of the cellular space has at most one predecessor. Such reversible systems have a close connection to physical reversibility, and have been known to play an important role in the problem of inevitable power dissipation in computing systems. In this paper, we investigate problems of logical universality and self-reproducing ability in two-dimensional reversible cellular spaces. These problems will become much more important when one tries to construct nano-scaled functional objects based on microscopic physical law. Here, we first discuss how logical universality can be obtained under the reversibility constraint, and show our previous models of 16-state universal reversible CA. Next we explain how self-reproduction is possible in a reversible CA.

1 Introduction

Cellular automata (CA) were first devised by J. von Neumann. He developed his theory of self-reproducing automata by using this framework [22]. Since then, studies on CA have been extensively done from several points of view. For example, they have been investigated as a model of a parallel processing system, models for describing and analyzing various space-temporal phenomena, and so on.

A "reversible" CA (RCA) is a special type of CA in which every configuration (i.e., the whole state) of the cellular space has at most one predecessor. Roughly speaking, it is a "backward deterministic" CA. The notion of reversibility were also defined for several other models of computing, such as reversible Turing machines, and reversible logic gates (see e.g. [4, 17, 18, 24] for general survey).

Reversible systems have a connection to physical reversibility, and in fact some RCAs can be considered as abstract models of reversible physical spaces. Recent studies on quantum mechanical models of computing, such as a "quantum Turing machine" and a "quantum CA" [6, 7, 25] have also a connection to reversibility. These models are in fact reversible because their state evolution is determined by a unitary transformation. Hence, a reversible computing system (in the former sense) can be regarded as a special type of a quantum computing system.

One interesting point of reversible systems is that it is closely related to the problem of energy dissipation in a computing process. It is a remarkable fact that, in an ideal situation, it is possible to construct a reversible computer that works without dissipating energy [2, 3, 8].

Another interesting point comes from a computational viewpoint. That is, several systems have universal computing ability even if reversibility constraint is added. Bennett [1] showed that any (irreversible) Turing machine can be simulated by a reversible one without leaving garbage symbols on the tape. Fredkin and Toffoli [8] showed that any logic circuit (even if it is irreversible) can be embedded in a circuit composed of only one kind of reversible gate called "Fredkin gate". As for RCA, it has also been shown that RCAs are computation-universal for both one-dimensional [14, 16, 19] and two-dimensional cases [13, 15, 23].

When one tries to construct, in near future, nano-scaled physical objects with some logical functions (rather than to design conventional VLSI circuits), it gets much more important to have new theories for logic design and computing that directly reflect microscopic physical laws, such as reversibility, quantum mechanical property, or like these. In particular, when developing a system with very low energy dissipation, reversibility problem seems very important. Furthermore, when manufacturing such a microscopic system, some self-assembling mechanism will be indispensable, since present techniques such as photoengraving will become useless.

In this paper, we investigate the problems how logical universality and self-reproducing ability can be obtained in a reversible space. We study these problems using reversible CA, an artificial model of a reversible space. In section 2, definitions of our framework of CA called "partitioned cellular automata" (PCA) and their reversibility is given. In section 3, we address the problem of designing a logically universal reversible CA. In section 4, the problem of self-reproduction in reversible CA is investigated.

2 Reversible Cellular Automata

In order to design an RCA, we use a framework of a partitioned CA (PCA) [14] rather than a usual CA. A PCA with von Neumann neighborhood (5-neighbor PCA) is a special type of CA whose cell is divided into five parts (Fig.1). The next state of each cell is determined by the present states of the center part of this cell, the lower part of the upper cell, the left part of the right cell, the upper part of the lower cell, and the right part of the left cell (not depending on the entire five cells). In PCA, injectivity of the global map is equivalent to injectivity of the local map [14]. Therefore, it makes easy to design an RCA.

Definition 2.1 A *deterministic two-dimensional partitioned cellular automaton* (PCA) with von Neumann neighborhood (5-neighbor PCA) is a system defined by

$$P = (\mathbf{Z}^2, (C, U, R, D, L), g, (\#, \#, \#, \#, \#))$$

where $C, U, R, D,$ and L are non-empty finite sets of states in center, up, right, down and left parts of each cell, $g : C \times D \times L \times U \times R \to C \times U \times R \times D \times L$ is a local

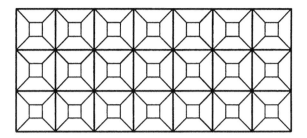

Fig. 1. Cellular space of a 5-neighbor PCA.

function, and $(\#, \#, \#, \#, \#) \in C \times U \times R \times D \times L$ is a quiescent state satisfying $g(\#, \#, \#, \#, \#) = (\#, \#, \#, \#, \#)$.

A *configuration* over the set $Q = C \times U \times R \times D \times L$ is a mapping $\alpha : \mathbf{Z}^2 \to Q$. Let $\mathrm{Conf}(Q)$ denote the set of all configurations over Q, i.e., $\mathrm{Conf}(Q) = \{\alpha \mid \alpha : \mathbf{Z}^2 \to Q\}$.

Let CENTER (UP, RIGHT, DOWN, LEFT, respectively) be the projection function which picks out the center (up, right, down, left) element of a quintuple in $C \times U \times R \times D \times L$. The global function $G : \mathrm{Conf}(C \times U \times R \times D \times L) \to \mathrm{Conf}(C \times U \times R \times D \times L)$ of P is defined as follows.

$$\forall (x, y) \in \mathbf{Z}^2 :$$
$$G(\alpha)(x, y) = g(\mathrm{CENTER}(\alpha(x, y)), \mathrm{DOWN}(\alpha(x, y + 1)), \mathrm{LEFT}(\alpha(x + 1, y)),$$
$$\mathrm{UP}(\alpha(x, y - 1)), \mathrm{RIGHT}(\alpha(x - 1, y)))$$

We say P is *globally reversible* iff G is one-to-one, and *locally reversible* iff g is one-to-one.

A *4-neighbor PCA* is a special type of a 5-neighbor PCA such that the state set C of the center part contains only one element. When depicting a figure of a cellular space or a rule of a 4-neighbor PCA, we simply omit the center part of each cell.   □

State transition of a cell by the equation $g(c, d, l, u, r) = (c', u', r', d', l')$ can be depicted as in Fig.2. We call this figure a *rule* of P. Besides such a figure, we also use an abbreviated notation

$$[c, d, l, u, r] \longrightarrow [c', u', r', d', l']$$

to denote a rule of P. In what follows, we regard the local function g as the set of such rules.

It is easy to show the following propositions on PCA. Proofs are omitted here, since analogous results for one-dimensional PCA are proved in [14].

Proposition 2.1 Let P be a PCA. P is globally reversible iff it is locally reversible.

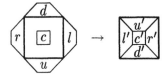

Fig. 2. A rule of a 5-neighbor PCA.

Proposition 2.2 For any PCA P, there is a CA A whose global function is identical with that of P.

By Proposition 2.1, globally or locally reversible PCA is called simply "reversible" and denoted by RPCA. Proposition 2.2 says that PCA is a subclass of CA.

By above, if we want to construct a reversible CA, it is sufficient to give a PCA whose local function g is one-to-one (note that the numbers of elements of domain and range of g are the same).

3   Logically Universal RPCAs

A CA is called *logically universal* if any logic circuit can be embedded in the cellular space by giving an appropriate configuration. Here, we discuss logically universal RPCAs, especially the problem how simple they can be.

In order to show universality of a given CA, it suffices to show that a universal set of logical elements (e.g. {AND, NOT}), signal routing, and signal delay are realizable in it. But in the case of "reversible" CAs, it is difficult to directly embed a non-one-to-one logical function (such as AND) in the cellular space. Instead, we employ a method to realize a Fredkin gate, a 3-input 3-output universal gate.

A Fredkin gate (F-gate) [8] is a reversible (i.e., its logical function is one-to-one) and bit-conserving (i.e., the number of 1's is conserved between inputs and outputs) logic gate shown in Fig.3. It has been known that any combinational logic element (especially, AND, OR, NOT, and fan-out elements) can be realized only with F-gates. Thus, any sequential circuit can be constructed from F-gates and unit delays.

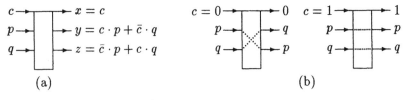

Fig. 3. (a) A Fredkin gate, and (b) its function.

Fredkin and Toffoli [8] further showed that an F-gate is embedded in the Billiard Ball Model (BBM). The BBM is a reversible and conservative physical model of computation, in which logical operations are performed by elastic collision of ideal balls. To realize an F-gate in the BBM, they introduced simpler logic gate called switch gate (S-gate) in Fig.4. An F-gate is then constructed from two S-gates and two inverse S-gates as shown in Fig.5. Note that an inverse S-gate is a gate that realizes inverse logical function of the former, and is obtained by reversing the direction of arrows of an S-gate in Fig.4.

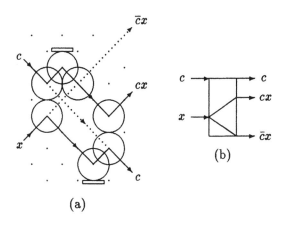

(a)

(b)

Fig. 4. (a) A switch gate in the BBM, and (b) its notation.

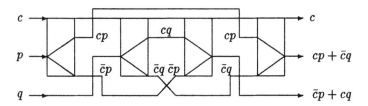

Fig. 5. Construction of an F-gate by two S-gates and two inverse S-gates.

Margolus [13] gave a very simple universal 2-state RCA in which BBM can be embedded. Though his model is interesting, it has a special kind of neighborhood, the "Margolus neighborhood", that varies depending on the parity of time, and thus is slightly non-uniform both in time and space.

On the other hand, Morita and Ueno [15] showed two models of 16-state 4-neighbor universal RPCAs, which can be regarded as special types of usual

CAs. Fig.6 shows the local function of the Model 1 (it is easy to verify its local reversibility). Fig.7 shows a reflection of a signal (or a "ball") by a mirror, where two consecutive dots form one bit of signal. Then by appropriately placing mirrors in the cellular space, we can construct an S-gate as shown in Fig.8. A whole configuration of an F-gate is given in [15].

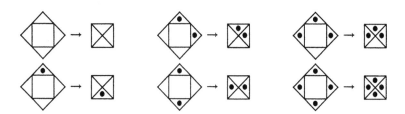

Fig. 6. The local function of a 16-state 4-neighbor universal RPCA Model 1. (Since the local function is rotation symmetric, rotated rules are omitted here.)

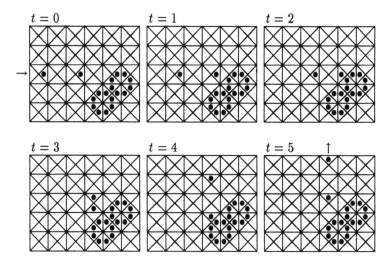

Fig. 7. Reflection of a signal by a mirror in the universal 16-state RPCA Model 1.

Figs.9 and 10 show the local function of another universal 16-state RPCA Model 2 [15] and its S-gate configuration.

If we consider the framework of PCA with square-shaped cells, 16-state PCAs are minimum ones except the one-state (trivial) PCA, provided that the local function is "isotropic" (i.e., rotation symmetric). Hence, the above two models are minimum-state universal two-dimensional RPCAs. However, it is un-

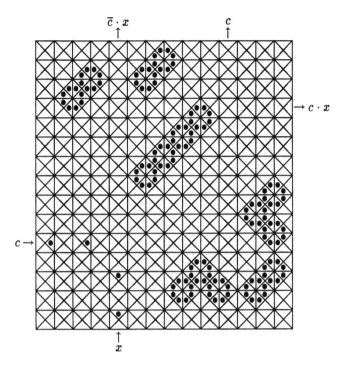

Fig. 8. Realization of an S-gate in the universal 16-state RPCA Model 1.

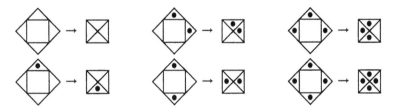

Fig. 9. The local function of a 16-state 4-neighbor universal RPCA Model 2. (Rotated rules are omitted here.)

known whether there is a non-isotropic universal RPCA having smaller number of states. It is also an open problem whether there exist a universal RCA (of a "non-partitioned" type) having 15 states or less.

As for a 3-neighbor "triangular" PCA (i.e., a PCA with triangular cells), Imai and Morita [10] investigate universality of an 8-state PCA with a local function shown in Fig.11. They showed that an S-gate can be embedded in its cellular space (but an F-gate configuration has not yet been obtained since its size seems to be large).

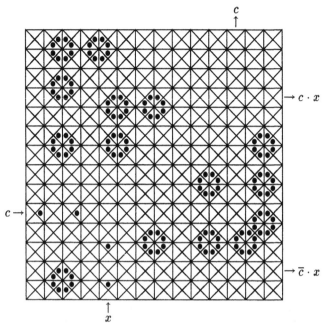

Fig. 10. Realization of an S-gate in the universal 16-state RPCA Model 2.

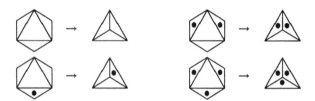

Fig. 11. The local function of a 8-state 3-neighbor triangular RPCA. (Rotated rules are omitted here.)

4 Self-Reproduction in RPCAs

Next, we discuss how self-reproduction is possible in a reversible cellular space. It is well known that von Neumann [22] first showed a self-reproducing machine using his 29-state CA model. When giving a definition of self-reproduction, von Neumann supposed the computation- and construction-universality as the condition for a self-reproducing machine to exclude passive or trivial replication of patterns in a cellular space. But, because of this condition, the early models [5, 22] needed huge configurations.

Instead of the above condition, Langton [12] posed a new criterion only requiring that the construction of a daughter configuration should be actively directed by the parent configuration, and that information stored in the configuration must be treated in two different manners, i.e., "interpreted" and "uninterpreted". Along this line, he designed a very simple self-reproducing CA in a modified 8-state model of Codd [5].

160

Although von Neumann's self-reproducing machine was very complex, his essential idea was elegant and useful. Hence, most models after von Neumann employed his method. In his model, a self-reproducing machine possesses its description as a "gene" besides the body itself. And self-reproduction is carried out by interpreting the description to construct a body, then copying and attaching it to the daughter.

However, if the machine can encode its shape into a description by checking its body dynamically, there is no need to keep the entire description. In fact, there proposed a few models that performs self-reproduction in such a manner [9, 11, 20, 21]. Ibáñes et al. [9] showed a 16-state model in which sheathed loops can reproduce by using a self-inspection method. Morita and Imai [21] also gave this kind of method independently. In the latter model, it is further shown that such a type of self-reproduction can be realized in a "reversible" cellular space.

We now describe the model called "SR_8" given by Morita and Imai [21], and explain how self-reproduction is possible in a reversible space. In the cellular space of SR_8, encoding the shape of an object into a "gene" represented by a command sequence, copying the gene, and interpreting the gene to create an object, are all performed reversibly. By using these operations, various objects called Worms and Loops can reproduce themselves in a very simple manner.

The RPCA "SR_8" is defined by

$$SR_8 = (\mathbf{Z}, (C, U, R, D, L), g, (\#, \#, \#, \#, \#)),$$
$$C = U = R = D = L = \{\#, *, +, -, A, B, C, D\}.$$

Hence, each of five parts of a cell has 8 states. The states A, B, C and D mainly act as signals that are used to compose "commands". The states $*, +,$ and $-$ are used to control these signals.

The local function g contains 765 rules, and is a one-to-one mapping. The complete listing of the rules is given in [21].

4.1 Signal Transmission on a Wire

A *wire* is a configuration to transmit signals A, B, and C. Fig. 12 shows an example of a part of a simple (i.e., non-branching) wire.

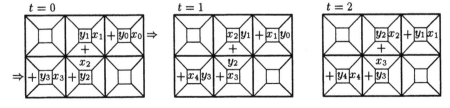

Fig. 12. Signal transmission on a part of a simple wire ($x_i, y_i \in \{$A,B,C$\}$).

A *command* is a signal sequence composed of two signals. There are six commands consisting of signals A, B and C as shown in Table 1. These commands are used for extending or branching a wire.

Command		Operation
First signal	Second signal	
A	A	Advance the head forward
A	B	Advance the head leftward
A	C	Advance the head rightward
B	A	Branch the wire in three ways
B	B	Branch the wire in two ways (making leftward branch)
B	C	Branch the wire in two ways (making rightward branch)

Table 1. Six commands composed of A, B, and C.

4.2 A Worm

A *Worm* is a simple wire with open ends that are called a *head* and a *tail*. It crawls in the reversible cellular space as shown in Fig. 13.

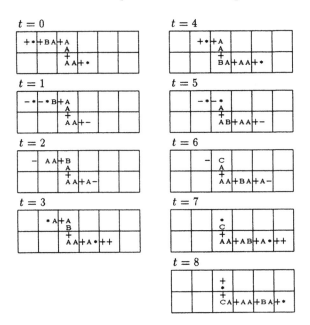

Fig. 13. Behavior of a Worm.

Commands in Table 1 are decoded and executed at the head of a Worm. That is, the command AA extends the head straight, while the command AB (or AC, respectively) extends it leftward (rightward). On the other hand, at the tail cell, the shape of the Worm is "encoded" into an advance command. That is, if the tail of the Worm is straight (or left-turning, right-turning, respectively) in its form, the command AA (AB, AC) is generated. The tail then retracts by one cell.

4.3 Self-Reproduction of a Worm

By giving a branch command, *any* Worm can self-reproduce indefinitely provided that it neither cycles nor touches itself in the branching process. Fig. 14 shows self-reproducing processes of Worms.

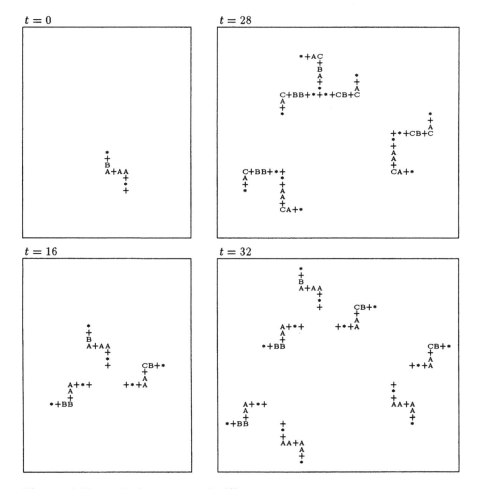

Fig. 14. Self-reproducing process of a Worm.

4.4 Self-Reproduction of a Loop

A *Loop* is a simple closed wire, thus has neither a head nor a tail as shown in Fig. 15.

Fig. 15. An example of a Loop.

If a Loop contains only advance or branch commands, they simply rotate in the Loop and self-reproduction does not occur. In order to make a Loop self-reproduce, commands in Table 2 are used.

Command		Operation
First signal	Second signal	
D	B	Create an arm
D	C	Encode the shape of a Loop

Table 2. Commands DB and DC.

Examples of entire self-reproducing processes of Loops are shown in Figs. 16 and 17. By putting a command DB at an appropriate position, *every* Loop having only AA commands in all the other cells can self-reproduce in this way.

5 Concluding Remarks

In this paper, we investigated the problems of logical universality and self-reproduction in RCA. These problems turned out to have relatively simple solutions, and we explained a few of them. However, there still remain several open problems whether it is possible to simplify these solutions. It is also left for the future study to design an RCA that supports both logical universality and self-reproducing ability in an elegant manner.

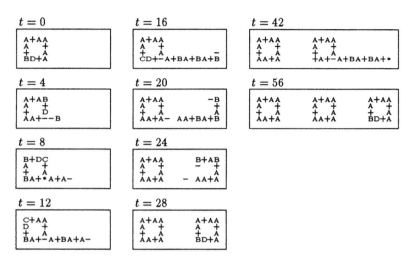

Fig. 16. Self-reproducing process of a Loop (1)

References

1. Bennett, C.H., Logical reversibility of computation, *IBM J.Res.Dev.*, **17**, 525–532 (1973).
2. Bennett, C.H., The thermodynamics of computation, *Int.J.Theoret.Phys.*, **21**, 905–940 (1982).
3. Bennett, C.H., and Landauer, R., The fundamental physical limits of computation, *Sci.Am.*, **253**, 38-46 (1985).
4. Bennett, C.H., Notes on the history of reversible computation, *IBM J.Res.Dev.*, **32**, 16–23 (1988).
5. Codd, E.F., *Cellular Automata*, Academic Press, New York (1968).
6. Deutsch, D., Quantum theory, the Church-Turing principle and the universal quantum computer, *Proc.R.Soc.Lond.A*, **400**, 97–117 (1985).
7. Feynman, R.P., Simulating physics with computers, *Int.J.Theoret.Phys.*, **21**, 467–488 (1982).
8. Fredkin, E., and Toffoli, T., Conservative logic, *Int.J.Theoret.Phys.*, **21**, 219–253 (1982).
9. Ibáñez, J., Anabitarte, D., Azpeitia, I., Barrera, O., Barrutieta, A., Blanco, H., and Echarte, F., Self-inspection based reproduction in cellular automata, in *Advances in Artificial Life* (eds. F. Moran et al.), LNAI-929, Springer-Verlag, 564–576 (1995).
10. Imai, K., and Morita, K., On computation-universal two-dimensional cellular automata (in Japanese), *Proc. LA Summer Symposium* (Kobe), 4·1-6 (1996). http://kelp.ke.sys.hiroshima-u.ac.jp/projects/rca/urpca/triangular/
11. Laing, R., Automaton models of reproduction by self-inspection, *J.Theor.Biol.*, **66**, 437–456 (1977).
12. Langton, C.G., Self-reproduction in cellular automata, *Physica*, **10D**, 135–144 (1984).
13. Margolus, N., Physics-like model of computation, *Physica*, **10D**, 81–95 (1984).
14. Morita, K., and Harao, M., Computation universality of one-dimensional reversible (injective) cellular automata, *Trans.IEICE*, **E-72**, 758–762 (1989).

$t = 0$

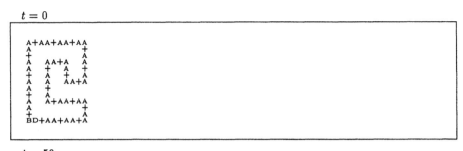

$t = 50$

$t = 100$

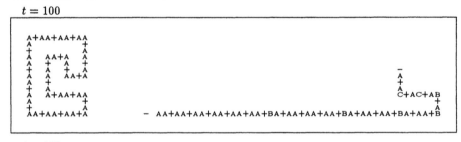

$t = 152$

Fig. 17. Self-reproducing process of a Loop (2)

15. Morita, K., and Ueno, S., Computation-universal models of two-dimensional 16-state reversible cellular automata, *IEICE Trans.Inf.& Syst.*, **E75-D**, 141–147 (1992).

16. Morita, K., Computation-universality of one-dimensional one-way reversible cellular automata, *Inform.Process.Lett.*, **42**, 325–329 (1992).

17. Morita, K., Reversibility in computation (in Japanese), *J.Inform.Process.Soc.of Japan*, **35**, 306–314 (1994).

18. Morita, K., Reversible cellular automata (in Japanese), *J.Inform.Process.Soc.of Japan*, **35**, 315–321 (1994).

19. Morita, K., Reversible simulation of one-dimensional irreversible cellular automata, *Theoret.Comput.Sci.*, **148**, 157–163 (1995).
20. Morita, K., and Imai, K., A simple self-reproducing cellular automaton with shape-encoding mechanism, *Proc. ALIFE V* (Nara) (1996).
21. Morita, K., and Imai, K., Self-reproduction in a reversible cellular space, *Theoret.Comput.Sci.*, **168**, 337–366 (1996).
   http://kepi.ke.sys.hiroshima-u.ac.jp/projects/rca/sr/
22. von Neumann, J., *Theory of Self-reproducing Automata* (ed. A.W.Burks), The University of Illinois Press, Urbana (1966).
23. Toffoli, J., Computation and construction universality of reversible cellular automata, *J.Comput.Syst.Sci.*, **15**, 213–231 (1977).
24. Toffoli, T., and Margolus, N., Invertible cellular automata: a review, *Physica D*, **45**, 229–253 (1990).
25. Watrous, J., On one-dimensional quantum cellular automata, *Proc. 36th IEEE Symposium on Foundations of Computer Science*, 528–537 (1995).

Engineering Applications of EHW

Data Compression Based on Evolvable Hardware

Mehrdad Salami*, Masahiro Murakawa** and Tetsuya Higuchi***

Computation Models Section
Electrotechnical Laboratory (ETL)
1-1-4 Umezono, Tsukuba, Ibaraki, 305, Japan.
e-mail: m_salami@etl.go.jp

Abstract. We have investigated the possibility of applying Evolvable Hardware (EHW) to data compression applications. One of the interesting area in data compression is Predictive Coding which we used for compressing block of data in the hardware configuration of EHW. The advantage of this approach is simplicity, adaptability, real time implementation for motion pictures and advantage of using non-linear prediction functions. Several configurations of EHW are tested to find the optimal system for data compression and the results show good performance compared with Neural Networks and JPEG approaches.

1   Introduction

Over the last ten years the idea of evolutionary systems have been attracted many people in engineering and computer science. Many systems in those fields are not optimized or don't have enough flexibility to work in different situations. Evolutionary systems claim to provide better performance for time-varying systems [1]. As many systems now working with high speed, a faster solution with high flexibility to deal with different conditions is highly required. Specialized hardware can be applied to these systems for speed improvement [2]. On the other hand specialized hardware doesn't have enough flexibility in time-varying systems and produce poor results in critical situations. Another approach is using reprogrammable hardware with the advantage of programming hardware for different situations [3]. The speed of reprogrammable hardware is lower than specialized hardware but gives more flexibility. The main question is how to find the best hardware architecture for a new situation in the system.

The approach that we choose here is based on a predefined hardware architecture which contains many non-linear functions and uses an evolutionary algorithm to find the best configuration for different situations. A system called Evolvable Hardware (EHW) is designed and will be fabricated in the near future. It contains floating point hardware blocks, programmable by a Genetic

---

    * The New Energy and Industrial Technology Development Organization (NEDO)
   ** University of Tokyo, 7-3-1 Hongo, Bunkyo, Tokyo, Japan
  *** Electrotechnical Laboratory, 1-1-4 Umezono, Tsukuba, Ibaraki. Japan

Algorithm (GA) software program. The software GA is capable of changing the architecture of EHW for different situations which bring adaptive behavior to the system. The system benefits from hardware speed and potential of applying to real time systems.

The rest of this paper is as follow. In the section 2 Evolvable Hardware will be explained in details. Section 3 explains the Data Compression (DC). Section 4 demonstrates how EHW can be applied to DC. Section 5 represents simulation results of data compression by EHW and compares it with other methods.

2   Evolvable Hardware

Research on EHW was initiated independently in Japan and in Switzerland around 1992 (for recent overviews, see [4] and [5]). Since then, the interest is growing rapidly (e.g., EVOLVE95, the first international workshop on evolvable hardware was held in Lausanne in 1995).

Most research on EHW, however, has the common problem that the evolved circuit size is small. The hardware evolution is based on primitive gates such as AND-gate and OR-gate, we call evolution at this level gate-level evolution. Gate-level evolution is not powerful for the use in industrial applications.

In order to solve this problem, we propose a new type of hardware evolution, function-level evolution, and a new FPGA (Field Programmable Gate Array) architecture dedicated to function-level evolution. Actually EHW can synthesize non-linear function genetically. This suggests that EHW may substitute Artificial Neural Networks (ANN) in industrial applications because EHW enables faster and more compact implementation than ANN, in addition to better understandability of learned results.

We use the FPGA model in Figure 1 to realize function-level evolution. The FPGA model consists of 20 columns, each containing five Programmable Floating processing Units (PFUs). Each PFU can implement one of the following seven functions: an adder, a subtracter, an if-then, a sine generator, a cosine generator, a multiplier, and a divider. The selection of the function to be implemented by a PFU is determined by chromosome genes given to the PFU. Constant generators are also included in each PFU. Columns are interconnected by crossbar switches. The crossbars determine inputs to PFUs.

In addition to these columns, a state register holding a past output is prepared for applications which deal with time-continuous data. This FPGA model assumes two inputs and one output. Data handled by FPGA are floating point numbers [6].

EHW can be uses in many kinds of applications where hardware specifications are not known. It may replace ANNs in industrial applications because of real-time response and compatibility with time continuous data. We are developing a prototype system board to demonstrate the on-line evolution of a hardware device.

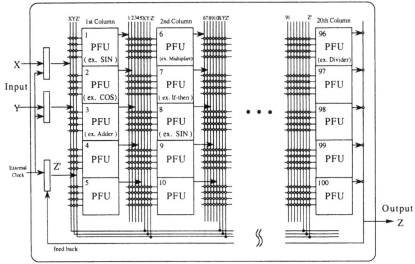

PFU : Programmable Floating processing Unit

Fig. 1. The FPGA model for function level evolution

3   Data Compression and Predictive Coding

Digital transmission is a dominant means of communication for voice and image. It is expected to provide flexibility, reliability and cost effective, with the added potential for communication privacy and security through encryption. The cost of digital storage and transmission media are generally proportional to the amount of digital data that can be stored or transmitted . While the cost of such media decreases every year, the demand for their use increases at an even higher rate. Therefore there is a continuing need to minimize the number of bits necessary to transmit images while maintaining acceptable image quality.

Normally, images show a high degree of correlation among neighboring samples. A high degree of correlation implies a high degree of redundancy in the raw data. Therefore if the degree of redundancy is removed, a more efficient and hence compressed coding of the signal is possible [7]. This can be achieved through the use of Predictive Coding (PC).

Figure 2 shows a block diagram of the predictive coding system. In the figure, $x(n)$ are the input values e.g. pixel values, $\hat{x}(n)$ are the predicted values, $e(n)$ are error values equal to $\hat{x}(n) - x(n)$, $y(n)$ are reconstructed values and $\hat{e}(n)$ is transmitted error value after quantization.

In predictive coding an image will be replaced by [8]

1 - A formula which predict each pixel by previous or neighboring pixels and

2 - The error at each pixel which is the difference between the predicted value and the original value.

In this coding repeating or correlated information can be saved in formula and uncorrelated information will be handled by the error values. This algorithm can be used for lossy or lossless compression. In lossy compression the error will be ignored or sent partially and in lossless compression the error fully quantised and recovered in receiver.

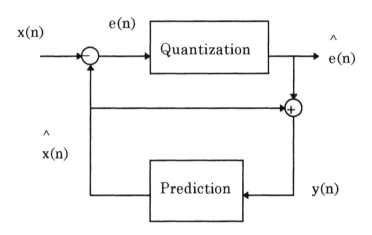

Fig. 2. Block diagram of the Predictive Coding.

Predictive coding formula may be linear or non-linear [9]. In linear Predictive Coding, a gray level $g(i, j)$ is predicted by the linear combination of four pixel values as shown in Figure 3. The coefficients of the linear prediction are determined by adapting the least square method to the neighboring area of $g(i, j)$.

In a non-linear approach, a function f will be selected and optimized to predict the pixel values. If the same neighboring as linear PC is used then each pixel can be estimated as

$$g(i, j) = f[g(i-1, j-1), g(i, j-1), g(i-1, j), g(i+1, j-1)]$$

JPEG (Joint Photographic Expert Group) uses a simple PC for lossless compression. The predictor function in JPEG combines the values of up to three neighboring samples ($g(i - 1, j)$, $g(i, j - 1)$ and $g(i - 1, j - 1)$) to form a prediction of $g(i, j)$ in Figure 3. This prediction is then subtracted from the actual value of $g(i, j)$, and the difference is encoded by entropy coding methods. Any one of the seven prediction functions listed in Table 1 can be used. Functions 1, 2 and 3 are used for one dimensional prediction and functions 4, 5, 6 and 7 are two dimensional prediction [10].

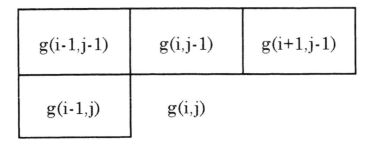

| g(i-1,j-1) | g(i,j-1) | g(i+1,j-1) |
| g(i-1,j) | g(i,j) | |

Fig. 3. Prediction of $g(i, j)$ from neighboring four pixels.

Table 1. Prediction function for JPEG lossless coding.

No.	Prediction Function f
1	$g(i - 1, j)$
2	$g(i, j - 1)$
3	$g(i - 1, j - 1)$
4	$g(i - 1, j) + g(i, j - 1) - g(i - 1, j - 1)$
5	$g(i - 1, j) + (g(i, j - 1) - g(i - 1, j - 1))/2$
6	$g(i, j - 1) + (g(i - 1, j) - g(i - 1, j - 1))/2$
7	$(g(i - 1, j) + g(i, j - 1))/2$

Most of the works so far in PC is based on linear functions, but this method is not always effective because it assumes the image obeys a linear model and the statistical properties of an image remains constant for the whole image [11]. In our work these two assumptions are removed and more flexible and effective algorithm based on EHW is applied. The EHW because of the structure of PFUs is designed to find a non-linear function for prediction. For compression, the image will be divided into smaller blocks and a non-linear function should be found for each block. It guarantees selection of a function for different parts of the image. The next section explains more details of this approach.

4 EHW for Predictive Coding

EHW can be considered as a predictive function for optimization. This function is not linear and changes depends on characteristic of the block for compression. In this approach one image is divided into a number of blocks (Figure 4) and then one function has to be found for each block. Selecting one function for the whole image will produce poor results. By using EHW for finding function for different region of the image we actually implementing an adaptive prediction coding for the image. Using non-linear function in EHW will enhance the performance because most blocks show a non-linear relationship.

Selection of the block size and the neighboring pixels are very important for our approach. If the block size is too small, then the EHW chromosome will be too long to represent block in compressed format. Assume a block size of 2 by 2 pixels is used for PC and the chromosome length for EHW is 250 on average then 32 (2*2*8) bits information of each block will be replaced by 250 bits! On the contrary, if the block size is too large, then the EHW cannot reflect the characteristic in a local area and that leads to poor performance. Practically finding a prediction function for the large blocks (say 64 by 64 pixels) is impossible even for non-linear functions. Besides the time required by EHW to find the function will be extremely high if the block size is very small or very large. If the block size is small then too many blocks should be compressed and if the block size is large then the computation time for finding function will be high. So there should be an optimum block size. For EHW we tested blocks of 8 by 8, 16 by 16 and 32 by 32 which are used in other works and we found 16 by 16 produces the best result regarding the quality of picture and the number of bits generated for each block.

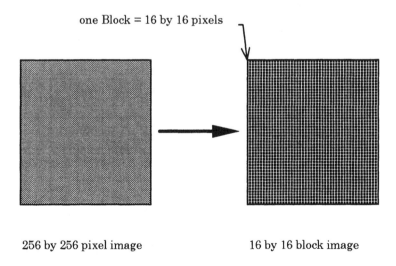

Fig. 4. Dividing a 256 by 256 pixel image to the blocks of 16 by 16 pixels.

If the number of neighboring pixels is small then finding an appropriate function to predict for all block will be difficult. On the other hand high number of neighboring pixels will take a long time to find the function. Not only number of neighbors is important but the location of neighbor pixel is also important. For EHW we select the neighboring which was shown in Figure 3. It is used in Predictive Coding for most of linear and non-linear functions, and we also follow that configuration.

5   Simulation Results

The system described above was used to generate compressed data for digital images. A 256 by 256 pixels image of Lenna was used for testing. As the image was gray each pixel has an intensity value between 0 and 255. The image was partitioned into 256 block each containing 256 pixels. For each block, one configuration was found for EHW to predict pixel values based on their neighboring pixels. Then each block could be compressed as one EHW configuration. If the quality of predicted pixels is not good (low signal to noise ratio) extra bits are required to send along the EHW configuration to increase the quality of picture.

Many EHW configurations are tested to find the best performance. First three block sizes of 8 by 8, 16 by 16 and 32 by 32 are tested to find the best one. The block size of 16 by 16, as mentioned in previous section, is produced the best results. Second we need to find what distribution of pixel values is matched with EHW. Each algorithm in data compression requires an appropriate distribution of pixel values. Three different distribution coding of pixel values are tested.

1 - The pixel value varies between -128 and 127.

2 - The pixel values varies between 0 and 255.

3 - Mixture of the above two coding. Originally pixel value varies between -128 and 127 and then negative parts is transferred to 128 to 255. So the code is finally varying between 0 to 255.

Table 2 compares the simulation results for the above codings.

Table 2. Signal to Noise Ratio (SNR) and bit per pixel (bpp) for different codings schemes (population=200 and generation=1000).

Coding	Signal to Noise Ratio (SNR)	bit per pixel (bpp)
Scheme 1	20.11 db	1.35
Scheme 2	19.90 db	1.33
Scheme 3	28.54 db	0.88

According to the table, the third coding produce the best results. Both the SNR and bpp are much better than the two other coding schemes.

In the next experiment the population size and the number of generations are tested to find out the performance for different execution times. Normally in real time operation the search must terminate as fast as possible and this simulation shows how much performance we will lose if the search is forced to terminate early. Table 3 demonstrate simulations when the population size changes and Table 4 shows the results of simulations when the number of generations is changed.

Table 3. Simulation results for two different population size (generation=1000 and scheme 3).

Population	Signal to Noise Ratio (SNR)	bit per pixel (bpp)
50	27.92 db	0.93
200	28.54 db	0.88

Table 4. Simulation results for two different generations(population=200 and scheme 3).

Generations	Signal to Noise Ratio (SNR)	bit per pixel (bpp)
100	27.45 db	0.97
1000	28.54 db	0.88

These two tables show by decreasing the amount of time to find the best configuration, the performance will be reduced but the reduction is not very large.

Most of algorithms in data compression provide an parameter for user to control the quality of image or SNR. In EHW approach we introduced a threshold parameter for SNR of a block. For each block if EHW can compress the block with SNR greater than the threshold value then no more error data is required for compression. On the other hand if the SNR of block is less than the threshold then error data is necessary for compression to have a target SNR. In the last experiment the threshold for sending extra bits for the block is examined. Two threshold values of 20 db and 25 db are tested and the simulation results are presented in the Table 5.

The table shows large difference in bit per pixel and suggests that 20db is a good choice but the SNR must be improved for that case.

Table 5. Simulation for two different threshold values (generation=1000 and population =200 and scheme 3).

Threshold	Signal to Noise Ratio (SNR)	bit per pixel (bpp)
20 db	25.60 db	0.53
25 db	28.54 db	0.88

Finally The compressed picture is compared with the result of compression by Neural Networks [12] and JPEG approaches [13] which is shown in Figure 5.

The figure shows that EHW is very competitive with NNs and JPEG compression and by further improvement in our system which is not very difficult we expect much better performance than NNs and JPEG approaches. We are working on variable block size for data compression and we find a much better performance which we will report it later.

6 Discussions

The previous sections show that EHW has a good capability for data compression and the simulation results predict a close performance compare with other methods in data compression. However these results definitely must be improved and the current investigations are aiming at produce results far better than before.

Another point is real time characteristic of this method. At the moment EHW is applied to static image pictures but later it will be applied to real time compression for motion image pictures. Using the ability of hardware compression by EHW and by establishing a mechanism for real time compression we should be able to compete with MPEG.

We are working on a new version of this approach based on variable block size. It has shown a good increase in the performance of our approach which we will report it later.

7 Conclusions

This research shows that the EHW can be applied to data compression and the results of simulations show good performance of EHW in compression. Several configurations are tested to find optimum system for EHW for data compression and more tests are required to produce a complete system with high performance. This system is capable of applying in real time for motion pictures and give us hope for a wide spread application of the evolvable hardware in current technology.

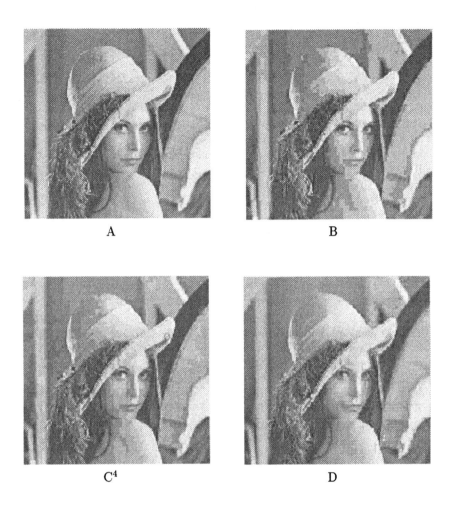

Fig. 5. Comparison between EHW, NNs and JPEG compression systems. A) Original Lenna image. B) JPEG compression (bpp=0.5, SNR=26.5 db). C) NN compression (bpp=0.7, SNR=26.95). D) EHW compression (SNR=28.54, bpp=0.88).

References

1. Hemmi H., Mizoguchi J., and Shimohara K., "Development and Evolution of Hardware Behaviors", Proceedings of Artificial Life IV, MIT Press, 1994.
2. Salami M. and Cain G., "Adaptive Hardware Optimization Based on Genetic Algorithms", Proceedings of The Eighth International Conference on Industrial Application of Artificial Intelligence & Expert Systems (IEA95AIE) , Melbourne, Australia, June 1995, pp. 363-371.
3. Higuchi T. et al., "Evolvable Hardware and its Applications to Pattern Recognition and Fault-tolerant Systems", Proceedings of the First International Workshop

---

[4] This is a copy from Lena image in [12]

Toward Evolvable Hardware, Lausanne, Switzerland, Lecturer Notes in Computer Science, Spring Verlag, 1995.

4. Higuchi T. et al., "Evolvable Hardware", in *Massively Parallel Artificial Intelligence*, edited by Kitano H. and Hendler J., pp.398-421, MIT Press, 1994.

5. Marchal P. et al., "Embryological Development on Silicon", Proceedings of Artificial Life IV, MIT Press, 1994.

6. Murakawa M. et al., "Hardware Evolution at Function Level", Proceeding of Parallel Problem Solving from Nature (PPSN) 1996.

7. Li J., and Manikopoulos C.N., "Nonlinear Prediction in Image Coding with DPCM", Electronics Letters, Vol. 26, No. 17, August 1990, pp. 1357-1359.

8. Kuroki N., Nomura T., Tomita M., and Hirano, K., "Lossless Image Compression by Two-Dimensional Linear Prediction with Variable Coefficients", IEICE Transaction on Fundamentals, Vol. E75-A, No. 7, July 1992, pp. 882-889.

9. Tekalp A.M., Kaufman H., and Woods J.W., "Fast Recursive Estimation of the Parameters of a Space-Varying Autoregressive Image Model", IEEE Transactions on Acoustics, Speech and Signal Processing, Vol. ASSP-33, No. 2, April 1985, pp. 469-472.

10. Wallace G.K., "The JPEG Still Picture Compression Standards", Communication of ACM, Vol. 34, No. 4, April 1991, pp. 30-44.

11. Dukhovich I.J., "A DPCM System Based on a Composite Image Model", IEEE Transactions on Communications, Vol. 31, No. 8, April 1983, pp. 1003-1017.

12. Parodi G., and Passaggio F., "Size-Adaptive Neural Network for Image Compression", Proceedings of the first International Conference on Image Processing 1994 (ICIP94), IEEE Computer Society Press, Vol. 3, pp. 945-7.

13. Lane T., The Independent Group Public Domain JPEG, Shareware Software, 1996.

ATM Cell Scheduling by Function Level Evolvable Hardware

Weixin Liu[1], Masahiro Murakawa[2] and Tetsuya Higuchi[3]

[1] The New Energy and Industrial Technology Development Organization (NEDO)
e-mail: w-liu@etl.go.jp
[2] University of Tokyo, 7-3-1 Hongo, Bunkyo, Tokyo, Japan
[3] Electrotechnical Laboratory, 1-1-4 Umezono, Tsukuba, Ibaraki, Japan

Abstract. In this paper, we study the possibility of using Evolvable Hardware (EHW) for scheduling real-time traffic in Asynchronous Transfer Mode (ATM) networks. EHW is hardware which is built on programmable logic devices and whose architecture can be reconfigured by using genetic learning to adapt to new environments. A novel design is the function-level EHW [5, 9] based on Field Programmable Gate Array (FPGA) chips, where a number of Programmable Floating processing Units (PFUs) are embedded in one chip. The selectable high-level hardware functions of each PFU make the function-level EHW to be suitable for a wide variety of applications in practice. In our experiment, a statistical multiplexer at an ATM node is modeled for the purpose of generating training data. Superposed bursty cell streams are applied to the model of the multiplexer. The EHW is trained by the collected data in the learning phase. After learning is complete, the best chromosome is tested with respect to various traffic characteristics and the Quality Of Service (QOS) requirements of the cell traffic. Simulation results show that the function-level EHW performs the control well for cell scheduling problem in ATM networks.

1 Introduction

In Broadband Integrated Services Digital Networks (B-ISDN), Asynchronous Transfer Mode (ATM) is a desirable transfer mode which multiplexes various types of traffic in a single transmission medium. Indeed, ATM provides the flexibility for carrying out diverse applications with different characteristics and the efficiency for allocating required bandwidth to a connection. However, ATM technique also introduces the challenge of traffic management within the networks. Efficient traffic control mechanisms are required for guaranteeing the quality of information transmission, and meanwhile for achieving high throughput of the network. Generally there are three control levels in ATM networks: network control level, connection control level and cell control level [6]. Network control level refers to the functions of optimal routing, link capacity assignment etc. Most important task in connection control level is connection admission control, which determines whether a new connection request can be accepted or not based on current traffic condition in the network. The functions in cell control level involves user parameter control and network parameter control, traffic shaping, congestion control, priority control etc. [7] Clearly, the lower the control level, the faster the control actions should be performed.

Due to the unpredictable behavior of the multiplexed traffic, it is difficult to implement an efficient and simple control strategy in ATM networks. With respect to the uncertain traffic characteristics in ATM networks, researchers have used Neural Network (NN) to deal with some control problems, such as connection admission control in [6] and traffic enforcement in [11]. However, the implementation of such an NN has not been announced yet. This task becomes more difficult from the fact of very fast transmission and switching in ATM networks.

A novel design is the function-level EHW [5, 9] based on Field Programmable Gate Array (FPGA) chips, where a number of Programmable Floating processing Units (PFUs) are embedded in one chip. The selectable high-level hardware functions of each PFU make the function-level EHW to be suitable for a wide variety of applications in practice. In comparison with the first generation EHW (i.e. gate-level EHW), the function-level EHW can attain much higher performances when applying to a series of applications [9]. Owing to its capability of real-time processing with high-level hardware functions and its capability of nonlinear functions learning based on Genetic Algorithms (GAs) [2], the function-level Evolvable Hardware (EHW) is a potential candidate for dealing with real-time and adaptive control problems in practice.

In this paper, we study the problem of scheduling real-time traffic in ATM networks and present an adaptive cell scheduling policy conducted by the function-level EHW. Cell scheduling is one kind of priority control performed in cell control level. In ATM networks, different classes of traffic, such as voice, video, data, and image, expect different Quality Of Service (QOS) from the networks. The real-time traffic such as voice and video is sensitive to delay, thus requires short transfer delay. If the delay is too large, the receiver would not be able to recover the sounds or pictures with good quality. Therefore, a scheduler should be able to reflect the QOS requirements of different classes of traffic, and should perform the function fast as well.

The proposed cell scheduling is based on the introduction of traffic classes. We classify incoming cells into two classes, class 1 cells from delay sensitive applications and class 2 cells from delay-insensitive applications. Under the condition of guaranteeing the required average cell delay for class 1 cells, the proposed scheduling policy will try to allocate as much bandwidth as possible to class 2 cells over the output link. The simulation results reveal that the flexible control can be realized by the function-level EHW. The rest of this paper is structured as follows: In section 2, the basic idea of EHW is reviewed first. Then we present the architecture and the operation principles of the function-level EHW. The simulator of the function-level EHW will be used for conducting cell scheduling function. Section 3 concerns cell scheduling problem in ATM networks. We present the model of an ATM multiplexer and the traffic source being applied, from which the training data are collected for the EHW simulator. Section 4 reports simulation results, and section 5 draws the conclusion of this paper.

2 Evolvable Hardware

EHW is hardware which is realized based on software-reconfigurable logic device such as Programmable Logic Device (PLD) and FPGA, and its architecture can be reconfigured by using genetic learning to adapt to the new environment [5]. EHW uses GA to search a suitable hardware structure. An optimal chromosome found by GA will be downloaded into PLDs or FPGAs.

Research on EHW was initiated independently in Japan and in Switzerland around 1992. Nowadays, more and more researchers realize its importance in industry and in laboratory. EHW has the following advantages [4]:

. The execution speed of the evolved system will be extremely fast (at least three orders of magnitude faster than a software implementation) because the result of adaptation is the hardware structure itself.
. EHW can implement finite state machines (FSMs) by genetic learning. This suggests the possibility that control programs of animates could be replaced with EHW implementing the real-time control of simple actions. Hence, microprocessors would simply not be required.
. Fault tolerant and flexible design is realized because EHW can change its own structure in the case of hardware error or environmental change.

Due to the primitive functions (e.g. AND-gate and OR-gate) within gate-level EHW, a difficulty arises when realizing large evolved circuit so as to apply gate-level EHW to industrial applications. To overcome the difficulty, the function-level EHW was proposed recently [5, 9].

2.1 The model of the function level EHW

The proposed function-level EHW is constructed by FPGAs, see Figure 1. The FPGA model contains 100 Programmable Floating processing Units (PFUs) that are arranged in a 20*5 grid. Each PFU may perform one of high-level functions such as addition, subtraction, multiplication, division, sine, cosine, and if-then. The selection of the function to be implemented by a PFU is determined by chromosome genes given to the PFU. Constant generators are also included in each PFU. Neighboring columns of PFUs are interconnected by crossbar switches. An output of a PFU will be fed into the input of an PFU located in the next neighboring column.

In addition to these columns, a state register holding the past output is prepared for applications which deal with time-continuous data. This FPGA model assumes two inputs and one output. FPGA is handling floating point numbers.

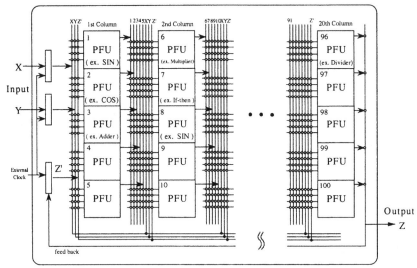

PFU : Programmable Floating processing Unit

Fig. 1. The FPGA model for function level evolutions

2.2 Genetic Learning of the function level EHW

PFU functions are selected by genetic learning at the function level evolution, so does the interconnection among PFUs. Genetic learning will find suitable hardware function for an application. Genetic operators include reproduction and the mutation: the crossover is not used yet.

Chromosome representation A variable-length chromosome representation scheme is used, which only encodes non-empty entries in the connectivity matrix of EHW (FPGAs). Such scheme enables faster GA execution and larger circuit evolution [8].

An allele in a chromosome may involve several elements, a template looks like:

$$(N, op, d1, d2)$$

where N is the identity of a PFU (expressed with a *number*) on which the *function op* is carried out, $di(i = 1, 2)$ can be an input value, a constant, or a PFU (identified by its number) from which the *input operand* is obtained. However, only is one operand necessary for the functions such as sine and cosine. $(8, +, 2, 5)$ is an allele, it implies that the hardware function '+' is executed at the 8th PFU in the second column, and two operands are the outputs from the second PFU and the fifth PFU, in the first column, respectively. It is assumed there are 5 PFUs in each column.

With such a representation, a chromosome may look like:

$$(2, sin, X), (5, cos, Y), \cdots, (8, +, 2, 5), \cdots, (98, \times, 93, 95)$$

which is a hardware function implemented by the whole FPGA.

A population for GA involves a number of chromosomes. The most desirable hardware function can be gradually synthesized for the environment through applying GA operations to the population repeatedly.

Reproduction The elitist strategy reproduction rule is used, and the chromosome with the best fitness value is always reproduced by the roulette wheel.

Mutation Mutation is applied to every chromosome after the reproduction. Due to the function level evolution, three kinds of mutations are adopted. Note that the second operand is not necessary for those single operand functions.

1. Mutation of an operand
   This is the mutation from $(N_i, op_i, d1_i, d2_i)$ to $(N_i, op_i, d1_j, d2_i)$, which means that the first operand will be obtained from the output of the PFU with number $d1_j$, rather than from the PFU with number $d1_i$.
2. Mutation of a function
   This is the mutation from $(N_i, op_i, d1_i, d2_i)$ to $(N_i, op_j, d1_i, d2_i)$, which means that the function op_j, instead of function op_i, will be carried out by the PFU N_i.
3. Insertion of a new allele
   If PFU N_j is not used in the forward column and if allele $(N_i, op_i, d1_i, d2_i)$ is mutated to $(N_i, op_i, N_j, d2_i)$, a new allele $(N_j, op_j, d1_j, d2_j)$ is generated and then inserted in the chromosome.

Readers may refer to [9] for detailed description about the function level EHW.

2.3 Simulator of the function-level EHW

Based on the design, the corresponding simulator has been implemented for the purpose of testing the functionality and the generality of the EHW. The simulator offers all functions of the specified function-level EHW. In fact, it has been successfully applied to a series of applications such as the two-spiral, the Iris data classification, 2-D image rotation, and synthesis of a finite state machine [9]. In this experiment, we use the same simulator to testify whether function-level EHW can perform cell scheduling function or not in ATM networks.

In the sequel of this paper, we will simply use term EHW to refer to the function-level EHW.

3 Cell Scheduling

In an ATM network, user messages are packetized into fixed-size packets called cells. Each cell consists of 48 bytes for user data and five bytes for the header. A fixed duration during which a cell is transmitted is called slot. The duration of a time slot depends on the physical link capacity. At most one cell is transmitted during each time slot on an ATM link. Due to the natural randomness of traffic in ATM networks, buffers are required for absorbing instantaneous bursty cells (i.e. a number of cells too close each other) at ATM nodes.

Cell scheduling in the context is to determine which cell (of different classes) in the buffer will be selected for output at each time slot. In other words, the function of a scheduler is to consider how to allocate the bandwidth of the output link to different classes of cells. Apparently, different scheduling policies will result in different service effects (e.g. cell delay) for each class of traffic.

3.1 Cell scheduling policies

First-In First-Out (FIFO) is a popular policy, in which cells leave the buffer according to their exact arrival sequence. Thus, all cells in an FIFO queue have the same priority. Although FIFO queue has the advantages of simplicity and neglected processing overhead, it cannot satisfy the QOS requirements of diverse types of traffic in an ATM network.

Static priority is another policy for cell scheduling. In static priority scheme, different priorities are assigned to different types of traffic, and a higher priority cell is always scheduled prior to a lower priority cell. Cells with the same priority comply with FIFO discipline. The priority for a traffic class is absolute and never changed.

Although it is effective and simple, static priority scheme is inflexible and do not contribute to overall performance in an ATM network. No doubt that real-time services such as packetized voice are quite sensitive to the transfer delay (which is mainly dominated by cell queueing delay in ATM networks), but an ATM node with static priority policy may result in too-good QOS (e.g. the delay that is far from their allowed limits) for delay-sensitive applications. If all queues adopt static priority policy and the proportion of each class of cells is reasonable, the network is able to deliver real-time applications to destination with much smaller delay than what the applications can tolerate. However, too-good service may not favor the real-time applications because receivers may have to buffer those earlier arrived cells to wait for their playback instant. In addition, lower priority cells would suffer larger delay if static priority policy is adopted.

A cell scheduling should be able to adapt to the change of traffic conditions and QOS requirements. This requires a scheduler to be able to select a cell according to necessity. Instead of using equal or absolute priority, we may assign a percentage of priority to each traffic class. Note that such a percentage is changable. Under this condition, the scheduling effect can shift between FIFO's and static priority's policy according to the QOS requirements of different classes. If the delay requirement is very strict, more bandwidth is allocated to the delay-sensitive cells; otherwise, the bandwidth will be allocated to other class of traffic. Our intention is to use EHW to perform such a control that an ATM node not only guarantee the QOS requirements for the real-time applications, but also reduce queueing delay of other classes of cells, thus improve overall performance of the network.

To study the possibility of using EHW to perform the cell scheduling function, we need to collect enough training data to teach EHW. For the purpose of obtaining the training data, we will give a model of an ATM statistical multiplexer in the next subsection.

3.2 Model for data collection

The aim of our experiment is to investigate whether EHW can provide adaptive cell scheduling so as to achieve efficient traffic control in an ATM node. As ATM is based on statistical multiplexing techniques, we first build a model of the statistical multiplexer in order to collect realistic data to train EHW. The model is shown in Figure 2, which is widely used for imitating ATM nodes. It involves traffic sources, input links, an output link, a transmission buffer and an associated server that selects a buffered ATM cell for further transmission. All input and output links may operate at the same speed to reflect the effect of cell statistical multiplexing, i.e. there is:

$$\sum C_{in} > C_{out}$$

where C_{in} is the input link capacity, and C_{out} is the output link capacity.

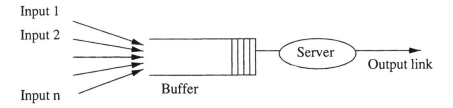

Fig. 2. ATM node model

3.3 Traffic source

For the model of a statistical multiplexer, a two-state Markovian representation of an ATM source has been used to generate bursts of fixed-size 53-byte cells via each input link. The source can be used to approximate a wide range of traffic flows [1, 3, 10]. At burst-state, cells are generated with the corresponding peak cell rate. No cells are emitted at the idle-state (i.e. silence time). At burst-state, the source has a probability α to transit to idle-state, and a probability $(1-\alpha)$ to stay at burst-state. At idle-state, the source has a probability β to transit to burst-state, and a probability $(1-\beta)$ to stay at idle-state.

For each individual traffic source, the mean burst-period M_{burst} is dependent on the peak bit rate and the average bursty length of the connection, whereas the mean idle-period M_{idle} is dependent on the peak bit rate, mean bit rate, and the mean burst-period of the connection. Given the values of the peak bit rate p, the mean bit rate λ, and the average bursty length B_l associated with a connection, it can be shown

$$M_{burst} = \frac{B_l \cdot S_{cell}}{p}$$

and

$$M_{idle} = \frac{(p - \lambda)B_l \cdot S_{cell}}{p \cdot \lambda}$$

where S_{cell} is the cell size in bits. The burst and idle lengths are exponentially distributed. Figure 3 gives the effect of such bursty traffic over a single input link.

Fig. 3. ON-OFF traffic pattern

Each input link has one traffic source attached. All sources independently generate cell streams on different links respectively. The traffic characteristic parameters (e.g. peak bit rate, mean bit rate) on one input link may differ from the parameters on another input link.

Two cell classes are assumed in this simulation, where a class 1 cell is more delay-sensitive than a class 2 cell. A generated cell can be either a class 1 or a class 2 cell, which is subject to the given value of the mixing rate m (e.g., 30 percent generated cells are class 1 cells if m is 0.3). Both classes of cells share a common buffer in the multiplexer. Table 1 gives the parameters used in the simulation.

3.4 Working conditions

Service rate s dominates cell scheduling effect to two cell classes. This parameter is used to determine the number of class 1 cells to be shipped in a control cycle

Table 1. List of parameters and their values used in the multiplexer simulation

Symbol	Definition	value
$R_{arrival}$	Average cell arrival rate	0.8~0.9
p	Peak bit rate of the source traffic	50 Mbit/s
λ	Mean bit rate of the source traffic	5 Mbit/s
B_l	Average length of bursts	10 cells
C	Capacity of input and output links	150 Mbit/s
T	Number of time-slots in a simulation run	10^6 time slots
m	Mixing rate of two classes of cells	0.3
d	Duration of one control cycle	100 time slots
s	Service rate for class 1 cells during a control cycle	0.29~0.4
L	Number of positions in the buffer	infinite

and may take a value ranging from 0 to 1. Given the control cycle of 100 time-slots and a service rate of 0.4, for example, 40 class 1 cells will be scheduled in a control cycle and the other time-slots will be allocated to class 2 cells. The scheduling effect becomes the same as static priority discipline when service rate takes the value of either 1 or 0, where class 1 or class 2 cells obtains absolute priority respectively.

The scheduling mechanism is work-conserving, i.e. it always schedules a cell for output as long as there are cells in the buffer.

To collect training data for EHW, identical and independent bursty traffic sources are applied to the model of the statistical multiplexer. The specific values for the operating parameters are given in Table 1.

We did not take the cell loss into account in this experiment. In fact, the scheduling policies we mentioned above do not affect cell input process; they only affect cell output process when a common buffer is shared by different classes of traffic. It is assumed that the buffer size is infinite in the simulation, thus cell loss never occur during the whole simulation process. In future work, we will study the possibility of using EHW for the control of cell selective discarding.

With this simulation, we got a group of training data for the EHW simulator. The data imply the relationships among cell arrival rate, cell mixing rate, cell delay constraint, and cell service rate.

4 Simulation Results

As mentioned in the previous section, an ATM multiplexer with a common buffer is considered in our experiment. Now we examine the control performance by EHW. Clearly, whether EHW can conduct correct control here depends on the value of service rate it estimates under a certain traffic condition and a delay constraint. The value of service rate should vary in time according to the change of incoming traffic streams and QOS requirements. A suitable value of service rate will ensure the delay bound of class 1 cells, but not overdo it. The delay of a class of cells can be controlled based on a given requirement as long as EHW can assign the correct values to the service rate during each control period.

With the training data achieved, EHW simulator starts to learn the suitable service rate with respect to the cell arrival rate and delay constraint. The purpose of this training is to let the EHW simulator be able to conduct adaptive cell scheduling so that more class 2 cells could be transmitted under the condition of guaranteeing the delay requirements of those class 1 cells.

For training EHW, the population size is 400 and each simulation run covers 2500 iterations. The EHW has 100 PFUs arranged in a 20*5 grid. Two inputs are applied to the simulator: cell arrival rate and average delay of class 1 cells. Assumed cell arrival rates are 0.80, 0.83, 0.87 and 0.90 respectively, whereas the values of the delay are from the multiplexer simulation. The reference values of the output are the corresponding service rates, which are used to calibrate the output values of EHW.

The training data involve 48 sets of triple-number, which are featured as four solid curves in Figure 4.

The fitness calculation is based on the variance in errors. The variance of each chromosome of the population is calculated, then it is normalized and used as the fitness value, i.e.

$$Fitness = \frac{1}{1 + Variance}$$

Figure 5 shows the fitness values of one simulation run during the learning process.

Table 2. Statistics from 30 simulation runs

	Learning (Fitness)	Testing (MSE)
Average	0.931529	1.99917×10^{-3}
Standard deviation	0.0509801	1.50608×10^{-3}
Maximum	0.984233	6.40248×10^{-3}
Minimum	0.80304	3.87618×10^{-4}

After learning is complete, the best chromosome is tested and the results of EHW simulator are compared with the original data. From this test, we can examine whether EHW really achieved the knowledge about the traffic streams and the delay requirement or not. Table 2 shows the statistics from 30 simulation runs, where MSE is the acronym of Mean Square Error. The results of one run of the EHW simulator are shown in Figure 4. Vertical bar is the values of the service rate, and horizontal bar represents the average cell delay of class 1 traffic.

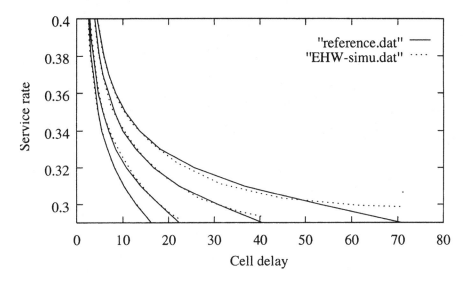

Fig. 4. Results from EHW simulator

The unit of time slot is used for the cell delay, where one time slot is about 2.8 μs. The original data are given as solid curves in the figure for comparison. The curves differ from each other by cell arrival rate of the superposed cell streams, which are 0.80, 0.83, 0.87, 0.90 respectively. The dotted curves were estimated by EHW simulator.

From the figure, it can be seen that the service rate values derived from EHW simulator are quite close to the data achieved from the simulation of an ATM statistical multiplexer. EHW is able to learn several non-linear functions simultaneously. Whenever a best chromosome is obtained, EHW will conduct the processing control very fast. Therefore, EHW can provide the required cell scheduling in ATM networks.

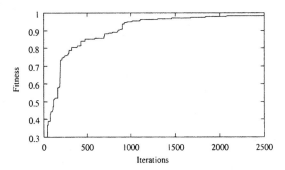

Fig. 5. Fitness during learning

5 Conclusion

In this paper, we presented our experiment and the results of using the function-level EHW to solve cell scheduling problem in ATM networks. Two simulators were involved in this experiment. One is the simulator of an ATM statistical multiplexer, and the other is the simulator of the function-level EHW. We obtained the training data from the multiplexer simulator, and then using the data to train EHW. The simulation results show that the function-level EHW can provide adaptive control well for real-time cell scheduling problem.

In addition to the adaptability, fast processing speed is another advantage of the function-level EHW. Indeed, the function-level EHW combines the advantages of both hardware and software, which makes it suitable for those applications that must be processed in very high speed, e.g. traffic control in ATM networks.

In future work, we will study the problem of cell selective discarding control by the function-level EHW. In the experiment, we will train EHW to learn the relationship between cell delay and cell loss ratio in an ATM node with finite buffer size.

References

1. Ani C. et al., "Simulation Technique for Evaluating Cell-Loss Rate in ATM Networks", *Simulation, Vol.64, No.5*, pp.320-329, May 1995.
2. Goldberg D., *"Genetic Algorithm in Search, Optimization, and Machine Learning"*, Addison Wesley, 1989.
3. Guerin R. et al., "Equivalent Capacity and Its Application to Bandwidth Allocation in High-Speed Networks", *IEEE J. Select. Areas Commun., Vol.9, No.7*, pp.968-981, Sept. 1991.
4. Higuchi T. et al., "Evolvable Hardware", in *Massively Parallel Artificial Intelligence*, edited by Kitano H. and Hendler J., pp.398-421, MIT Press, 1994.
5. Higuchi T. et al., "Hardware Evolution at Gate and Function Level", in *Proc. of International Conference on Biologically Inspired Autonomous Systems: Computation, Cognition and Action*, Durham, NC, USA, 4-5 March 1996.
6. Hiramatsu A., "ATM Communications Network Control by Neural Networks", *IEEE Trans. Neural Networks, Vol.1, No.1*, pp.122-130, March 1990.
7. ITU-T: Recommendation I.371, *"Traffic Control and Congestion Control in B-ISDN"*, Geneva, 1993.
8. Kajitani I. et al., "Variable length Chromosome GA for Evolvable Hardware", in *Proc. of the 1996 IEEE International Conference on Evolutionary Computation (ICEC'96)*, pp.443-447, May 1996.
9. Murakawa M. et al., "Hardware Evolvable at Function Level", in *Proc. of the International Conference on Parallel Problem Solving from Nature (PPSN'96)*, Berlin, Sept. 1996.

10. Nagarajan R. et al., "Approximation Techniques for Computing Packet Loss in Finite-Buffered Voice Multiplexer", *IEEE J. Select. Areas Commun., Vol.9, No.3,* pp.368-377, April 1991.
11. Tarraf A. et al., "A Novel Neural Network Traffic Enforcement Mechanism for ATM Networks", *IEEE J. Select. Areas Commun., Vol.12, No.6,* pp.1088-1095, August 1994.

Evolutionary Robotics

An Evolutionary Robot Navigation System Using a Gate-Level Evolvable Hardware

Didier Keymeulen[1], Marc Durantez[2], Kenji Konaka[3], Yasuo Kuniyoshi[1], Tetsuya Higuchi[1]

[1] Electrotechnical Laboratory
Tsukuba, Ibaraki 305 Japan
[2] Ecole Nationale Superieure des Telecommunications
46, rue Barrault, Paris, France
[3] Logic Design Corp.
Mito, Ibaraki 305 Japan

Abstract. Recently there has been a great interest in the design and study of evolvable systems based on Artificial Life principles in order to control the behavior of physically embedded systems such as a mobile robot. This paper studies an evolutionary navigation system for a mobile robot using a Boolean function approach implemented on gate-level evolvable hardware (EHW). The task of the mobile robot is to reach a goal represented by a colored light while avoiding obstacles during its motion. Using the evolution principles to build the desired behaviors, we show that the Boolean function approach using gate-level evolvable hardware is sufficient. We demonstrate the effectiveness of the generalization ability of EHW by comparing the method with a Boolean function approach implemented on a random access memory (RAM). The results show that the evolvable hardware system obtains the desired behaviors in twice fast time and that the EHW generates a robust robot behavior insensitive to the robot position and the obstacles configurations.

1 Introduction

Robotics has until recently developed system able to automate mostly simple, repetitive and large scale tasks that did not require any decision. These robots, e.g. arm manipulators, are mostly programmed in a very explicit way and the environment of the robot and all the operations are described in minor details. But for mobile robot applications, the environment must be perceived via sensors and is far well less defined. It implies that the mobile robot must take decisions in a complex and dynamic environment in order to achieve its autonomy. This is one example of the "nouvelle" AI and Artificial Life research the aim of which is to build autonomous system which can adapt to the world it is embedded in, this world being changing and often unpredictable [13] [20].

To answer this new domain of robotics, Brooks has proposed a new designing methodology for mobile robots called "subsumption architecture" [2]. His approach consists to build behavior-based robots. To deal with the autonomous arbitration of the multiple behaviors, Brooks outlines an evolutionary approach based on Koza's genetic programming techniques [12]. He proposed a high level language, GEN, which could evolve, and then compiled down into behavior language and further on down on the actual robot hardware. Koza successfully used the technique of Genetic Programming to develop subsumption architectures for simulated robots engaged in wall-following and box moving tasks. Since then a lot of researchers have explored this new approach. Beer explores the evolution of module controlled by continuous-time recurrent neural networks for the locomotion-control of a six-legged insect-like agent [1]. Parisi et al. evolve back-propagation neural networks which are the "brains" of agents collecting food in a simple cellular world [17]. Reynolds uses genetic programming to create control programs which enable a simple simulated vehicle to avoid collisions [19]. Hoshino et al. analyze the manifestation of neutral genes in the evolution of neural network as well as the robustness of navigation systems [8]. Cliff et al. have developed an incremental evolution of arbitrarily recurrent neural networks [3].

While in mobile robotics, the evolutionary approach allows to design robot behaviors by auto-organization of a great number of modules, in technology new hardware devices allow to implement *evolvable hardware*. It is a hardware device which modifies its own hardware structure using genetic learning according to the rewards received from the environment [5]. In this way the evolutionary principles applied to mobile robotics can be implemented directly into hardware evolution as suggested by Higuchi and Manderick [6].

In this paper we tackle the navigation task for a mobile robot which has as target to reach a goal in the environment while avoiding the obstacles. The robot has no knowledge of his own system such as its geometry and the positions of its sensors. The robot has also no knowledge of its environment such as the geometry of the obstacles and their position. In this robotics application, the robot is considered as a *reactive system*, that is, without internal representations, so that the robot behavior is based only on the current sensory inputs and is described by a Boolean function. The Boolean function can be represented by a *functional form* easily implemented on a gate-level evolvable hardware (EHW) or a *truth table form* easily implemented by a random access memory (RAM).

We show that the gate-level evolvable hardware (EHW) is able to find the Boolean function under its functional form, describing the desired reactive behavior of the robot, from a set of observable sensor and action pairs. We show also that the generalizing ability of the gate-level evolvable hardware is necessary to obtain the desired behavior due to the small number of observable data obtained by the robot interaction with its environment. This generalizing ability of EHW is made effective through the exploitation by the EHW of the symmetries in the sensors-motions mappings representing the desired behavior [2].

Because of the small input space, the reactive behavior can also be represented by the truth table form of the Boolean function implemented by a random access memory (RAM). In this case the sensor input is an address and its content is the action associated with this sensor input. We demonstrate the effectiveness of the generalization ability of the functional approach by comparing EHW with an evolvable truth table approach. It shows that the functional approach is at least twice time faster and that it implements a more robust robot behavior than the truth table approach.

The paper presents first the robot task and its environment. In the second part, it describes the reactive navigation system based on a Boolean function controller represented by a truth table form and by a functional form. In the third part we present the evolution mechanism applied to both representations to find the desired robot behavior. Finally we illustrate the advantages of the functional approach and compared it to the truth table approach.

Fig. 1. Real Robot

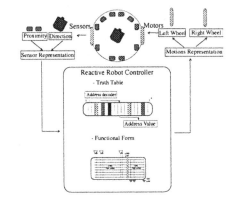

Fig. 2. Reactive Navigation System

2   Robot Environment and Task

The robot was built by the research group of Dr. Kuniyoshi of the Autonomous System Section of ETL. Its shape is circular with a diameter of 25 cm (Fig.1). It has 10 infra-red sensors of 1 bit resolution situated at its periphery and equally distributed. It is equipped with an intelligent vision system able to identify and track a colored object. The intelligent vision module returns the direction in which the colored object is located and the distance from the robot to the colored object deduced from the tilt angle and the zoom of the camera. The robot moves using two motor wheels at both sides driven independently. It allows the robot to perform motions such as translation, rotation or any combination of rotation and translation. The robot is controlled by a transputer board dealing with the infra-red sensors, the intelligent video camera sensor and the motor wheels and connected to the EHW board. The environment is a world with low obstacles such that the colored target can always be detected by the robot.

We have built a robot simulation to test and evaluate the performance of the gate-level evolvable hardware. From the 10 infra-red sensors, 5 returns a boolean indicating the presence of objects at a distance of 30 cm performing obstacle avoidance. The 5 other detect an object at a distance less than 5 cm implementing a bump sensor to detect when the robot hits a wall. The intelligent video system is simulated by a direction and a distance sensor to the target. The direction sensor returns one of the 8 sectors, covering 45 degree each, in which the goal is located. The simulated robot is placed in an artificial world with a square shape partitioned with obstacles. The obstacles shape is such that using a reactive system the robot will not be stuck. For example it doesn't include obstacles with the horseshoe.

The task assigned to the robot is to reach the target without touching the walls within a minimum number of motions and that from any positions in the virtual world. To perform its task, the robot must acquire a set of navigation strategies such as obstacle avoidance, turning at in the "right" direction at the obstacles and going to the target. To realize these navigation strategies, we chose to control the robot with an evolvable boolean function.

3   Reactive Navigation System

With new robotics challenges issued from mobile robotics, many new ideas of control system for mobile robots have been proposed. In the introduction we mention the "subsumption approach" which decomposes the control system into a set of interacting behavior modules. Each interacting behavior simulating a augmented finite state automaton, i.e. finite state machines with timing elements [2]. Other researchers in the domain of autonomous agent have proposed to use artificial neural networks of some variety as the basic building blocks for the control system of the robot due to its generally smoother search space and its working with very low-primitives avoiding using preconceptions about the properties of the systems [14]. Finally Luc Steels proposes to use a dynamical systems, called a process network, inspired by the couple map latticed to control the robot behaviors [20].

Boolean Function Controller. In our approach, we assume that the robot behavior is described by a boolean function. It is more simple than the set of behavior modules implemented by augmented finite state machines, although the boolean function represents in an implicit way all the knowledge to perform the task. Furthermore it doesn't assume any knowledge of the necessary behaviors and of the high level primitives of each behavior. It is able to perform the navigation task in a reactive way and is well suited for an evolutionary search algorithm. Finally it is easily implemented in hardware. But for performing more complex tasks such as a navigation task in an environment with obstacles of any shape or when the target is not always visible, a finite state automaton may be needed and probably even a stochastic finite state automaton. This boolean function approach was applied already to navigation tasks by Hoshino who uses a compressed form of the boolean function to deal with the huge input space due to the large number of proximity sensors increase [7].

The robot controller follows the following line (Fig. 2). For each proximity sensors, the controller extracts in a synchronous manner the distance to the nearest obstacle and the direction of the target. It transforms the distance into a boolean giving an indication of the presence of the obstacle on that side of the robot. The target direction indicates if the goal is in front, at the back or on the sides of the robot. The proximity and direction sensors form a binary-code furnished to the controller. The boolean function controller is represented in two ways: (i) a *truth table controller* implemented with a Random Access Memory and (ii) a *function controller* implemented with a Programmable Logic Device. The output values of the controller is the binary representation of the motion. The motion encodes the motor speed for the left and right wheel. In a synchronous way, the controller determines the speed of both wheels and then repeats the cycle of the operation.

Mathematical Expression. The target of the boolean function approach is to find a function $\mathcal{F}$ of n boolean variables which represents the desired reactive behavior. The domain of the input variable is $(0, 1)^n$ where 2^n are the possible world robot situations and the domain of the output variable is $(0, 1)^m$ where 2^m are the possible motions. The input variable is encoded by 5 bits for the proximity sensor and by 3 bits for the direction sensor and represents 256 robot world situations. The output variable is encoded by 3 bits to represent the 8 possible motions. Each point of the input space represents a world robot situation and the associated output represents the action it will perform in this situation.

$$y = \mathcal{F}(x_0, \ldots, x_{n-1})$$

where $x_i \in \{0, 1\}$ codes the proximity and direction sensors and $y \in \{0, 1\}^m$ codes the 2^m possible motions.

4 Evolvable Reactive Navigation System

The goal is to identify the unknown function $\mathcal{F}$ from a given set of observable input and output pairs, representing respectively the sensor inputs and the corresponding motions to guide the robot to the goal while avoiding the obstacles.

If we suppose that each pair can be evaluated separately, the task can be seen as a *Boolean concept learning* [16]. For this task, there are algorithmic approaches which, although complete, suffer from computational complexity. We have chosen a stochastic method, the evolutionary method, which use probabilistic search and is more efficient than algorithmic approach but does not assure completeness[10]. They are also well adapted to robotics task for which many Boolean functions are solutions.

Unfortunately evaluating the pair sensor-motion at each step of the robot motion suffers from the problem of local maxima. The learned behavior settles into a very suboptimal policy that prevents the robot to reach the colored target. In our approach the robot performance is measured when the robot fails as its task to track the target while avoiding the obstacles. By consequence our approach evaluates globally a set of pairs of sensor-motions resulting from the robot interaction with its environment until it fails. This learning task is called the *credit assignment* task: figure out what behavior of the robot (pair sensor-motion) is defective and how to fix it [4].

Another difficulty is that the learning is done in an *unsupervised mode* without active guidance of a teacher. The robot has to collect its own training data opening interesting possibilities where the robot can actively choose to explore and experiment with certain parts of its environment. For this learning mode, the generalization ability of the learning method is essential because there are so many combinations of sensory input and motion that is unlikely that the robot will ever experience most of them.

As mention in the introduction many researchers have proposed, besides reinforcement learning algorithms [11], supervised neural networks [18] or knowledge based system [4], to use an evolutionary approach using genetic algorithm [5]. We follow the same line that we resume briefly here. The GA operates on a population of bit strings each of them representing a function $\mathcal{F}$, and determining the behavior of a robot. The performance of all the behaviors for the tracking task is evaluated and the resulting performance measures the fitness of the corresponding function $\mathcal{F}$. The GA gives the better bit strings a higher chance to survive and to recombine. This way we get a new generation of bit strings which might incorporate good features of their parents. Each bit strings of this new generation is then used to represent the functions $\mathcal{F}$s and make behave the robots accordingly. Their performance is again evaluated and this process continues that way.

We have applied the evolutionary method to two representations of the function $\mathcal{F}$: one based on truth table form and the other based on the functional form. Although both descriptions give the same input-output and thus behavior of the robot, they differ in terms of their transformations by evolutionary algorithms and by consequence the representation influences considerably the learning performance as well as the generalization ability of the learned controller as explained in the following paragraphs.

4.1 Truth Table Controller

The truth table controller is implemented by a random access memory (Fig.3). The address is the binary representation of both the proximity and the direction sensors: $x_0, \ldots, x_{n-1}$ and represents an address space of $2^8 = 256$ words. At each address is memorized one of the $2^3 = 8$ possible motions as a integer value. The motion is further interpreted into the speed for the two wheels.

$$\mathcal{F} \overset{\text{def}}{\equiv} y \overset{RAM}{\longleftarrow} x_{n-1} \cdots x_0$$

The basic idea of evolvable truth table controller, is to regard the content of the table as a chromosome for a Genetic Algorithm. We use a asexually reproducing population in which genetic variation is produced only by mutation. The probability to change the value at a memory address is given by the mutation rate. The probability of transition of the memory value to one of the 8 possible values is set equal. In this way no particular motion is favored.

When looking at the mutation operator as a transformation of the sensors-motions mapping, the mutation operator changes the value at some point of the input space in an independent way, letting the other association sensors-motions unchanged (Fig.5).

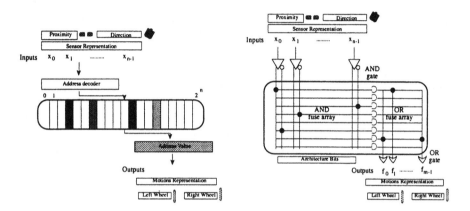

Fig. 3. Truth Table Controller (RAM).   **Fig. 4.** Function Controller (EWH).

4.2 Function Controller

The function controller represents the Boolean function $\mathcal{F}$ by m Boolean functions in their disjunctive form. The input variables of the Boolean functions are the binary representation of the sensors. The output value of the function $\mathcal{F}$ is the possible motions associated with the input variables. The function $\mathcal{F}$ is defined by:

$$\mathcal{F} \stackrel{def}{\equiv} y \stackrel{PLD}{\equiv} f_{m-1}(x_0, \ldots, x_{n-1})\, 2^{m-1} + \cdots + f_0(x_0, \ldots, x_{n-1})$$

where $f_u(x_0, \ldots, x_{n-1}) = (x_{i_u} \wedge \cdots \wedge \overline{x_{j_u}}) \vee \cdots \vee (x_{k_u} \wedge \cdots \wedge \overline{x_{l_u}})$

Hardware Implementation. The function $\mathcal{F}$ described by m Boolean functions in their disjunctive form is implemented by a PLD (Programmable Logic Device) to increase considerably the speed of execution. A PLD consists of logic cells and a fuse array. The logic cells performs logical functions such as AND or OR gates. The fuse array determines the interconnection between the input devices and the logic cells. The input device creates for each input variable its value and its inverse. If a column on a particular row of the fuse array is switched on, then the corresponding input signal is connected to the row and thus to the entrance of the logic cells. The fuse array is programmable. So in addition to the input bits and the output bits of the logic circuit simulating a Boolean function, the architecture of the PLD is determined by the architecture bits. Each link of the fuse array corresponds to a bit in the architecture bits [6].

The PLD structure suited for the implementation of the $\mathcal{F}$ function consists of an AND-array and an OR-array (Fig.4). Each row on the AND array calculates the product of the inputs connected to it, and each column of the OR-array calculates the sum of the results of the output of the AND-gates if connected. The number of columns of the AND array and the OR array is determined respectively by the number of inputs and outputs of the PLD device. The number of rows of the arrays depends on the complexity of the disjunctive form of the boolean function f_i. In the worst case (f_i is a xor of $n - 1$ boolean variables) the disjunctive form contains 2^{n-1} terms and the PLD needs $n - 1$ lines. In our experimental set up the PLD needs $2 * n + 3 = 19$ columns and a maximum number of $2^{n-1} = 128$ rows. But due to the symmetries of the desired Boolean functions, the number of rows is reduced to 50 lines.

Evolvable Function Controller. The basic idea of EHW is to regard the architecture bits of the PLD as a chromosome for a Genetic Algorithm[5]. If these bits change then the architecture and the function of the logic device also change. The basic idea is to let the architecture bits evolve in order to adapt the function of the logic devices to perform the tracking task.

Chromosome Representation The architecture bits are built in the following way. For each link in the AND array one bit is used. It represents that the input is either connected (directly or inversely) or not connected. Each link in the OR array is represented also by 1 bit: connected or not connected. We can interpret the PLD representation in the following way (Fig.5). Each term of the disjunctive form (PLD row) defines a hypercube of the input space where the value of the term is 1. It is called the support of the term. The support of each boolean function (PLD output) is the union of the hypercubes defined by each term of the disjunctive form (implemented by the rows connected to the PLD output).

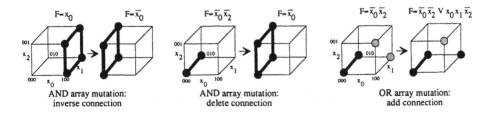

Fig. 5. Mutation of a Boolean function F of 3 variables using EWH.

Mutation Operator We use only mutation operator to change the architecture bits. There are five kinds of mutation operators: removing or setting a connection on the OR array, removing a connection on the AND array, setting directly and inversely a connection on the AND array.

The effect of mutation on the Boolean function is different if it happens on the OR array or on the AND array (Fig.5). In the OR array the mutation means removing or adding one hypercube to the union that defines the support of one output. The change in the fitness generated by the mutation depends on the overlapping of the support of the mutated line with the other hypercubes that define the support of the PLD output. If it is completely overlapped there will be no change in the fitness of the individual and the mutation will be neutral.

If the mutation occurs in the AND array it means a change of the shape of the hypercube defined by the mutated line. If the mutation removes/adds one connection then the size of the hypercube grows/decreases along the dimension associated with the added/removed variable. The third kind of mutation turns a direct into an inverse connection or the contrary. This mutation "translates" the support of the line along the dimension of the mutated connection.

Mutation Rate The mutation rate must be low to allow the selection algorithm to do a selection of the good lines. If it is too high it introduces noise to the evaluation of the lines and it causes genetic unstability. The mutation rate used is inferior to the inverse of the length of the chromosomes so there is few mutations at each generation.

The different mutations have different effects on the lines. We changed the relative probabilities between them to favor full connected lines. The mutation that enlarges the support of one line is rarer than the mutation which translates the support. The reason is to direct the evolution of the lines from small supports to larger ones. Once a line is fixed its support can grow.

We are fairly tempted to try to evolve the lines and then build the function by selecting the best combination of lines. However we do not evaluate a line alone but a set of lines. In order to evolve lines for a Boolean function that needs severals lines, the technique which can be used is to use the cross-over between the chromosomes. In that way the structures that remain unchanged between the generations are the lines. So the pressure of selection will efficiently operate on the lines even if the fitness uses a complete individual.

Truth Table and Function Evolvable Controller. One may observe that both approaches are identical if we could find the genetic operators on the truth table transforming the Boolean function in the same way as the genetic operators applied on the function controller (EHW). For example by transforming one mutation on the architecture bit of the PLD into multiple mutations on the memory (RAM). But the efficiency of the gate-level evolvable hardware will still be greater because the number of hardware access needed for the RAM mutation is of the order of the truth table dimension $O(2^n)$ while for the evolvable hardware it will be of the order of $O(1)$.

5 Genetic Algorithms

Genetic Parameters. In genetic algorithm a point to be careful with is the selection algorithm. In this problem we considered that we could not afford the loss of a useful mutation by the hazard of a selection algorithm as the roulette wheel. We have preferred to use a deterministic tournament selection with tournament size $s = 3$. In the deterministic tournament selection, s individuals are picked at random from the old population, and the fittest of them is put in the mating pool. This is repeated until the mating pool is as large as the old population. The tournament size s can be used to tune the the selection pressure. The population size value is 20 individuals.

We use an elitist algorithm to be sure of keeping the state of the memory and the lines of the above-average individuals but we also restrain the number of offsprings of each individual to avoid an excessively fast convergence to a local maxima, using as explain a tournament selection.

Fitness Evaluation. Each robot is evaluated in the simulated environment starting always from the same initial position. Its fitness Φ is evaluated when it fails at the task of reaching the color target. It can fail for two reasons:

- it hits an obstacle: it is detected by computing the intersection between the environment walls and the robot geometry. In the real world it is obtained from the bump sensors.
- it reaches the maximum number of steps it is allowed to perform in the environment. This situation occurs when the robot is stuck in a loop.

The fitness Φ is represented by a scalar number in the interval 0 (worse) and 1 (best). There are two contributions to the fitness.

- the distance to the target $D(robot, target)$: it forces the robot to reach the target. This distance is obtained using the tilt and the zoom information of the camera and is normalized using the dimension of the simulated world L.
- the number of steps used to reach its actual position from its initial position. It forces the robot to chose a shorter path and to avoid to be stuck in loop.

The fitness is a balance between the distance and the number of steps. In our experiment both evaluation has an equal contribution to the fitness:

$$\phi = 0.5 \left(\frac{D(robot, target)}{\sqrt{2}L} \right) + 0.5 \left(1 - \frac{Nbr.\ Steps}{Maximum\ Nbr.\ Steps} \right)$$

6 Experiments

In this section we present the advantages of the evolvable function controller by comparing its performance and its robustness with the evolvable truth table controller.

6.1 Faster Convergence with Evovable Function Controller

We have conducted experiments with an environment containing 9 obstacles of different shapes. The distance between the obstacles is such that the robot can only perceive one obstacle at the same time, to limit the possible sensor inputs encountered in the environment. The target is situated in the bottom right corner (Fig.6).

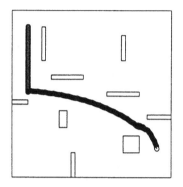

Fig. 6. Path obtained after 24 generations by a function controller.

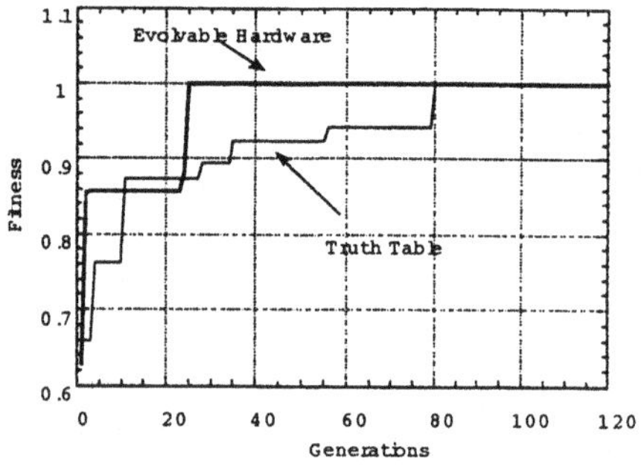

Fig. 7. Fitness throughout generations.

Function Controller. The robots are always placed at the same initial position in the environment: at the left upper side of the environment. At the beginning of the evolution, the controller is built in a random way. The behavior of each robot is then simulated until it hits an obstacle or is stuck in a loop.

Figure 6 shows the result of the function controller. It represents the behavior of the best individual of the population after 24 generations. In this approach, the controller is completely reactive taking into account only the actual values of the sensor to guide the robot motion. We observe, although this reactive behavior, the robot finds effectively a short path between the obstacles using its direction sensor to guide its motion to the target. Its behavior also integrates an obstacles avoidance strategy to avoid the horizontal and square obstacles.

Evolution of Best Individuals. Figure 7 shows the evolution of the fitness with generations. The maximum value of the fitness is obtained after 24 generations. They are five jumps in the evolution before reaching the individual able to perform the task, respectively at the first, second, third, twenty third and twenty fourth generations. The best individuals of the first three generations have failed because being stuck in a loop of small dimension. The best individual at the 23rd generation hits the wall but has accumulated the necessary navigation strategies discovered by the previous best individuals to find the desired behavior at the next generation.

Function vs. Truth Table. The same experiment was conducted for a truth table controller. The initial position of the robots are exactly the same. Figure 7 compares the evolution of both controllers. We see first that the function controller converges nearly three time faster than the truth table controller which takes approximately 80 generations to give a satisfactory behavior. Second we observe that the evolution of both systems is different in terms of the jumps in the evolution. The evolution is much more smooth for the truth table controller: it reaches the satisfactory behavior in 15 jumps in place of 5 for the evolvable hardware.

The reason is that the functional approach discovers the symmetries of the boolean function solutions: boolean functions for which the output, the robot motion, is independent of some of the input variables. In the functional approach, the mutation of the architecture bits which enlarges the boolean function support results in an individual not only valid for the situation where the robot just failed but also in situations that the robot has not yet encountered. It allows the robot to react in a proper way in the next motion steps. In contrary, for the truth table approach the mutation of the controller results in an individual only valid for the particular situation it failed, not allowing the robot to anticipate the right motion for similar future situation.

6.2 Better Robustness with Evolvable Function Controller

A robust controller is obtained by designing carefully the controller which results in satisfactory performance over wide variations in the environment [15]. The structure of the controller is then fixed and its parameter are known. The robustness is needed because there are many combinations of sensory input and motions that are unlikely that the robot will ever experience most of them. The robustness of the control system is intimately related to generalization. Indeed by generalization it can extrapolate what it has seen so far to similar situations which may arise in the future.

To increase the robustness of navigation systems, some researchers either include noise in the environment either use a particular way to evaluate the fitness [9] [7]. In our approach we guide the evolution during the learning process to obtain robustness. We show by experiments that the function controller presents a generalization capability and by consequence is robust.

Guided Evolution To build robust evolutionary controller at one side we force the robot, during its evolution, to encounter many different situations and on the other side we use the generalization capacity of the evolvable hardware able to profit of the symmetries of the boolean function solutions.

We proceed exactly in the same way than in our previous experiment, but in this case every time the robot reaches the target position, we replace its initial position by another initial position from where it has again to reach the target and increase the fitness by 1. There are 64 initial positions which are distributed equally in the environment. The important point is that during the evolution, the robots start always from the same initial position and that the sequence of initial positions is always the same for all the robots.

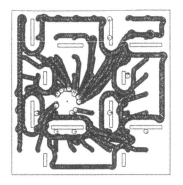

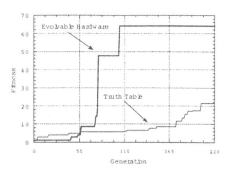

Fig. 8. Robust Navigation System

Fig. 9. Fitness throughtout generations.

Robustness Once the robot reaches 64 times the target being passed by the 64 initial positions we test the robustness of the system for three perturbations on the environment: the modification of the initial condition, the modification of the shape and the position of the obstacles and the modification of the target position. The results of the experiment, shown in Figure 8 concludes that the robot is able to reach the target from different initial positions in different environments.

Figure 9 shows the fitness evolution with generations. Once more we observe in the evolution of the function controller a small number of big jumps due to the discovery of symmetries in the boolean function solutions. The evolvable truth table controller was even not able, after 200 steps, to find the controller to reach the target passing by the 64 initial conditions.

7 Conclusion

In this paper we have presented an evolutionary navigation system for mobile robot. The navigation task consists to track a colored target in the environment while avoiding the obstacles. The mobile robot is equipped with infra-red sensors and an active vision system furnishing the direction and the distance to the colored target. The navigation control system is completely reactive and is based on a gate level evolvable hardware implementing a Boolean function in its disjunctive form. We have compared the approach with a evolutionary system where the robot behavior is described by a truth table.

We demonstrate the feasibility to implement an evolvable Boolean function controller at the hardware level using PLD, a programmable logic device. We have shown that a complete reactive navigation system was able to perform the tracking task. We have also shown and demonstrated that the gate level evolvable hardware was able to take advantage of the numerous symmetries in the desired behaviors. It has as consequences that the evolvable function controller implemented on a gate level evolvable hardware converges two to three times faster than the evolutionary truth table system and that although the small numbers of interaction of the robot with the environment, the generalization ability of the evolvable hardware allow to build a highly robust control system which is insensitive to the initial conditions, the shape of the obstacles and even the position of the target.

Currently the gate-level evolvable hardware is being design to work as a stand alone navigation system on a mobile robot and opening the door for *online* evolvable hardware adaptation to real world applications.

Acknowledgements

The authors would like to express great thanks to Prof. Hoshino of Tsukuba University for his valuable discussion and his visionary thinking on artificial life as well as Prof. Manderick for his valuable comments. We want to thank Mr. Mitsumoto, Mr. Nagano and Mr. Hito of Tsukuba University for the graphical interface design and many suggestions for the implementation of the robot simulation and Dr. Iwata from ETL who has helped us to design the evolvable hardware applied to robotics.

This research was partially supported by the RWCP (Real World Computing Partnerschip) and NEDO Grant (New Energy and Industrial Technology Development Organization). Part of this research was done when Didier Keymeulen was supported by the Post-Doctoral fellowship grant of the Japan Society for the Promotion of Science in collaboration with the National Belgian Science Foundation (NFWO) at Tsukuba University.

References

1. R.D. Beer and J.C. Gallagher. Evolving dynamic neural networks for adaptive behavior. *Adaptive Behavior*, 1(1):91–122, July 1992.
2. R. Brooks. Artificial life and real robots. In F. J. Varela and P. Bourgine, editors, *Proceedings of the First European Conference on Artificial Life*, pages 3–10, Cambridge, MA, 1992. MIT Press / Bradford Books.
3. D. Cliff, I. Harvey, and P. Husbands. Explorations in evolutionary robotics. *Adaptive Behavior*, 2(1):73–110, July 1993.
4. Jonalthan H. Connell and Sridhar Mahadevan. *Robot Learning*. Kluwer International Series in Engineering and Computer Science. Kluwer Academic Publisher, 1993.
5. D.E. Goldberg. *Genetic Algorithms in search, optimization and machine learning*. Addison Wesley, 1989.

6. T. Higuchi and B. Manderick. Evolvable harware with genetic learning. In *Massively Parallel Artificial Intelligence*, pages 398–421. AAAI Press / The MIT Press, 1994.

7. T. Hoshino, D. Mitsumoto, T. Nagano, and D. Keymeulen. Loss of robustness in evolution of robot navigation. In *Artificial Life V*, Sciences of Complexity. Addison-Wesley, 1996.

8. T. Hoshino and M. Tsuchida. Manifestation of neutral genes in evolving robot navigation. In *Artificial Life V*, Sciences of Complexity. Addison-Wesley, 1996.

9. Philip Husbands, Inman Harvey, Dave Cliff, and Geoffrey Miller. The use of genetic algorithms for the development of sensorimotor control systems. In F. Moran, A. Moreno, J.J. Merelo, and P. Chacon, editors, *Proceedings of the third European Conference on Artificial Life*, pages 110–121, Granada, Spain, 1995. Springer.

10. Hitoshi Iba and Masaya Iwata. Machine learning approach to gate-level evolvable hardware. In *Proceeding of the First International Conference on Evolvable Systems: from Biology to Hardware*. Springer Verlag, 1996.

11. Leslie Pack Kaelbling. *Learning in Embedded Systems*. Bradford Book, MIT Press, Cambridge, 1993.

12. J. Koza. *Genetic Programming: On the programming of computers by means of natural selection*. MIT Press, 1992.

13. Pattie Maes. Behavior-based artificial intelligence. In J-A. Meyer, H.L. Roitblat, and S.W. Wilson, editors, *From Animals to Animats 2: Proceedings of the Second International Conference on Simulation of Adaptive Behavior*, pages 2–10. MIT Press, 1993.

14. O. Miglino, H.H. Lund, and S. Nolfi. Evolving mobile robots in simulated and real environments. *Artificial Life*, 2(4):417–434, summer 1995.

15. Kumpati S. Narendra and Anuradha M. Annaswamy. *Stable Adaptive Systems*. Prentice Hall, Englewood Cliffs, New Jersey, 1989.

16. Balas K. Natarajan. *Machine Learning: a theoretical approach*. Morgan Kaufmann Publisher, 1991.

17. D. Parisi, S. Nolfi, and F. Cecconi. Learning, behavior and evolution. In *Proceedings of the First European Conference on Artificial Life*, pages 207–216, Cambridge, MA, 1992. MIT Press / Bradford Books.

18. D.A. Pomerleau. Neural network based autonomous navigation. In *Vision and Navigation: The CMU Navlab*. Kluwer Academic Publishers, Boston, MA, 1990.

19. C.W. Reynolds. An evolved, vision-based model of obstacle avoidance behavior. In *Artificial Life III*, Sciences of Complexity, Proc. Vol. XVII, pages 327–346. Addison-Wesley, 1994.

20. Luc Steels and Rodney Brooks, editors. *The Artificial Life Route to Artificial Intelligence: Building Embodied, Situated Agents*. Lawrence Erlbaum Assoc, 1995.

Genetic Evolution of a Logic Circuit
Which Controls an Autonomous Mobile Robot

Taku Naito, Ryoichi Odagiri, Yutaka Matsunaga,
Manabu Tanifuji, Kazuyuki Murase

Department of Information Science, Fukui University,
3-9-1 Bunkyo, Fukui 910, Japan.
e-mail: naito@synapse.fuis.fukui-u.ac.jp

Abstract. In this paper, we propose a new approach to evolve controllers of autonomous robots, and experimental results of its application to a real mobile robot are described as well. It is based on two concepts: Firstly, behavior of a system in environment is generated by combinations of multiple sensory-motor reflexes, and secondly, the system behaves and evolves under the direct influence of its environment, thus the system is expected to adapt well for its environmental situations with flexibility.
The sensory-motor reflexes were realized by logic circuits connecting sensors to motors, which are composed of AND, OR and NOR elements. The genetic evolution in the environment was employed to determine the connections among the elements, thus types of reflexes and the ways of their combinations being obtained automatically by the interactions between the environment and the system itself.
This algorithm was implemented in a real miniature mobile robot, Khepera, and several experiments were performed. Khepera successfully learned to navigate and avoid obstacles in test fields. In comparison to a conventional algorithm, the acquired behavior scored higher in the values of fitness functions.

1 Introduction

1.1 Evolutionary robot controllers

In the last few years, several approaches to apply evolutionary robotics to real robots have been described. Various forms of control systems have been proposed for evolutionary robots such as explicit programs with a high-level language, classifier systems and neural networks. Among them, neural networks seem to be used most frequently in the applications to real robots, because of the advantages in learning and in robustness. That is, neural networks could generate most of necessary nonlinear input-output relations by learning. And also, they are resistant to external disturbances probably because of its distributed nature of information storage and of the continuity in input-output relations. (See Nolfi, Floreano, Miglino & Mondada, 1994 for references and further arguments).

In neural network controllers, however, some essential behaviors of robots in real situations do not seem easy to generate, or it is necessary to have special arrangements. One of the examples is a flip-flop operation. A robot in real environment often needs to turn around or to go backward whenever it senses a certain object (or situation) by the front sensors. In neural network controllers, this seems to be achieved by adding recurrent connections in the networks. The learning process of recurrent neural networks, however, is often unstable and thus time-consuming.

Neural network controllers could deal with continuous signal levels. However, considering if the tasks given to robots are relatively simple such as navigating and avoiding obstacles, input signals with fewer levels, or even all-or-none input signals, might be sufficient as the sensory inputs. For example, the sense of touch with obstacle, or the sense of adjacent wall could be expressed in all-or-none forms.

The information is stored in a distributed manner to all the weights among cells in neural networks. However, most of elementary behaviors could be described by simple logic's, which may be called reflexes. For example, if a wall is getting closer to the left sensor in a forward-moving robot, it should turn right to avoid a collision.

1.2 Logic circuits as the evolutionary controller

Considering above arguments, we started to think that more direct representation of sensory-motor behavior could lead to a better design of the controllers. That is, we assume as follows: Behavior is generated by a combination(s) of simple reflexes. The reflexes are described by logic circuits. Any logic circuits can be constructed if we allow to use AND, OR and NOR elements because they consist a complete set. Sensory input signals are expressed in a all-or-none form. If different levels of the signal are essential to express a certain sensory information, the signal could use several signal lines with different threshold or could be digitized.

Evolution of behavior in environment is achieved by evolving the logic circuits. That is, the genetic algorithm is used to determine types of logic elements as well as the targets of input terminals. The inputs could be connected with outputs of sensors and other logic elements. Thus, it can be expected to form flip-flop circuits, and even memory elements, in the evolution process, if necessary to achieve certain tasks. Elements to evoke basic behaviors, such as going forward, going backward, and so on, are also included in the gene.

1.3 The logic evolutionary controller in a real mobile robot

In order to test the above idea of the logic evolutionary controller, whether or not it is feasible, we have implemented the mechanism in a real robot, the miniature mobile robot, Khepera (Mondada, Franzi & Ienne, 1993, Lami/EPFL). As the results, the robot evolved to perform simple tasks, navigating and avoiding obstacles.

In this paper, we describe the detail of the methods used to control the mobile robot Khepera, as well as some results obtained in the experiments.

2 Implementation of the logic controller in Khepera

2.1 Khepera, a real mobile robot

Fig. 1. Khepera

The structure and function of the miniature mobile robot Khepera has been well described elsewhere (Mondana et al., 1993, Fig.1). In short, it features 55mm of diameter, 30mm of height, and 70g of weight. It is supported by two wheels, each controlled by a DC motor with an incremental encoder. It has 8 infra-red proximity sensors placed around its body, 6 in front and 2 in rear as seen in Fig.2. It can be attached to a external workstation by means of a lightweight cable and rotating contacts.

Khepera is installed a microprocessor (MC68331, 16MHz). The processor can execute programs written by the user and downloaded from the external workstation. In the program, sensory information form the infra-red sensors can be acquired and used to control the motors. In addition, multiple processes can be executed in parallel by time-sharing.

2.2 Logic circuits in Khepera

In this study, each logic element of the evolutionary controller was realized as one process in the microprocessor, not as a hardware. All the logic elements in the controller worked in parallel by time-sharing. Output value of each infra-red proximity sensor was coded to 0 or 1 at a certain threshold.

Two examples of logic elements are illustrated in Fig.2 and 3. In Fig.2, an AND element obtains its input from proximity sensors at both sides, and sends the output to both motors. Therefore, the robot moves forward if walls exist at both sides. In Fig.3, the robot rotates right to avoid the obstacle if there is an obstacle in the front or left-front direction.

In practice, we used 8 logic elements in this series of experiments though expandable. Each element could be AND, OR or NOR. Each could have up to

15 input terminals because the input could be obtained from all of 8 proximity sensors as well as output of 7 other logic elements.

Among 8 logic elements, outputs of 4 elements were assigned to have motor functions; 1st and 2nd ones are to rotate the left motor in the forward and backward direction, respectively, and 3rd and 4th ones to rotate the right motor in the forward and backward direction, respectively.

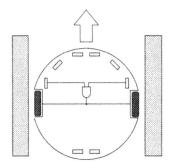

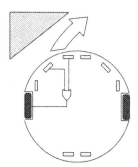

Fig. 2. Example of simple behavior (Forward)

Fig. 3. (Rotate)

3 Genetic evolution of the logic controller

3.1 Encoding

The phenotype was the logic controller circuit, and the genotype was defined as a sequence of characters. The chromosome consisted of 8 parts, each of which represented one logic element.

One logic element was coded in 18 characters as shown in Fig.4. The initial one character represented the type of functions, NOR, OR, AND, and NIU (not in use), as the numbers, 0, 1, 2, and 3, respectively. The following 8 characters represented the presence of connections from 8 proximity sensors, and the last 8 characters from outputs of other logic elements. Getting input from the output of the same element was prohibited, however.

The logic element placed at the 1st and 2nd parts in the chromosome had the motor functions; rotating the left motor forward and backward, respectively. The 3rd and 4th are for rotating the right motor forward and backward, accordingly. The rests, from 5th to 8th parts, had no direct motor functions.

3.2 The fitness function

The fitness function was designed to select the individuals who (1) moved a long distance (2) as straight as possible (3) without getting close to any obstacles.

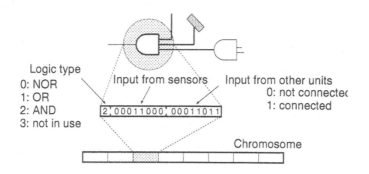

Fig. 4. Genetic encoding

That is, the behavior is basically evaluated by the following function,

$$E = D \times (1 - V) \times (1 - S)$$

Here, D, V and S represent, the total mileage of both left and right motors, the difference of mileage between both motors, and the total of all the sensor outputs. Some additional rules such as thresholds and limits were used in order to tune up. The state of the robot was observed at every 100ms by itself. V and S were detected at every observation, and D was obtained as the distance from the last observation made at 100ms before.

Individuals were allowed to move in a test field for 10 sec, and each individual was evaluated by the total sum of E's obtained at every 100ms during the 10 sec period.

3.3 The genetic rules

We employed a standard genetic algorithm with one-point crossover, bit-reverse mutation, fitness scaling, and elite preservation. The parameters used in experiments were listed in Table 1.

Table 1. GA parameters

Population	20
Gene Length	136
Cross Rate	0.9
Mutation Rate	0.03
Normalize Rate	1.5
Elite Rate	0.10

4 Experimental results

The test field was approx. 60cm square with some obstacles as shown in Fig.5.

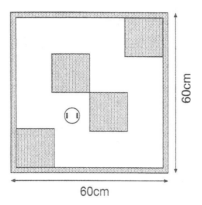

Fig. 5. Test field

The robot started to move from the final point of the last trial. The robot first made a small random movement which was programmed to get away from difficult situations such as heading toward corners, and then performed the next trial. An example of the fitness curves is shown in Fig.6. Fig.7, Fig.8 and 9 show an example of behavior obtained during the evolution, and a logic circuit which was reconstructed from the gene that performed well to avoid collisions.

In this series of experiments, the starting point of the robot was not fixed, and varied from one trial to another. Therefore, the values of fitness also fluctuated from one trial to another for the same individual. The upper and middle raws in Fig.11 show the distributions of fitness values, E, D, 1–V and 1–S, for the best individuals at generations of 30 and 247, obtained in 100 trials. We can see that the behavior is getting better after generations although the experience of each individual in a generation was very much limited.

The bottom raw in Fig.11 was obtained by the robot in which had the identical mechanism to the Braitenberg's vehicle. In comparison to the middle raw, our algorithm scored better or similar to the Braitenberg's vehicle.

5 Discussions

In the experiment, the starting point was different in each individual. Therefore, an identical behavior of the same individual may get evaluated differently,

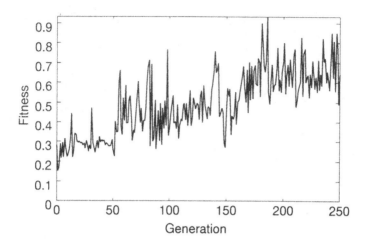

Fig. 6. Best individual fitness in each generation

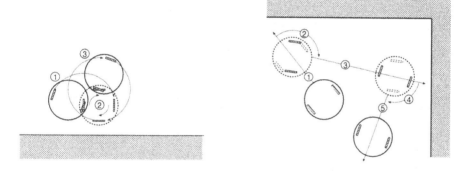

Fig. 7. Behavior of the best individual in generation 30

Fig. 8. Behavior of the best individual in generation 247

because the environment was different, and thus we could expect that the behavior be robust. Some individuals after generations acquired such robustness in experiments.

The life time of one generation, 10 sec, might be thought to be too short. However, the purpose of this series of experiments was not let the robot learn the path in the field and get the map, but acquire various useful reflexes in a variety of situations. If the life was long, individuals ought to make complex behaviors, and did not get high scores even if they make good simple reflexes, such as avoiding left obstacles, going straight in a path, and so on. By setting the life time short, one can expect that individuals with simple good behaviors

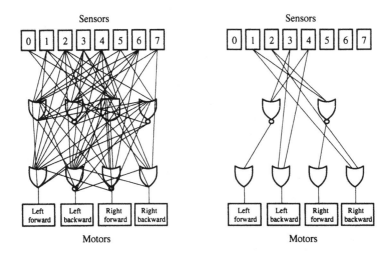

Fig. 9. Circuit of best individual in generation 247

Fig. 10. Circuit of Braitenberg's vehicle

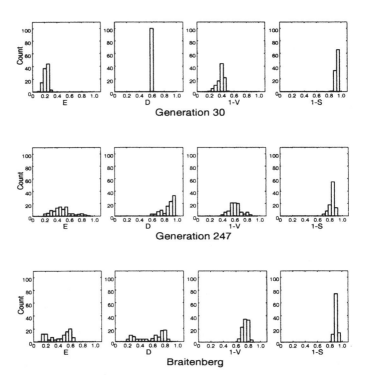

Fig. 11. Histograms of fitness values

could get high scores, and thus, individuals with combinations of such simple behaviors would appear by evolution. The best individual in generation 30 was only rotating (Fig.7). As shown in Fig.11, D and 1–V were not so high, because almost all the time it drove only a motor at a side. At generation 247, however, both motors were activated (Fig.8), and thus, the fitness values became higher.

It is essential to think that intelligent and well-organized behavior is generated by two mechanisms. One is the response of the system to the external environment, and this may called reflexes. The second is the response to the internal triggers. The internal triggers are generated within the system, and not caused through sensory mechanisms. The examples include oscillatory behavior such as running on a circle without sensory stimuli, and behavior based on memory. One may think that the more complex both internal and external response mechanisms are, the more intelligent is the behavior.

In this study, we demonstrated that an evolutionary controller based on logic circuits could generate a variety of behavior. One could argue that the logic circuitry is too simple as the controller, because, in real biological systems, the controller is the complex neuronal connections that consist of neurons with complex structures and functions, and that are constructed under complex principles. However, in order to understand intelligence of behavior, analysis of the neuron network is not sufficient, or we may say that such analysis is even useless. It is most important and essential that the system be evaluated by whether or not it has the suitable external and internal response mechanisms to the environment, not by the kinds of material that the system is constructed of.

In this experiment, we have seen not only the logic controller generated the right reflexes, but also behaviors which, it seemed, based on internal triggers. One such example is the behavior based on flip-flop. The robot rotated until it sensed a wall in left-front. And then, it went forward straight. During both rotating and forward moving periods, the sensory information should be identical, but the robot evoked two types of behavior, rotating and moving forward. The second example we observed was running on a circle with a radius of 3 times of the width between wheels. Because the motor ran in a on-off manner, the radius should have been the width between wheels if one motor was on and another off. In practice, a controller contained a oscillator to generate a periodical triggers that made another wheel rotate slowly in a on-off manner.

In this study, we demonstrated that the logic circuit controller was able to generate behaviors based on both external reflexes and internal triggers. It seems that the logic circuit controller is suitable to express both mechanisms if it is further refined, and we are currently trying to improve the method.

References

Braitenberg V. 1984. Vehicles. Experiments in Synthetic Psychology. MIT Press, Cambridge, MA.

Mondada F., E. Franzi, P. Ienne. 1993. Mobile Robot miniaturisation: A tool for investigation in control algorithms. In: Proceedings of the third International Symposium on Experimental Robotics,Kyoto,Japan.

Nolfi S., D. Floreano, O. Miglino, F. Mondada. 1994. How to evolve autonomous robots: different approaches in evolutionary robotics. In: Proceedings of the Fourth International Workshop on the Synthesis and Simulation of Living Systems, 190 – 197.

Autonomous Robot with Evolving Algorithm Based on Biological Systems

Jun YAMAMOTO[1] and Yuichiro ANZAI[1]

Department of Computer Science, Keio University, 3-14-1, Hiyoshi,Yokohama 223,
Japan

E-mail:{yamajun,anzai}@aa.cs.keio.ac.jp

Abstract. This paper describes an evolvable system that has a computational model of basal ganglia and hippocampus which have close relation to the generation and management of memory and movement faculty in our brain. We also describe its application to the design of learnable autonomous mobile robots. From a view point of Evolvable System, first we argue how the particular functions of basal ganglia and hippocampus work for memory and movement. Next a computational model of basal ganglia and hippocampus is presented as a specific type of evolvable systems, and how this model can be used for the design of learnable robots is described. Then we make our actually constructed robot learn, using the implemented model, four fundamental behavior tasks such as those insects learn in their baby stages. It follows with the analysis of the learned behavior. Finally, we discuss the significance of our research.

1 Introduction

The hippocampus has been implicated in the processes of learning and memory based on both its structure and function. Also, the basal ganglia and cerebellum play significant roles with the motor cortex in voluntary motor behavior. From the biological point of view, hippocampal pyramidal cells showed a remarkable property during spatial behavior [3] — individual cells fire robustly when an animal enters a circumscribed region of space with different cells firing in different locations. Furthermore, at each location within an environment, a unique pattern of neuronal activity is generated which may be thought of the "representation" of that particular place [6]. The reason why we applied such mechanism of the brain to robot leaning tasks, is that animals, from humans down, have excellent ability of learning behavior patterns.

On the other hand, there are various approaches to the design of robot behavior systems. Broadly, it can be divided into three approaches. The first approach is *The algorithmic approach* that is based on functional or behavior based decomposition of the overall control task such as obstacle avoidance and others. Brook's subsumption architecture[2] is a representative one. The next one is *The dynamics approach*[4] that is based on the notion of behavior systems as continuous dynamical systems instead of discrete computational systems as in the algorithmic approach. The third approach is *The connectionist approach* that is based on direct coupling between a system's sensory inputs and action parameter outputs through artificial neural networks, especially recurrent networks [1] [5].

Different from these approaches, we provide a biological approach to the learning behavior patterns of mobile robots.

2   Building a Computational Model

In real brain, most of sensory messages from the body are received by cerebrum, especially in neo-cortex. This part is a kind of sensor data handler which integrates a number of nerve signals, but it does not analyze or evaluate patterns of nerve information.

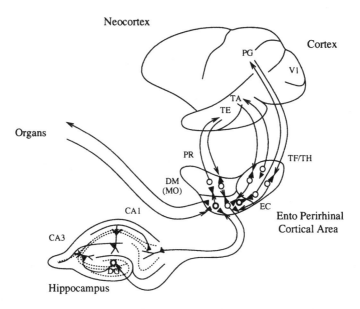

Fig. 1. A Model of Hippocampal-Cortical Memory System

Those integrated sensor data are submitted to hippocampus and basal ganglia where the sensory data are analyzed and evaluated. Between the hippocampus and basal ganglia, some sensory data which are well evaluated may put into short term memory called Long-Term-Potentiation (LTP) cycle. This cycle is a kind of what is called reinforcement learning, and forms a specific type of neural activating pattern. Afterwards, those special patterns gradually change into behavior patterns, so called long-term memory.

A schematic drawing of the hippocampal-cortical memory system is shown in Figure 1. The neural networks in the neocortex show representation of visual object quality in area TE, object location in area PG and information for auditory recognition in area TA, as long-term memory. The entorhinal cortex(EC), perirhinal cortex(PR) and para-hippocampal gyrus(TF/TH) are an interface between the neocortex and the hippocampus, and receive projections from the neocortex and are the major source of projections to the hippocampus. The entoperirhinal cortices also receives projections from the decision making area(DM) or the motivational area(MO). All these projections are reciprocal. The hippocampus consists of dentate gyrus(DG), CA1 and CA3 regions and plays an important role at the time of learning in establishing long-term memory[7].

Basal ganglia and hippocampus have close relation to the generation and management of memory and movement faculty in the brain. Figure 2 shows a summary of brain information processing, being concentrated on learning behavior tasks.

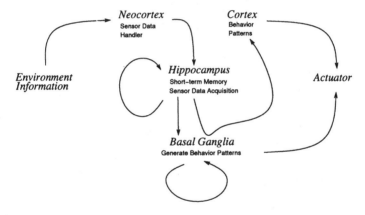

Fig. 2. A Computational Model for Learning Robot

2.1 Learning Algorithm

Let us introduce how our computational model of brain acquires or learns behavior pattern from it's environment. The primary goal of learning is to build a good feed-forward model of behavior tasks through several trails of experiment so that the robot can predict next sensor image to make a new action. According to our biological review of brain information processing, we constructed a feed-forward model shown in Figure 3.

In our model, the notation m_n and m_{n+1} imply motor values, s_n and s_{n+1} current and predicted sensor images, respectively, and r_n and r_{n+1} are internal state values that represent the condition of the robot.

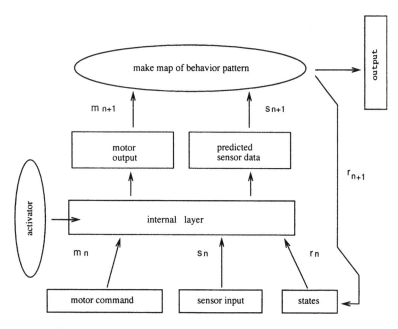

Fig. 3. Internal Configuration of Computational Model

One cycle of learning proceeds as follows. At first, sensor data are temporarily stored and observed by an activator that examines the stream of raw data whether they are candidate of LTP or not. Is is done in a sensor handler which corresponds to neo-cortex and hippocampus. Then those sensor data are put into the feed-forward model corresponding to basal ganglia that repeats the creation and prediction of motor value and sensor images to minimize the error evaluating function E shown below :

$$E = \epsilon e_g + \gamma e_c \tag{1}$$

$$e_g = 1/2(s^{r*} - s_r^r)^T(s^{r*} - s_r^r) \tag{2}$$

$$e_c = 1/2 \sum_{n=1}^{\tau} (s_n^l)^2 \tag{3}$$

where e_g means the error between actual sensor image s_r^r from the environment and predicted image s^{r*} from the computational model. e_c means the evaluated cost of trajectory. s_n^l means traveling cost of the robot. The model tries to minimize the factor E, so that the robot can acquire proper behaviors. where ϵ and γ are constants.

2.2 A diorama

When we considered variations of learning tasks for our robot, **Obstacle Avoidance** and **Target Detection** were selected, since the computational model can manage sequential input data, and also robot returns sequential values of their sensors such as IR distance sensors, CDS light intensity sensors and so forth. The robot were given such kind of diorama as shown in Figure 4.

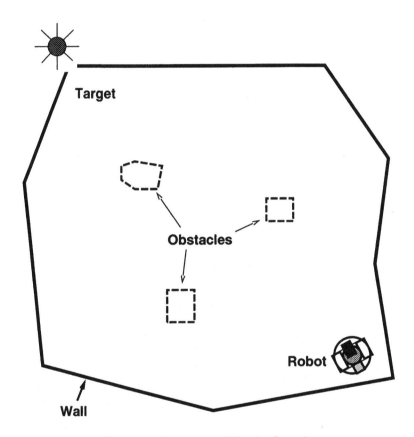

Fig. 4. A Diorama for Behavior Learning

The features of our experimental diorama are as follows. The diorama is separated with walls from surroundings. The walls are required to be neither of fixed size nor with regular polygonal outlines. Moreover, as the learning proceeds, the size, form or number of obstacles does not much influence on the learned result. After several trials, the skill of the robot gradually increases, and it can move along walls and obstacles. Finally it accomplishes the task when value of the light sensor exceeds a certain value.

3 Robot in detail

The autonomous robot that we constructed are shown in Figure 5. The dimensions are $(W) : 150mm(6in.) \times (L) : 150mm(6in.) \times (H) : 240mm(9.5in.)$. Several sensors and actuators are packed in it.

To accomplish our learning experiment, three kinds of sensors, infrared range sensors (Figure 6), CDS light intensity acquisition sensor (Figure 5) and collision detect touch sensors (Figure 6), were implemented in the robot.

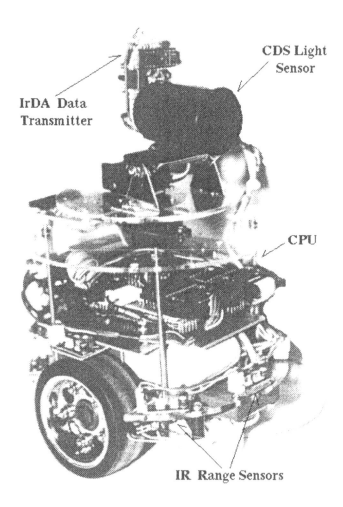

Fig. 5. A Portrait of the Robot

The infrared range sensors are to detect walls or obstacles so that the computational model can evaluate and generate the next behavior pattern. In the experiment, the sensors are mounted toward eight directions at even intervals and, set to react precisely against a certain threshold level of distance.

Another sensor is what we call the CDS cell light intensity acquisition system that measures the light intensity and is used to find its target in the diorama. The acquisition system is composed of a CDS cell unit, an operational amplifier and A/D converters. Converted data are scaled into 0 to 1 analog value so as to input those values into the brain model.

Those two sensors are directly used for learning, but that alone would not complete the learning cycles. Since our brain model requires such signal that terminates feed-forward data acquisition and switches the learning phase to the back-propagating phase. The two touch sensors (Figure 6), mounted in the front side of the robot, enables the termination.

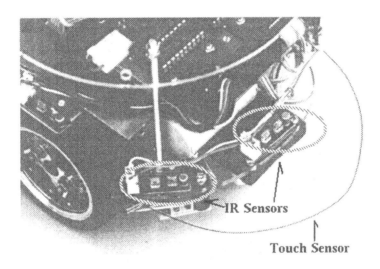

Fig. 6. The IR Sensor and Touch Sensor

Figure 7 shows an internal configuration of the learning robot. Infrared range sensor images and other data are combined by a on-board CPU into small data packets. The CPU transmits those packets in regular intervals to a host computer by using the IrDA infrared data communicator. The intervals are approximately $200msec$.

The learning system configuration is shown in Figure 8. At first, robot sends sensory data packets at regular intervals while explores the diorama. Then a front-end computer (shown as Sensor Data Handler in Figure 8), corresponding to neocortex, receives sensory messages from the robot and reconstruct a history of brief data stream.

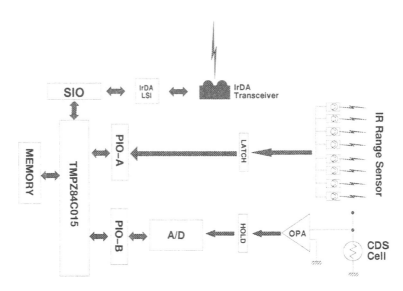

Fig. 7. A Block-Diagram of Robot

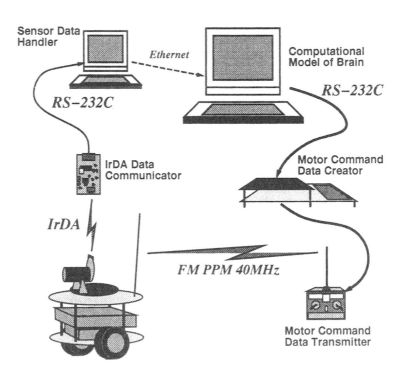

Fig. 8. A Robot Controlling Configuration

The short data streams are down-loaded from Ethernet to a main computer (workstation) where the biological brain model is executed. Next, the output of actuator data are sent to "Motor Command Data Creator" that converts the protocol of output data to linear actuator values for the robot. Finally, the data from the brain model are transmitted by radio to the robot. This cycle takes $700msec$ approximately.

4 Learning Experiment

We carried out two types of experiments, each with two cases. The experimental setup is shown in Table 1.

Table 1. Experimental Setup

Variation	Light Attractor	Obstacle
Exp. A-1	No	No
Exp. A-2	No	Yes
Exp. B-1	Yes	No
Exp. B-1	Yes	Yes

The experiment items are "Wall/Obstacle Avoidance" and "Target Detection ". The first experiment, Exp. A-1 and A-2, implements an avoidance task which we considered as a most fundamental task that insects or animals are engaged to learn in their baby stages. The next experiment , Exp. B-1 and B-2 is for a target detection task.

At the beginning of the experiment, there are no limit of variations of movements. So the learning did not converged. But we gave some limit such as, "move backward when something straight ahead" and so on, that accelerate the learning speed and get rid of falling into local minimum problem.

4.1 A-1:Wall with no obstacles

As noted above, this experiment was considered to be the most primitive and fundamental behavior for robot learning task. A snapshot is shown in Figure 9. The results of what the robot acquired from the environment summarized below.

– If no limitation is imposed at all, the learning cycle would not converge.
– There is such a moment that the limitation directly affects it's avoiding behavior.
– At acute corner, the robot had a tendency to fall into a local minimum.
– The robot tended to move away from the wall.

4.2   A-2:Wall with obstacles

The next experiment is to verify its robustness and validity of what the model learned from the environment. Needless to say, that number and location of obstacles are not given to the model.

A snapshot with a trajectory is given in Figure 11. The results are summarized as follows.

- By the behavior pattern learned in experiment A-1, the robot could avoid the obstacles in most cases.
- But there were a few special cases difficult for the robot to deal with: e.g., when there is not enough space to go through(Fig. 10 and Fig. 12).
- After 25 to 30 of learning trial, the robot could go through the path with much shorter time.

4.3   B-1:Target detection with no obstacles

The latter half of experiment consists of "Obstacle Avoidance" and "Target Detection". The target source of light was placed at the edge of the environment. The objective of this experiment was to inspect whether previously acquired behavior patterns were effective against a new environment or not.

Figure 13 shows an experimental scene. The results are summarized as follows.

- The learning does not work well at the beginning. Movement of the robot was too unstable to accomplish the task. The reason was considered that there was no priority between the two tasks.
- So we explicitly gave the priority of "avoidance" being higher than "detection".
- After 15 to 20 of repetitive trials, the robot could find the target more stably than before.

4.4   B-2:Target detection with obstacles

The final experiment included whole items of Table 1. In this experiment, the priority was set to the condition. The results showed that the previously learned behavior was applied effectively.

A snapshot is shown in Figure 14, and the results are summarized below.

- There were no stuck problems like Exp. A-2, but when the target of light went hidden behind an obstacle, the robot's movement went slightly unstable.
- We attributed the fact to the lack of adaptability to rapid change of the environment, supposed in the brain, for example, by eye movement. We suspected that the priority which we explicitly set was too affective to the change it's internal states.
- Another cause, may have been the redundancy of the robot's eye movement. That is, our robot has a panning mechanism that seeks the light source around it. So the eye movement and its motor movement gave redundant factor.

5 Discussion

The above experiments confirms the possibility of applying the biological information processing mechanism to the learning behavior mechanism of autonomous robots. We have made the model as faithful to the biological mechanism as possible so that the robot can learn specified tasks in real time. In the Exp. A-1 and A-2, the model provided a test whether it could acquire the appropriate task from unknown environment. From the results of the experiment, it could gradually acquire the avoiding task in about 30 to 40 trials. This leaning speed was accelerated by limiting some conditions. But if there are any limitations, the trial would over hundreds of trials. We considered the limitation as a genetic inheritance underlying human faculties so as not to be diverged.

On the other hand, the Exp. B-1 and B-2 took much time to converge their learning cycles. As we described before, the redundancy movement of the robot's eye directly affected to the learning speed. This phenomenon reflects our coordinated movements of muscle when we are learning a new behavior.

6 Conclusion

This paper presents an application of some specific biological information processing to learning of autonomous robots. We constructed a computational model of hippocampus and basal ganglia to find a way of robot learning problem. As we described, the experiments worked well, and the robot acquired two behavior from unknown environments. Our results can be a kind of "feed back" from biological science to engineering science.

References

1. Lisa A.Meeden. Towards planning:incremrmtal investigations into adaptive robot control. *PhD dissertation*, Indiana University, 1994.
2. Rodney A. Brooks. A robust layered control system for a mobile robot. *IEEE Journal of Robotics Automation*, RA-2:14–23, 1986.
3. O'Keefe J and Conway DH. The hippocampus as a spatial map. preliminary evidence from unit activity in freely-moving rat. *Brain Res*, 34:171–175, 1971.
4. Luc Steels. The artificial life roots of artificial intelligence. *Artificial Life,MIT Press*, 1:75–100, 1994.
5. Jun Tani and Naohiro Fukumura. Learning goal-directed sensory-based navigation of a mobile robot. *Neural Networks*, 7:553–563, 1994.
6. Matthew A. Wilson and Bruce L. McNaughton. Dynamics of the hippocampal ensemble code for space. *Science*, 261:1055–1058, 1993.
7. Minoru Tsukada. A Proposed Model of The Hippocampal-Cortical Memory System and Temporal Pattern Sensitivity of LTP in Hippocampal Neurons. *Neuroscience*, Vol. 3, No. 2,213–224, 1992.

Fig. 9. An Experiment with No-Light, No-Obstacles

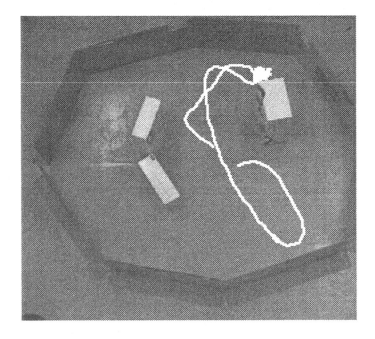

Fig. 10. An Experiment with Obstacles, but No-Light Attractor (1/3)

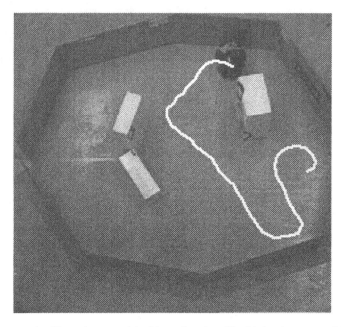

Fig. 11. An Experiment with Obstacles, but No-Light Attractor (2/3)

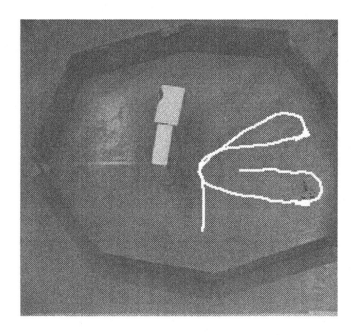

Fig. 12. An Experiment with Obstacles, but No-Light Attractor (3/3)

Fig. 13. An Experiment with Light Only, but No Obstacles

Fig. 14. An Experiment with Light and A Few Obstacles

Memory-Based Neural Network and Its Application to a Mobile Robot with Evolutionary and Experience Learning

Hidetaka ITO[1,2], Tatsumi FURUYA[2]

[1] Industrial Research Institute of Kanagawa
705-1 Shimo-Imaizumi, Ebina-shi, Kanagawa, 243-04, JAPAN
E-mail: ito@kanagawa-iri.go.jp
[2] Department of Information Sience,Toho University

Abstract. Use of the neural network in pattern recognition problem has many beneficial aspects, including advantages in learning, generalization, and robustness. However, the use of neural networks also has drawbacks. Problems encountered during the use of neural networks include an extended learning period and the inability of the users to process data. In an attempt to overcome these disadvantages, we propose a memory-based implementation of neural networks. Our method realizes neural network-like properties such as learning, generalization and robustness and is free from the weak points of the neural network. In our approach, training data are stored in a memory in the form of distributed manner by the use of several random number tables. On-line learning can be realized easily in our approach. This method was applied to a behavior-learning mobile robot. This robot acquires instinctive behavior by evolutionary method.

1 Introduction

In pattern recognition problems, layered neural networks show superior abilities in learning algorithms, such as the back propagation methods. However, neural networks need a long execution time for training. And also, if we want to add or eliminate trained patterns, we have to prepare new training data sets. In order to overcome these problem, we proposed a memory-based approach called "Probabilistic Distributed Memory" a few years ago.[2] This memory stores input-output pattern pairs in the form of distributed manner by the use of hash functions. The memory based approach to memorize training data instantly. The memory-based approach realizes robustness and generalization. Furthermore, it is easy to modify the stored patterns. However, the only problem was a tuning of a hash function. In this paper, we propose a new version of the memory-based method using several random number tables instead of the hash function. We applied this method to the behavior learning of simple mobile robot. This mobile robot is assumed to have two pairs of drive wheels, and several sensors which detect the direction of the goal and the distance to the obstacles. In our simulation, a robot learns behavior patterns to reach to the goal based on the sensor information. First, the robot learns fundamental behavior in an obstacle free environment by the use of an evolutionary technique (section 3.1). Then, the elite robot selected in the evolutionary learning repeats trial and error to acquire an obstacle evading behavior by its experience(section 3.2).

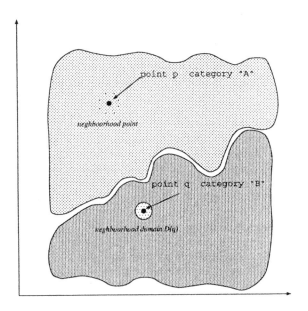

Fig. 1. Pattern space

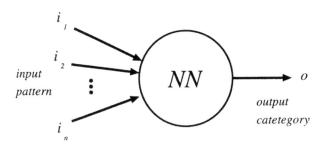

Fig. 2. Memory-based Neural Network Module

2 Memory-based Implementation of Neural Network Properties

A layered-type neural network is applied to various pattern recognition problems in combination with learning methods like back propagation. The neural network allows superior abilities in learning, generalization and robustness, and it enables the use of flexible information processing. However,two major problems are encountered during the use of neural networks:(1) It requires long period for leaning and (2) the partial addition or deletion of data is done with difficulty. Therefore, the use of such networks in systems which are situated in dynamic environment is not possible. Learning in dynamic environments requires on-line training capabilities. We have devised mechanisms which overcome such problems. We have tried to make a mechanism which not only has conventional neural network properties, such as robustness and generalization, but also on-line training capabilities. In order to improve upon our previous work in this area, we are proposing a new memory-based neural network. The mechanism uses random number tables instead of the hash function. A description of the Probabilistic Distributed Memory method comes first. Then we will demonstrate the new random number table approach.

2.1 Probabilistic Distributed Memory

We demonstrate the probabilistic distributed memory method by the use of a simple category classification problem. In this problem, input pattern space is classified into two categories: "A" and "B" as shown in 1. Input patterns are multi-dimensional vectors and output patterns are put into the category "A" or "B". (cf. Fig.2) The Probabilistic Distributed Memory consists of a hash memory which stores output patterns.

(a) Training procedure

A procedure to train a point p in an input pattern space belongs to category "A" is as follows. An input pattern is used to determine to an address "add" in which the corresponding output pattern is stored. At first, The input pattern is input to a hash function $f(p)$. The output of the hash function is a memory address where the corresponding output pattern is stored. The category "A" (output pattern) is written on the hash memory. At the same time, some points which are closed to the input pattern p are generated. The points are also applied to the hash function, and the output pattern "A"s are stored in those addresses (cf. Fig.3).

(b) Recall procedure

The recall procedure is as follows: One point on an input pattern space is described as q. To determine a category of q, many points nearby to (neighborhood) the point q (domain D in Fig.1) are selected. The hash addresses of the points are used to recall the output pattern. By referring to the hash

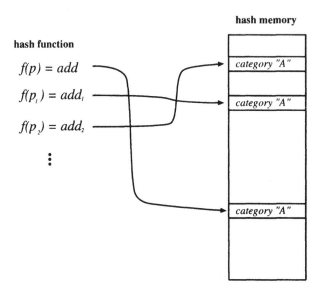

Fig. 3. Procedure store data with hash memory

memory, many output patterns are read from the addresses. The final output (category) is determined by the majority decision. The procedure realizes some properties of the neural network, such as robustness and generalization ability. In addition, this method also has the advantages in additional learning and partial elimination learning. Users can process learned data because the training data are stored in a form easy to understand. In the case of the neural network however, users cannot understand learned data which are saved as weights of many synapse combinations. One problem of this approach was the optimization of hash function.

2.2   From the hash function to the random number table

The new approach we proposes is to determine the address by using random numbers tables. We propose to determine the address by using a random numbers tables instead of the hash function. The outline of this method is shown in Fig.4. At first, a table of random numbers (address map) is prepared for each field of input. The table is replacement a hash function. Each address map outputs a partial address for the input value. Memory address is made by concatenation of the partial addresses. By using random number tables, the stored addresses are expected to be uniformly distributed on the memory. When constructing a neiborhood address, one partial address should be replaced by a neighborhood partial address.

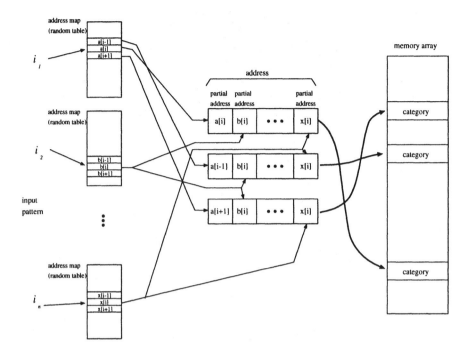

Fig. 4. Address determined by using random number

2.3 An application to two spiral problem

In order to examine the performance of our method, we applied it to the two-spiral problem. The problem is to classify the points into categories "Black" and "White" judging by its coordinate (x, y). Training data are spiral points as shown Fig.5. This problem is said to be difficult for the two-layered neural network. Input value (x, y) ranges from 0 to 255. The points in the neighborhood are selected as shown as Fig.6. and each point has a weight value which shows the distance to the original input. A training datum is written into the memory M, by accumulating the next formula.

$$M[add] = M[add] + C * W \tag{1}$$

Here, M is a memory array, add is address ,C=1 (if Black) or C=-1(if White), W is weight. during the recall phase, the data on the memory located inside of the circle 10 radius around the point are summed up. If the result is a positive value, its category is "Black" . Otherwise the category is determined to the "White". Fig.7 to 14 show retrieved results, memory sizes are 65536=256*256, 36864=192*192, 16384=128*128, 9216=96*96, 4096=64*64, 2304=48*48, 1024=32*32, 512=16*16. These results show that the proposed method can separate a complicated nonlinear space. We can observe the neural network-like

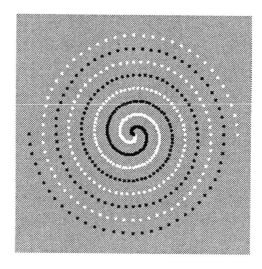

Fig. 5. Training data at two spiral problem

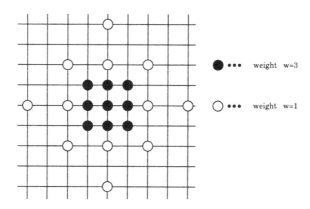

Fig. 6. Neighborhoot point and those weight

properties of generalization and robustness in those figures as well. These results are better than the hash function method.[2]

Fig. 7. Retrived result memory size 65536=256*256

Fig. 9. Retrived result memory size 16384=128*128

Fig. 8. Retrived result memory size 36864=192*192

Fig. 10. Retrieved result memory size 9216=96*96

Fig. 11. Retrieved result memory size 4096=64*64

Fig. 13. Retrieved result memory size 1024=32*32

Fig. 12. Retrieved result memory size 2048=48*48

Fig. 14. Retrieved result memory size 512=16*16

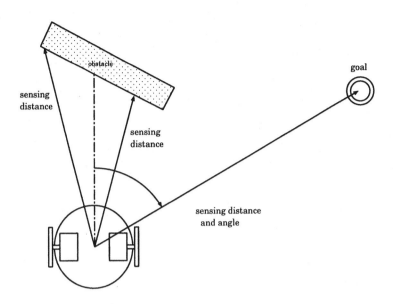

Fig. 15. Mobile robot

3 Application to behavior learning of mobile robot

We carried out the simulation in which a simple mobile robot learns a behavior pattern by applying the neural network module the memory-based approach. The mobile robot is assumed to have a structure as shown in Fig.15. It has four sensors to detect (1) distance and (2) direction to a goal, and (3) (4) distance to an obstacle of the two front directions. The robot has two drive wheel and each wheel is controlled independently by a motor. These motors can select one of three states in unit time. Those three states are 1) standstill, 2) slow, and 3) fast. Each motor both the right and left is controlled by a neural network module. The input data to the neural network module are received from four sensors (cf.Fig.16).

3.1 An acquisition of fundamental behavior by evolutionary method

First, we examined a simple, obstacle-free case. In this case, the front direction sensors (sensors (3)and (4)) do not need to react since there is no obstacle. The robot behaves on the basis of information received from only the distance and direction sensors (sensors (1) and (2)).We describe the algorithm which allows the robot learn how to arrive at a goal. We prepare a square region in x-y plane and put a goal.

(i) We prepare an initial population of robots. These initial states of robots (positions and directions) is determined by the random numbers.

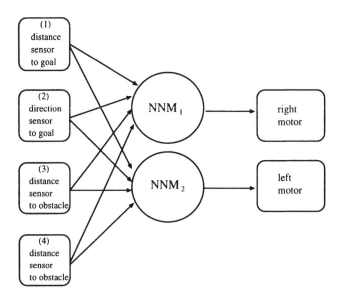

Fig. 16. Block diagram of mobile robot

(ii) The system obtain distance (L_0) to a goal, angle (θ_0) between the directions of robot and the direction of the goal.

(iii) A robot is moved by a random number. When a robot moved out of the region, the system returns to (i). The system receives information on distance(L_1) and difference(θ_1) of direction in the same way. In case of $(L_1 < L_0$ or $|\theta_1| < |\theta_0|)$, behavior pattern is succeeded to by a next generation.

(iv) System returns to (ii) (several times).

(v) The system creates a next generation from selected robots.

After this procedure, we select an elite robot that acquired the best fundamentals behavior. And, we let learn a behavior evading an obstacle to the robot by experience learning as shown in the next section. Fig.17 shows some behavior patterns acquired after learning.

3.2 Behavior learning with an obstacle

The robot learns a new behavior evading an obstacle on the basis of data earned in section 3.1. At first, an obstacle is put between the robot and the goal.

(i) The robot proceeds until it detects an obstacle, a learnd response as described in section 3.1.

(ii) Next, the robot walks randomly several times.

(iii) When the robot evaded an obstacle by chance, the behavior pattern stored in the memory.

Fig. 17. some behavior pattern Without obstacle

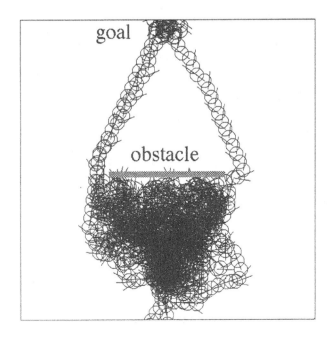

Fig. 18. Situation of learning with an obstacle

245

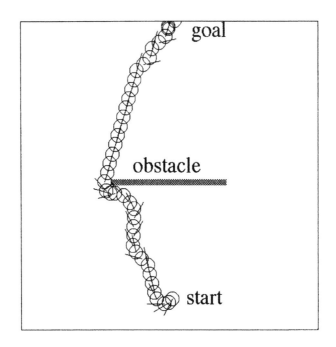

Fig. 19. Evasion behavior pattern 1

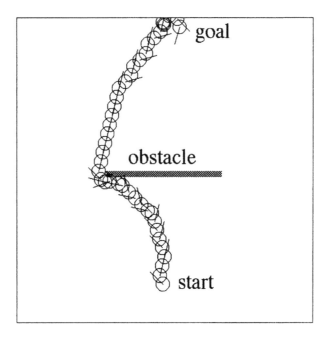

Fig. 20. Evasion behavior pattern 2

The robot acquires behavior patterns for evasion by repeating this procedure. Fig.18 shows how the robot learns the appropriate behavior. Some evasion behavior patterns acquired are shown in Fig.19 and 20.

4 Conclusion

We proposed a method which employs several tables of random numbers instead of the hash function in the memory-based approach. Tuning of hash function is not required in our new method. Learning data are scattered on the memory ideally by using tables of random numbers. In the two-spiral problem, clear results were obtained in the whole region, even when the memory size is made to 1/64th of full size. In the behavior learning of a robot, additional learning is important. In our case, we first trained the robot to respond to an obstacle-free course. Next the robot was able to maneuver a course with obstacle. Our memory-based approach enables the reinforcement learning, which is essential to adaptive behavior.

References

1. P.Kanerva, *Sparse Distributed Memory*, MIT Press,1988.
2. T.Furuya, *Probabilistic Distributed Memory*, IJCNN,1992,Beijing
3. T.Furuya, *Self-Programming Network(SPN)*, IJCNN,1993,Nagoya
4. R.A.Brooks, *New Approaches to Robotics*, Science, vol.253,1991.
5. H.Ito, *Intelligent mobile robot*, ICNN,1995,Perth

Innovative Architectures

Multiple Genetic Algorithm Processor for Hardware Optimization

Mehrdad Salami*

The New Energy and Industrial Technology Development Organization(NEDO)
Computation Models Section
Electrotechnical Laboratory (ETL)
1-1-4 Umezono, Tsukuba, Ibaraki, 305, Japan.
e-mail: m_salami@etl.go.jp

Abstract. A hardware description of Genetic Algorithms is presented to handle optimization problems. Genetic Algorithm Processor (GAP) is a reliable and fast processor for emulating genetic algorithms in hardware. Following that the multiple genetic algorithm processor configurations have been described based on the GAP. The simulation results show that multiple genetic algorithm processor configurations work better than single configuration with lesser complexity. It is possible to apply multiple configurations to more complex problems.

1 Introduction

Genetic algorithms represent a class of adaptive search techniques which can be applied to optimization problems or in artificial intelligence. A genetic algorithm (GA) is a robust global optimization technique based on natural selection. The basic goal of GAs is to optimize functions called fitness functions. GA-based approaches differ from conventional problem-solving methods in several ways. First, GAs work with a coding of the parameter set rather than the parameters themselves. Second, GAs search from a population of points rather than a single point. Third, GAs use payoff (objective function) information, rather than auxiliary knowledge. Finally, GAs use probabilistic transition rules, not deterministic rules. These properties make GAs robust, powerful, and data-independent.

Due to its evolutionary nature, a GA will search for solutions without regard to the specific inner workings of the problem. Much of the interest in genetic algorithms is due to the fact that they provide a set of efficient domain-independent search heuristics without the need for incorporating highly domain-specific knowledge [1]. There is now considerable evidence that genetic algorithms are useful for global function optimization and NP-Hard problems [2].

This paper starts with an explanation of how a GAP can be applied to an optimization task. Then the multiple GAP is introduced. Finally the model is applied to two applications to compare the performance of the Multiple GAP with single configuration.

---

* The work was done by the author when he was a Ph.D. student in the Electrical Engineering Department - Victoria University of Technology - Melbourne - Australia

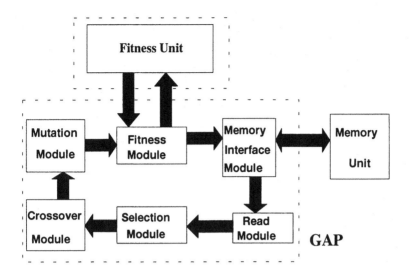

Fig. 1. Internal architecture of the GAP

2 The Genetic Algorithm Processor

A hardware genetic algorithm can be constructed to directly execute the operation of a genetic algorithm as shown in Figure 1 [3]. Such a processor can be used in situations where high throughput is required and where the logic of the genetic algorithm is expressible in simple units which can be synthesized in hardware. This is generally the case as genetic algorithms are inherently simple and contain only a few logic operations. Such speed enhancement is important for many applications such as adaptive filters.

The GAP maintains a population of bit strings in its memory. Each of these member strings represents a tentative solution to the problem at hand. In execution the GAP delivers a member string to the Fitness Unit (FU) which evaluates its performance and returns a response known as a *fitness* value which is stored in memory. The fitness value is used by the Selection Module (SM) to select suitable members for mating and reproduction to obtain new member strings for the next generation.

In the GAP, mating involves switching substrings from two selected parent strings in simple operation which takes place in the Crossover Module (CM). New child strings formed in this fashion may then be subjected to a simple random bit-flipping process in the Mutation Module (MM). After mutation the newly formed strings are delivered to the Fitness Unit for evaluation. The FU returns a fitness value which is written to memory with the new member string.

The above steps continue until the FM determines that the current GAP run is finished. The whole GAP section was written in VHDL (VHSIC Hardware Descriptive Language) and simulated and synthesized by Mentor Graphics tools. After synthesizing the GAP Modules, the NeoCAD FPGA Foundry^{TM} tools was used to convert the GAP modules to the FPGA (Field Programming Gate Array) technology [4].

3 Multiple Genetic Algorithm Processor

For real applications of the GAP a means of splitting the bit string of a member between multiple GAPs is needed [5]. One practical way of doing this is by dividing the full member into bit slices. Figure 2 shows how a 32 bit member string can be distributed over four GAPs. Each 8-bit slice is assigned to a separate GAP which can be implemented on one FPGA chip. The Fitness Unit works with a full 32 bit member and delivers an 8 bit fitness value to all GAPs.

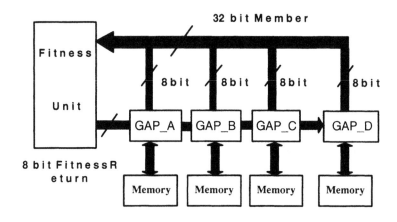

Fig. 2. Splitting one member between four GAPs.

All GAPs in Figure 2 operate concurrently but at any time only one of them is waiting for a response from the Fitness Unit. Each GAP produces new members and tries to access the FU to calculate the fitness value. Whenever one new member (bit slice) is ready in one of the GAPs (say GAP_A) then GAP_A tests the FU to see if it is free. If so then the GAP_A provides a new member slice on its output.

The FU receives the new member which is made up of the new slice from the GAP_A and the current (unchanged) slices from the other GAPs. The FU calculates the fitness value and returns the 8-bit result on the buss where it is received by GAP_A. This configuration splits the bit string and enables the GAP to handle large bit strings.

The operation of the multiple GAPs can be demonstrated with an example. Suppose the fitness function is $F(x, y) = x + y$. The x value is represented as 4 bits and the y value is also represented as 4 bits. Two GAPs can be used to solve this problem. One of them contains the x values (GAP_X) and the other one contains the y values (GAP_Y). The fitness value is represented in 4 bits. If there is a change to be made in x then GAP_X delivers its new string. The string for y remains unchanged at the last value delivered by GAP_Y. Table 1 demonstrates how the changes in x and y values affect the fitness values.

Table 1. A simple example for the multiple GAP configuration.

Action	x (4 bits)	y (4 bits)	Fitness value (4 bits)	Fitness Unit busy signal
Initial Values	0	0	0	'0'
Change in x Value	2	0	—	'1'
Return to GAP_X	2	0	2	'0'
Change in y Value	2	10	—	'1'
Return to GAP_X	2	10	12	'0'
Change in x Value	7	10	—	'1'
Return to GAP_X	7	10	17	'0'
Change in y Value	7	14	—	'1'
Return to GAP_X	7	14	21	'0'

In the next two sections the GAP in multiple configuration will be compared with single configuration.

4   Inverse Matrix

An inverse matrix problem is selected to compare the multiple configurations with the single configuration. The objective is to find the reverse matrix (A^{-1}) of the following matrix:

$$A = \begin{bmatrix} 1 & 4 \\ 5 & 6 \end{bmatrix}$$

which is equal to:

$$A^{-1} = \begin{bmatrix} -0.4286 & 0.2857 \\ 0.3571 & -0.0714 \end{bmatrix}$$

During simulations, each member of A^{-1} matrix is varied between -1 and +1 and defined as 8 bits. It means all configurations need to produce 32 bits for the Fitness Unit. The Fitness Unit receives 32 bit and returns the fitness value as an 8 bits. The Fitness Unit multiplies the input matrix (M) by matrix A to calculate the B matrix (A*M=B). Then The B matrix subtracted from unit matrix (I) to produce the C matrix (C=B-I). The fitness value then calculated as:

$$Fitness = \sum_{i=1}^{2} \sum_{j=1}^{2} abs(c_{ij})$$

and scaled to fit in 8 bits. Following parameters are selected for all GAPs:

population size = 64
generations = 256
fitness value = 8 bits
crossover rate = 90
mutation rate = 2
member size = 32 bits.

Three GAP configurations are examined:

1. Single GAP: The member length for the GAP is equal 32 bits.
2. Two GAPs: Each GAP delivers 16 bits to the Fitness Unit.
3. Four GAPs: Each GAP delivers 8 bits to the Fitness Unit.

Figure 3 shows the result of GAP simulations for the three configurations averaged over 10 individual runs and Table 2 summaries the results.

The table and figure show that in the multiple configurations, diversity in population is higher and the final error value is lower than the single configuration. The four GAP configuration generally produces better performance than two GAP configuration.

5   The PID Controller

In PID control we attempt to derive a plant in accordance with a given reference signal (Figure 4). If a mathematical model of the plant can be derived, then it is possible to apply various design techniques for determining parameters of the controller that will meet the transient and steady state specifications of the closed loop system [6]. However, if the plant is so complicated that its mathematical model cannot be easily obtained, then analytical approach to the design of a

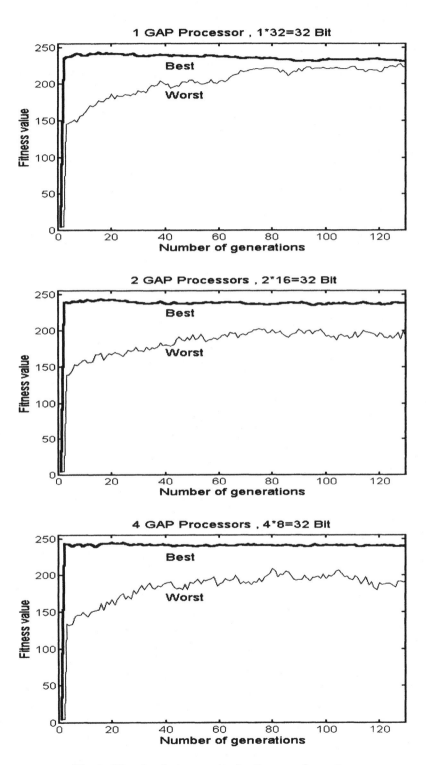

Fig. 3. The simulation results for three configurations.

Table 2. Comparing the results for the three configurations.

Characteristics	Single GAP =32 bits	Two GAPs =16*2 bits	Four GAPs =8*4 bits
Best Matrix	243.5	244.0	245.5
Average Best Matrix	233.8	237.7	240.0
Worst Matrix	210.0	189.0	106.2
Average Worst Matrix	228.0	208.3	208.9

PID controller is not possible and we must resort to experimental approaches. The process of selecting the controller parameters to meet given performance specifications is known as controller tuning [7].

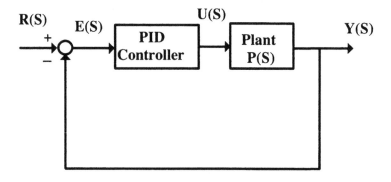

Fig. 4. A typical PID controller system.

The three gain parameters (K_i, K_p, K_d) of the PID control law interact with the plant parameters $P(S)$ in a complex fashion when the designer attempts to achieve the specified roots of the closed loop equation. These roots are chosen in order to obtain the desired transient response of the closed loop, while taking the resultant zeros into account.

The controller introduces a new pole at the origin of the s-plane and shifts the original roots of the closed loop system to new positions. PID controllers increase the order of the closed loop equation by one. In addition to these effects, PID controllers introduce a pair of zeros, usually a complex conjugate pair, which will normally have a significant effect on the transient behavior of the compensated system.

Designing PID controllers, even for low order plants such as a robot arm, can be a difficult problem. Consider the system illustrated in Figure 4 where the PID controller obeys the following control law:

$$U(S) = (\frac{K_i}{S} + K_p + K_d S)E(S) \tag{1}$$

where $Y(S)$ is the output of the controller system, $R(S)$ is the reference signal, $E(S)$ is the error signal equal to $Y(S) - R(S)$ and $U(S)$ is the output of the PID controller. Using the equality $S = (1 - Z^{-1})$, (1) can be expressed in the Z domain as:

$$U(Z) = (\frac{K_i}{1 - Z^{-1}} + K_p + K_d(1 - Z^{-1}))E(Z) \tag{2}$$

The goal of PID controller design is to determine a set of gains, (K_i, K_p, K_d), of the control law such that the set of roots of the closed loop equation chosen by the designer are obtained.

The efficiency of the system can be measured by calculating the integral of the time multiplied by the error for the unit step response during $[0, T]$:

$$error = \int_{t=0}^{T} t|e(t)|dt \tag{3}$$

The problem confronting the designer, therefore, is to calculate the three gains of the PID controller while ensuring that transient response specifications (minimum error, overshoot, rising time, settling time and steady-state error) are met.

6 Simulation Results for PID controller

Genetic Algorithm Processor simulations have been conducted for the PID controller system in Figure 4. The simulations measures the response $Y(t)$ to the reference signal $R(t)$ which is a unit step function. The transfer function for a typical plant is described by Hwang and Thompson [8] as:

$$P(S) = \frac{2}{s(s-1)s+5)} \tag{4}$$

To calculate the fitness value, all transfer functions have to be converted from the S domain to the Z domain, and to the discrete domain. Then 2000 points are selected between 0 and 10 seconds ($T = 10/2000 = 0.005$), and the integral of time, multiplied by the absolute error values, is defined as the fitness value.

$$Fitness = \sum_{k=1}^{2000} (kt)|e(k)| \tag{5}$$

The following parameters have been selected for the GAP model:

population size = 64
generations = 128
fitness value = 8 bits
crossover rate = 90
mutation rate = 2
member size = 24 bits.

Figures 5 shows the result of simulations for configurations of two, three, four and six GAPs averaged over ten individual runs for the PID controller system. In the figure the normalized error value is shown after each generation. All three configurations deliver 24 bits to the objective function. Table 3 shows the best final values for the multiple configurations. Note that the final error values are close, but the K values are very different. This means the problem surface has many local optimum points and each configuration merges to one of them.

Table 3. The best final values for the five configurations.

Best Values	Single GAP	Two GAPs	Three GAPs	Four GAPs	Six GAPs
Normalized Error Value	7e-2	5.8e-2	4.8e-2	5.7e-2	6e-2
K_i	5	3	13	3	40
K_d	63	42	105	45	45
K_p	95	83	185	95	103

The normalized error value column in Table 3 shows that all multiple processor configurations work better than the single processor. This is because they work on a smaller search space with the same resolution for the objective function. On the other hand single processor configurations give a smoother curve graph than the multiple configurations. The graphs show that in the 2 and 6 GAP configurations the convergence speed is higher than 3 and 4 GAP configurations but the curves are smoother and the error result in the final generation is worse. In the 3 GAP configuration there are no correlation between bit strings as each GAP is optimizing one K value. This is the reason why the 3 GAP configuration achieves the lowest error result.

7 Conclusions

This research shows that the multiple GAP configurations have potential to be used in real applications. All simulations show this configuration is capable of competing with the single GAP. In two applications the multiple GAP appeared to optimize the application better than the single GAP but it has not been proven that the multiple GAP is always better than the single GAP. To investigate the operation of the multiple configurations further mathematical analysis is required.

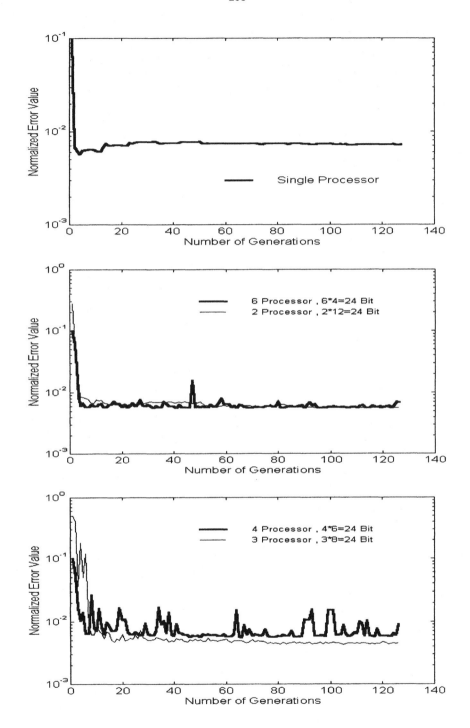

Fig. 5. The results of the PID controller simulation with single, 2, 3, 4 and 6 processors (Error Value).

References

1. Holland, J.H., "Adaptation in Natural and Artificial Systems", 2nd Edition, The MIT Press, Cambridge, Massachusetts, 1992.
2. Davis, L., "Handbook of Genetic Algorithms", International Thomson Publishing, New York 1991.
3. Salami, M., Cain, G., "Adaptive Hardware Optimization Based on Genetic Algorithms", Proceedings of The Eighth International Conference on Industrial Application of Artificial Intelligence & Expert Systems (IEA95AIE) , Melbourne, Australia, June 1995, pp. 363-371.
4. Salami, M., Cain, G., "Implementation of Genetic Algorithms on Reprogrammable Architectures", Applications Stream Proceedings of The Eight Australian Joint Conference on Artificial Intelligence (AI'95), The University of New South Wales, Canberra, Australia, November 1995, pp. 121-128.
5. Salami, M., Cain, G., "A PID Controller Based on a Multiple Genetic Algorithm Processor", The Proceedings of Control 95 Conference (Control'95), University of Melbourne, Melbourne, Australia , October 1995, pp. 359-362.
6. Dorf R.C., "Modern Control Systems", Addison-Wesley Publishing, Reading, MA, 6th Edition, 1991.
7. Ogata K., "Modern Control Engineering", 2nd Edition, Prentice-Hall, Englewood Cliffs, NJ, 1990.
8. Hwang W.R. and Thompson W.E., "An Intelligent Controller Design Based on Genetic Algorithms", Proceedings of the 32nd Conference on Decision and Control, San Antonio, Texas, 1993, pp. 1266-7.

NGEN: A Massively Parallel Reconfigurable Computer for Biological Simulation: Towards a Self-Organizing Computer

John S. McCaskill [*][*], *Thomas Maeke* [◊], *Udo Gemm*[°],
Ludger Schulte [*], *Uwe Tangen* [*]

[*]Institut für Molekulare Biotechnologie, Beutenbergstr. 11, D-07745 Jena, Germany.
[◊] Resedaweg 3, Göttingen-Nikolausberg, Germany.
[°] Max Planck Institut für Biophys. Chemie, am Faßberg, D37077 Göttingen, Germany.
[*]Author to whom correspondence should be addressed. Email: jmccask@imb-jena.de

ABSTRACT

NGEN is a flexible computer hardware for rapid custom-circuit simulation of fine grained physical processes via a massively parallel architecture. It is optimized to implement dataflow architectures and systolic algorithms for large problems. High speed distributed SRAM on the chip-to chip interconnect enables a transparent extension of problem size beyond the limits posed by the number of available processors. For simulated evolution tasks for example, this takes the effective population sizes up into the range of millions of strings without computational bottlenecks. Using FPGA technology, multiple processors per chip may be configured down to the level of individual gates if need be. 144 agent FPGAs are grouped in blocks of 4 and connected with one another via one of several possible broad band electronic frontplanes (36 channels per chip) which implement 2D, 3D or higher geometries. The communication of the parallel computation with a UNIX host workstation via VME-bus is mediated also by configurable interface FPGAs allowing problem specific communication needs to be respected. A separate 100 Mhz clock card frees the machine from the 16 MHz VME clock and allows designs to run at their optimum speed. Configuration files may be downloaded in series or parallel from the host workstation in less than a second. They may be created by user programs or commercial schematic entry or VHDL products. A run-time library for writing simulations in C which use the configurable hardware has been completed including a graphical interface allowing parallel symbolic debugging and display. The machine is a logical consequence of the shift of programming effort to effective communication in massively parallel applications. Its flexible structure also admits applications to real-time intelligent data acquistion tasks.

1. Introduction.

NGEN is a hardware platform for low level user-configured massively parallel computation. Based on FPGA (Field Programmable Gate Array) technology, it is reconfigurable for each problem by the user in software right down to the level of the elementary digital processing and communication. Up to 10000 processors may be configured by the user with connection topologies ranging from simple linear arrays to high dimensional hypercubes. NGEN arose from the desire to perform long-time simulations on large populations of interacting molecules to study basic principles of creative information processing as found in natural evolution[1][2][3][4]. Compared with FPGA arrays employed as logic simulators, NGEN is distinctive in its use of massively parallel memory access in conjunction with the wide inter-FPGA connections and the fine grained configurability of the global interconnection topology.

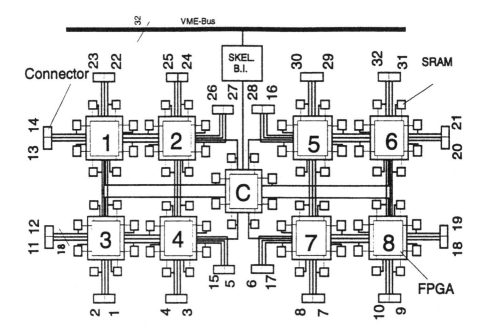

Fig. 1 Functional design of the NGEN processor cards.

The control FPGA C and agent FPGAs numbered 1-8 are each surrounded by 8 SRAMs. The control FPGA is connected to the VME bus via the skeleton bus interface. The agent FPGAs are connected in blocks of 4 to 32 connectors for alternate architectures.

The design of NGEN stems from an appreciation of the natural information processing capabilities of genetic systems and their special relevance to parallel processing. An understanding of biological information processing accompanied von Neumann in his work on the serial computer, prior to the unravelling of the double helical structure of DNA. It appears now to be equally important in the design of the self-organizing parallel computer. Self-organization is needed in at least three points for further progress in the development of high speed computation:

- in the manufacturing of hardware components exploiting the third dimension
- in the organization of data communication in parallel processing
- in learning and optimization of algorithms for specific problems

The vision of programmable matter, especially as found now in the artifical life community, was implicit in the initial conception of NGEN. Earlier work with networks of interacting molecular Turing machines [2,5] had shown the close relationship between the proliferation and interconversion of genetic information, both in and out of the cell, and parallel information processing. Experience in chemistry has shown that molecular processing is decomposable into a combination of binary recognition events and unary processing so that it suffices to bring together arguments for a computation in a pairwise fashion, cf [6]. The conventional cellular automata view of programmable matter fails to capture the natural movement of information packets implicit in molecules. A molecular architecture should ideally allow smooth interpolation between perfect random mixing of molecules between reactive collisions and spatial diffusion, which determines a lower bound to the intercommunication bandwidth. A key problem addressed here is how to extend the size of interacting processor simulations smoothly beyond the limit posed by available processing resources (i.e. without incurring data bottlenecks to memory storage).

The remainder of this contribution is structured as follows. In section two, we present an overview of the NGEN hardware. Section three discusses the multiprocessor design process with FPGAs, delineating the special properties of the FPGAs in the NGEN hardware, and then presents a multiprocessor example based on the Game of Life. In sections four and five we describe the special features of the control and agent FPGAs and their configuration, discussing the application to evolutionary designs. In the final section, we discuss the performance of NGEN and draw some conclusions about its potential fields of application.

2. System Overview

The chief design goals for NGEN were:
(1) a large number of configurable processors (10^4 or more);
(2) configurable interface to host workstation for rapid design iteration;
(3) interconnect between processors transparent to chip and board boundaries;
(4) large distributed memory to extend problem sizes without bottlenecks;
(5) flexible global interprocessor communication for high dimensional architectures.

The basic hardware consists of 18 copies of a custom designed VME-Bus Board (6U) together with a RISC workstation board (HP743i) in a 21 Slot VME-Bus backplane. An 1800 Watt power supply drives the boards using high current bars and 24 of the user definable pins on the bus. The chassis is air-cooled using a battery of conventional fans. Optionally a custom designed VME-BUS clock card may be used to generate the design clock at variable frequency up to 100 MHz. The 18 configurable boards containing 162 FPGAs may be connected via ribbon cable with one another to form the desired global architecture, with 288 external data lines per board. A relatively fine grained unit of 9 data lines per cable allows high dimensional architectures to be wired. Currently, the cables have been connected in parallel to a shielded multilayer front plane (currently 2 and 3 dimensional architectures have been implemented in the front plane) to improve performance. A key feature of the architecture is the placement of high speed SRAM on almost all the interchip connections.

Functionally, the 18 processor boards comprise, as shown in Fig. 1:
- a skeleton VME-bus interface
- a control FPGA which may be configured via the skeleton bus interface and then takes over control allowing customization of the bus interface and intelligent control of clocks and data to and from the agent FPGAs.
- agent FPGAs (8) which may be independently configured via the control FPGA from the host workstation and are usually used to implement an array of problem specific processors (with multiple processors per chip).

- high speed RAM units both dedicated to the 8 agent FPGAs (4 per FPGA) and placed on the communication paths between FPGAs (4 per agent FPGA and 8 on the control-agent connections)
- connectors (4 * 9 data bits per agent FPGA) for broad band communication between FPGAs.

The skeleton bus-interface is implemented with bus drivers, jumpers and two programmable GALs. With almost all its logic specified in two GALs, it is sufficient to allow the host to load the *control* FPGA with configuration data and thereby bring up the rest of the VME-Bus interface. This approach ensured maximum flexibility in the configuration and a minimization of irreparable design-errors. A VME-bus was chosen as likely to offer the least unexpected problems in definition and practice and allow us to concentrate on the problems of massively parallel configuration. Pairs of processor boards respond to 32 bit wide access with the host which is bus master. A LED indicates the selection of the board with an address matching that chosen by the jumpers and one of the two appropriate address modifiers. Interrupt processing and read and writes of varying word length are supported by the skeleton interface.

The control FPGA (a Xilinx XC4008 Logic Cell Array[7]) interacts directly with the VME-Bus and with each of the 8 *agent* FPGAs via 9 specific data and 8 global clock lines. Although the chips are identical, and all 9 FPGAs drive their own SRAMs, the pin configurations for control and agent FPGAs are different. The control FPGA supports a 16-bit data access to the VME-bus, and takes over the bus communications from the skeleton logic in the 2-GALS once it is loaded and active. The control FPGA communicates with the VME-bus, once it has been configured, via 32 16-bit registers using 5 address lines. It supports readback by the host, interrupts and a 32-bit loadable counter for high-speed multi-stepping operation. It can communicate with the 8 agent FPGAs in parallel using the 8 clock lines and 64 data-lines (8 per agent FPGA which run via the SRAMs of the control FPGA). Maximally affordable pin-counts of 208 (144 user) for the control LCA dictated the achievable 16 bit host access and the 8 bit communication to each agent LCA after allowing for global clock distribution, configuration, memory and control. Boundary scan for diagnostic of the agent FPGAs is supported reusing 32=4*8 of the 64 data-lines.

The 8 agent FPGAs (also XC4008) on each card also each control 8 SRAMs via a single address, as shown in Fig. 2. The single address choice was dictated by the limited pin count on the agents (144 user pins in the 208 pin package). The 8 SRAMs per agent are divided into two classes, the 4 exclusive SRAMs with data pins on dedicated lines and the 4 interconnect SRAMs with data pins on 4*8 of the interconnect lines to neighbouring chips. The neighbouring agents' SRAMs lie on the other 4*8 interconnects. Additionally, 4*2 lines lead directly to four neighbouring agents. The two different classes of SRAM have separate write enable and chip enable allowing independent operation. The remaining pins include the 8 global clock lines from the control FPGA and the 8 bidirectional data lines plus control lines for configuration and readback.

The SRAMs have the advantage over DRAMs of not requiring a dynamic refresh cycle. The only form of shared RAM which could be simply addressed must lie on the agent

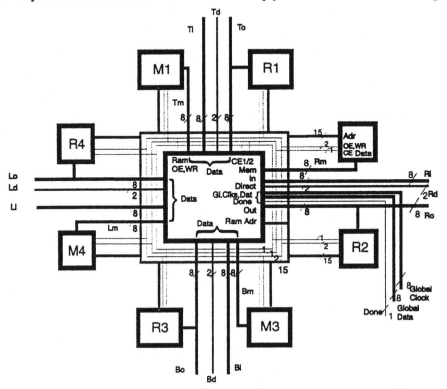

Fig.2 The connection scheme for an agent FPGA and its eight SRAMs.
The agents communicate with the control FPGA via 8 global clocks and 8 independent global data lines; with their neighbouring agents via 18 data lines per side; with their own 8 SRAMS via 16 data lines per side and a single address. Control via output enable (OE) and write (WR) distinguishes the 4 exclusive (M) from the 4 connecting R SRAMs.

interconnect data lines. A division of the 26 pins available for local data transmission into 16 SRAM mediated IO, 8 exclusive memory access, and 2 non-memory IO was found to provide the most balanced resource utilization under different connection geometries and gave an SRAM count of eight 8 bit wide SRAMs per agent. This results in half the memory on the interconnects and half as exclusive to the agent. We have employed 15nsec chips in order to give as much scope as possible for synchronizing the RAM read-write cycle with that required for the internal memory in the agents. The 64-bit wide memory access per agent was also employed for the control FPGA, and the control memory data-lines were coopted as 8x 8-bit bidirectional lines between the control and 8 agent FPGAs. The 8 SRAMs dedicated to each FPGA in NGEN provide 64*32k (i.e. a total of *ca* 40 megabytes). The 1296=18*9*8 SRAMs required for NGEN are of type Micron MT5C2568 or NEC 43258-15-E2, wiith almost identical specifications. The chips consume 500mW when running at the minimum cycle period, or 125mW at 16MHz. When disabled, the power consumption sinks to 30mW, with memory retention.

A flexible geometry for the development phase of work with the new machine, required that the board interconnect allow arbitrary chip to chip connections and the largest practicable fan-out. In addition, inter-agent connections should run symmetrically via SRAM data pins of SRAMs belonging to both the FPGAs. A compromise of 8+1=9 data-lines per cable was adopted, with 2 connectors for the 18 data lines for each edge of the agents. Two lines go directly without involving SRAM data, while the other sixteen go via the SRAMs belonging to the connected agents. The eight data pins of each interconnect SRAM are split equally between the two cables to allow symmetric designs in higher dimensions with only a single interconnecting cable. It was further decided to encompass the additional few connectors necessary to split the 8 agent FPGAs into two groups of 4 to improve the symmetry and flexibility of the geometry of routing in higher dimensions. The requirement of minimizing crosstalk on ribbon-cable dictated using alternate pins, with alternate GND lines, and thus 20 pin connectors. In total there are then 2*4*4=32 such connectors which give 32*9=288 signal lines connecting block edges via flat-band cabel, which may be used to connect different boards. This allows for the test development of both planar and non-planar overall architectures.

Examples of connection architectures which can be realized on NGEN include the Linear(Ring), Planar(Torus), 3D and N-Cube whose implementation diagrams are shown in Fig. 3. The Ring, 2D Torus and 3D architectures can all be realized with a single flat band cable connection scheme, although then only 2/3 of the connections are utilized for the 3D case compared with a 100% utilization in the specialized 3D connection scheme. Other configurations are available on request. Two front planes have also been developed to implement the two and three dimensional architectures. The front planes greatly simplify the wiring task and its verification, for commonly used architectures. The front planes are in 14 layer technology, with 576 micro-connectors (20-pin Speedy) on the inside face and a single row of test connectors on the outside.

All designs are intended to run synchronously with a common clock. Synchronous design is not essential for the FPGAs, but greatly enhances the stability and verifiability of the computations. In the initial phase the clock was extracted from the 16MHz host bus, but a clock card has now been developed which distributes a tunable clock frequency with stop and start and synchronization signals via the VME backplane. This allows the frequency to be chosen in software tailored to the needs of individual designs. The card is realized as a complete slave VME-bus agent. The card may be addressed with a standard read cycle. The given address determines the frequency the card generates. Two different base frequencies can be used 100 MHz and 66 Mhz. These base frequencies can be divided by powers of two up to a division of 32768 which gives a minimum frequency of about 2 KHz. The maximum frequency usable is 100 MHz. A simple handshake protocol is provided to synchronize the actions with the cards of NGEN. Only if the master card of NGEN is ready does the clock card begin operation. Counting is provided by the control LCAs in NGEN. The clock card is a 6 layer PCB with high speed (100 MHz) optimization.

The processor boards required a two-sided device placement and 8-layer technology (layers 2 and 7 being Vcc and ground sheets) and were routed using Cadence Allegro software. They were manufactured by Brockstedt GmbH and the SMD parts, FPGAs

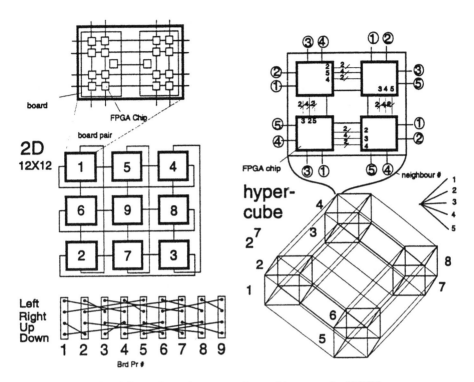

Fig. 3 Some alternative connection architectures for NGEN.
Two of the many possible connection architectures for NGEN are shown: a two dimensional architecture with 12 by 12 agents and toroidal boundaries and a hypercube with 2^7 agents. Multiple processors per agent give more fine grained architectures.

(MQ208) and SRAMs (SOJ), were machine-soldered by Paragon Electronic GmbH. The NGEN boards are equipped with two 3×32 pin VME-bus connectors. Additional VCC and GND signals are required on pins A1-12 and C1-12 of P2 if the board is to be clocked at speed. At a clock frequency of 25Mhz, the individual boards will sink up to 11 amps of current with the SRAMs and FPGAs running in parallel. Careful distributed capacitative loading was employed to reduce ground bounce and signal spikes from the synchronous operation of thousands of flip-flops.

3. The Field Programmable Gate Arrays as Multiprocessor

In this section, we provide a brief introduction to the configurable logic resources of FPGAs used in NGEN, including a simple multiprocessor example. Using high levels logic design tools, the user need not be concerned with the available hardware resources, since powerful automatic partition, place and route software is available. On the other hand, it is important to be aware of the type and abundancy of hardware resources available[7]. The Xilinx 4008, with 8000 gate equivalents, is employed both for the control and agent FPGAs and contains an 18x18 array of Configurable Logic Blocks (CLBs). The choice of this family of FPGAs as processors was determined by their on-chip memory and the convenient complexity of the CLBs. The compromise between more or larger chips was made in 1993 to optimize price-performance. Using the same chip for control and designs has meant that the detailed experience gathered in processor design could then be used directly to improve performance of the interface.

The FPGAs are configurable both in terms of the logic blocks and their interconnections. Dedicated logic on the FPGA provides for the user functions of configuration, startup, readback and boundary scan. Configuration involves loading a serial binary file (147,504 bits) into the configuration memory of the FPGA while operation is suspended prior to startup. Partial reconfiguration is not supported in these FPGAs. Readback and boundary scan enable interactive verification of logic and board connections.

3.1 Configurable Logic and IO Blocks

The CLBs (Configurable Logic Blocks), may be configured as combinatorial functions or as RAM. In each CLB, two combinatorial functions of 4 inputs may be configured (F and G function generators). These functions may be combined with a further input (H function generator) enabling combinatorial functions of 5 variables to be realized. Two D flip-flops provide outputs with clock enable and set/reset control and inputs from F,G,H or external data. Two additional outputs can be linked to the combinatorial functions or data inputs in each CLB. Each CLB may also be used to provide up to 16x2 or 32 bits of RAM. This is achieved by opening up the configuration memory, underlying the combinatorial functions F and G, to data input and control by the user signal level. The four inputs to the combinatorial functions (F and G and for 1x32bit also the outside input to H) then serve as addresses, while write control and data enter via the four CLB control inputs. In addition, the CLBs support dedicated carry logic allowing them to be linked in columns for high speed arithmetic operations.

Input-Output Blocks (IOBs) connect internal chip logic with the pins of the external chip package. They contain configurable pad control in terms of pullups, pulldowns, switching speed and delays, configurable tri-state for outputs to the pads, two configurable inputs and both input and output D-flip-flops with separate clocks. Of the 208 package pins (MQ208), 33 are not connected, 16 are GND and 8 VCC. Of the remaining 151 pins, 144 are available as general purpose user I/O after configuration. There are also 144 IOBs, i.e. two per row and column of CLBs per chip edge. In addition to this direct IO control, the FPGAs contain configurable decoder logic. The wide fast decoders are a powerful resource for implementing wide combinatorial AND logic of inputs or their complements.

3.2 Routing Resources

There are a hierarchy of routing resources in the XC4000 family with varying fan out, range and delay characteristics as shown in Table 1. The basic device used to implement configurable routing is the PIP (programmable interconnect point), an array of transistors under the control of loadable configuration memory bits which allows signal propagation to be permitted or prevented under software control. While the routing is usually performed automatically, there is low level routing control down to the level of single PIPs. In addition to these resources, tri-state buffers (TBUFs) are useful for data buses (for example controlling the flow of data between agent and control FPGAs), with two invertable TBUFs per CLB available.

Routing Resource	Number	Fanout (guide)	Extent
Primary global nets	4/chip	All CLBs, IOBs	Entire FPGA
2ndary global nets	4/chip	All CLBs, IOBs	Entire FPGA
Horiz/Vert longlines	6/row,col	20	Row or Column
Double length lines	4/CLB	4	Two CLBs
Single length lines	32/CLB	4	Next CLB
Switching matrix	2/CLB	8*8 matrix	Between CLB
PIPs (see text)	ca 200/CLB	1-2	Local

Table 1 Routing Resources of 4008 FPGAs

3.3 Configuration, Readback and Boundary Scan

The way in which these operations are induced differs for agent and control FPGAs. See ref.[7] for detailed specifications. Here we describe only those aspects relevant to the current implementation. Configuration is implemented in a serial fashion (bitwise) for each FPGA chip. The XC4008 requires 147,504 configuration bit, each controlling either a function look-up table bit, a multiplexer input or an interconnect pass transistor. This data (.bit file) is generated from the symbolic output (.lca file) of partition, place and route software. While the control FPGAs are configured directly (separately or in parallel) from the VME-Bus in NGEN, the agents are configured in parallel from the control FPGA over separate data lines, each in the bit-serial slave mode. The external configuration clock option is used (CCLK) and potential problems during configuration are monitored (using the INIT pins). Successful startup is indicated by a dedicated line, the DONE pins of all agent FPGAs on a processor board forming a wired AND input to the control FPGA.

Readback allows the serial readout of configuration data and the state of each flip-flop. This includes the bits of the function generators (which may be in use as RAM). The format is similar to the configuration file. Symbolic access to individual signals is provided via a symbolic table (.ll file) created during compilation of the configuration file. This is supported at a high level via a software library. Readback proceeds in parallel on NGEN with either two control or all agent FPGAs being read out at once. If the SRAM on the control FPGAs is used to store the data from the agents, readout can occur in parallel for the entire machine. Readback may proceed concurrently with processing, but then the values of the CLB RAM bits will be undefined since they may change during operation, and as opposed to the values of flip-flops, are not latched on initiation of readback.

Boundary scan is a low level *in situ* device test facility via a standard 4 pin interface. The XC4000 chips implement IEEE 1149.1 Boundary Scan. The agents have been physically routed on the board to allow boundary scan testing via the control FPGA. The boundary scan allows the IOBs to be linked in a single serial shift register with IOB signal levels being read out or set to particular values. Extensive support for boundary scan of the agent FPGAs has been implemented in a control FPGA design.

3.4 Multiprocessor Applications: Flow Processor for the Game of Life.

Space limitations prevent us from describing the evolutionary applications for which NGEN was designed here and these will be reported elsewhere. The example of Conway's "Life"[8] is chosen because it is familiar to many and because, although the capabilities of NGEN go significantly beyond cellular automata (CA), it shows how the distributed memory resources of NGEN can allow large simulations to be performed without data bottlenecks. It also allows one aspect of the computational power of NGEN to be assessed. Of course, this application could be run on a dedicated cellular automata machine such as CAM6, but as we shall see, the simulation size on NGEN is very large (a field of $32383 \otimes 9216$ cells). It has not yet been resolved whether there is a critical size to "life" above which suitably constrained random configurations generate patterns with asymptotically increasing complexity in time. While these issues will not be explored further here, this may serve to indicate that even this well studied system is by no means resolved.

We utilize a slightly unusual statement of the rule of life which operates on a two state CA with the Moore neighbourhood. Let N be the number of cells in the Moore neighbourhood (including the central cell) which are white (or 1). If N is 3 then the center cell turns white and if N is 4 then if it is white, it stays white. In all other cases the new cell state is black (or 0). This reformulation of the usual rule is slightly simpler from a logical point of view and allows a denser packing of the logic. In the current implementation of this rule, N mod 8 is used (which can be represented in 3 bits) instead of N. This makes no difference since the outcome is the same for N=0,1,8,9.

Instead of setting up a two dimensional array of processors, each performing the game of life calculation in one cell, we have chosen a data-flow architecture in which the processors form a linear array corresponding to one axis of the two dimensional field. The second axis is formed by the address space of the high speed distributed SRAMs of the agent LCAs which are accessed so as to implement a large parallel shift register. The details of how this can be realised will become apparent in section 6, where the agent FPGA design process including memory access is presented. The rows of the CA are presented one at a time to the array of processors, with one processor per column over the entire width of the CA field. The new states will be shifted back into the memory with an appropriate delay.

The calculation of the number of white cells can then be performed using horizontal 3-neighbourhoods with three inputs (left, right, middle) in a pipeline fashion as shown in Fig. 4. A new state is computed every clock step and the processor incurs an insertion delay of two clock cycles between the incoming center state and the computed new state. This means that the data will be written back to the shift register at a different location to that from which it is read, with the intervening values only being stored in the processor flip-flops. The entire files of the CA then drifts with respect to the SRAM address cycle during the update process. This schematic also demonstrates the hierarchical nature of the schematic design process, resembling programming in terms of subroutines or functions. The simple encoding function for example is defined as a symbol and used twice in the design. The hierarchical naming scheme for nets allows structured debugging of designs in hardware.

The mapping between schematic and the FPGA design is performed automatically by the firm software. For multi-processor applications such as this example, it is useful to pack the processor into a contiguous area of the CLB array which may be packed regularly to form multiple processors per FPGA. The processor for life above can be packed into a 2x2 array of CLBs, allowing 64 processors per FPGA. This also matches the number of SRAM data channels. Using hierarchical relative location constraints and logic mapping, the 64 processors were successfully routed in the interior 16×16 array of CLBs.

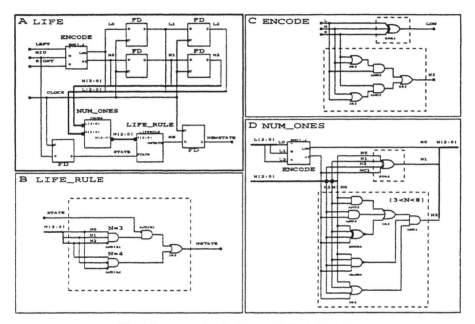

Fig. 4 Schematic for the Game of Life in Dataflow.
The top level hierarchical design for the pipeline processing of Life shown in A involves flip-flops FD and three user defined symbols whose schematics are shown in B-C. The dashed boxes enclose logic which is mapped onto a single 4-input combinatorial function in a CLB. The entire processor shown can be mapped to 4 CLBs. Further explanation in text.

4. Control Design

4.1 Configuration and Readback
Configuration of the control FPGAs may be performed independently by addressing individual cards or in parallel using the address 0. In the former case the control FPGAs on different boards may be configured differently from oneanother. Configuration is performed in the "slave mode" with a maximum clock frequency of 8MHz. Reconfiguration is initiated by a low signal on the PROGRAM pin which is produced by the skeleton interface when data is written to the Program Clear address. The configuration clock for the control FPGA is generated when the VME-Bus writes to the Configuration Address. The configuration data is written to the control FPGA on the lowest bit of the data word (DATA0). Reading at the Status Address gives the

current values of the control DONE and INIT signals monitoring the progress of configuration. The DONE signal (DATA1) is held low during configuration and goes high to indicate the completion of configuration and startup. The INIT signal (DATA0) gives information about the progress of configuration: during reprogram prior to reconfiguration it is low, during configuration it is high unless a configuration error occurs.

Readback of the control FPGA requires that it has been correctly configured as a VME-Bus interface. The readback can only be implemented through a register in the control FPGA (Readback, also responsible for agent FPGA readback). Since the connections of the 4 readback signals (trigger, clock, data and rip) are user configurable in the Xilinx 4000 series, these can be connected to the chosen register in the control FPGA itself and controlled from there.

4.2 Registers
The boards have been designed to allow flexible communication with the control FPGA via registers. 32 addresses are available for this purpose, corresponding to the address bits 2-6 from the VME-bus as determined by the skeleton bus interface. Bit 1 may be used to specify which of two paired boards is accessed as a short word, while bit 0 specifies the individual bytes. The addresses with bit 7 high are used for status, reprogram and configuration of the control FPGA as described above.

A top level schematic of the basic design commonly used in simulations is shown in Fig. 5. The control FPGA registers are reconfigurable so the design shown is rather a description of the classic mode of operation of NGEN rather than a hard and fast limit of the hardware. Only a fraction of the available registers are used in basic controlling designs. See reference[9] for a more detailed account. The 8 global clocks may be configured by the control FPGA for specific functions. In the above example, this involves 3 different clock functions, a bit-serial address cycle, reset and 3 different global signals which may be set high or low from the host. Reset is a signal line used to reset the flipflops in agent CLBs and IOBs to their default state

When using the VME-bus 16MHz clock, rather than the custom clock card, synchronization is done via dedicated user available pins on the P2 VME-bus connector. Setting a certain register in the control FPGA determines whether it acts as a master or a slave. With each access to the custom boards, the master drives down the synchronization bus and detects via the dedicated input pin the ongoing event itself. This active low signal is also received at all other control FPGAs. The maximum allowable timespan to realize a synchronization event is about 62nsec. The actual design needs about 25 ns delay in the control FPGAs and a further 10 to 20 ns delay for removing the charge from the synchronization bus. The synchronization bus is terminated at both ends with a 10kΩ pullup resistor. With a counter implemented on the custom clock card, the synchronization task for this variable clock frequency is made much easier.

4.3 Host Software Interface
Based on a custom VME-driver which is integrated into the host unix kernel, and allows read and write operation to the registers of the processor cards, an interactive run

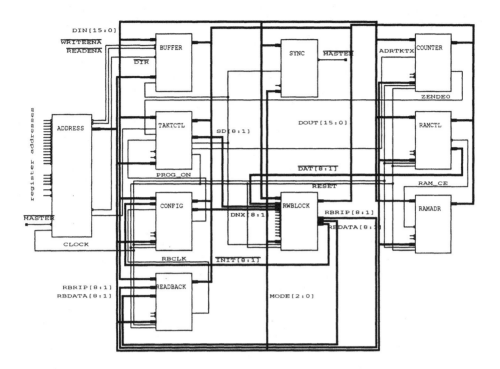

Fig. 5: Basic Control FPGA Design used in simulations. The individual logic blocks control the behaviour of the design at the various registers of the interface and are labelled accordingly. A brief description of their function follows. Each register consists of two 16 bit words, one for the upper and one for the lower card of a pair. The addresses on the figure are in hexadecimal and relative to the chosen card pair, while those below are in 4 byte multiples and decimal for clarity.

Clock Control (0, "taktctl") for the eight clocks and control lines linked to the agent FPGAs;
Configuration (1) controlling the configuration process of the agents;
Readback (2) controlling readback of the control and agents;
Counter (3) specifying the state of a 32-bit clock counter for stepping the agents;
Agent Data (4,5,6,7, "rwblock") each with 4x8 bits data for a pair of agent FPGAs;
Interrupt (8) allowing interrupt control (not shown in the figure) ;
Ram Control (9) controlling the use of the control FPGA memory;
Ram Address (10) the address for the control SRAM;
Common Agent Data (11) allowing identical data to be radiated to all agents;
Synchronisation (12) controlling the synchronisation of the clocks on various cards.

time environment for NGEN has been developed in C++. This is complete with a graphical interface and allows users to communicate with the machine at various levels using either low level register read-write commands or high level configuration, execution, symbolic debugging and data processing with graphic support (UIL). Shells and hierarchical files of interactive commands are supported. In addition a user library is available in C++ so that users can integrate all functions of NGEN into their own programs.

5. Agent Design

5.1 Configuration and Readback

Configuration of the agent FPGAs may be performed independently or in parallel and in both cases the agents may be configured differently from one another. Configuration is performed in the "serial slave mode" (Xilinx) with a maximum clock frequency of 8MHz. Reconfiguration is initiated by a low signal on the individual *program* line for each agent. The signals of each agent are joined in a wired AND and indicate the completion of configuration and startup. The *init* signal gives information about the progress of the configuration: during clearing of the device prior to reconfiguration it is low, during configuration it is high unless a configuration error occurs. Analogously to *done*, it may also be used as an active-low input to hold the agent in an internal *wait* prior to configuration. The *init* signals are available in parallel for each of the 8 agents.

Readback of the agents is accessed through the readback register of the control FPGA. Parallel readback for two boards with 8 bits of readback data per board, lower byte, from the 8 agents has been implemented. The individual RIPs, for each agent FPGA which indicate the status during the readback, are also available as the upper byte on each card. The readback data and RIP status are transferred across the general purpose 64 control-agent data lines. The readback is triggered by a dedicated control line to each agent.

5.2 Global Signals

8 clock lines, 4 primary and 4 secondary are available for high fan-out signals such as clocks, write enable and reset in each agent. Every CLB in the FPGA can be connected to the global clock signals, but only at specific pins. While the primary global clocks have the least skew, their connections are more constrained than the secondary global clocks. It is also possible to use the global clock signals with sources internal to the agent FPGAs by connecting the internal signal to the IO Block output of the corresponding pins. However, since all of the 8 agent FPGAs share a common net for each of the 8 different clock signals (excepting two which are wired independently to each of the two blocks of 4 FPGAs), designs must avoid sending conflicting signals to these pins from different agent FPGAs. This does however limit the autonomous activity of the agent FPGAs and we may at a future date remove one of these 8 pins on each agent FPGA to allow for more autonomous operation. The function of the global clock signals used externally is determined by the control FPGA configuration (see above).

5.3 SRAM Control

Because the control of the SRAMs are mediated by the corresponding control or agent FPGA, user designs must respect the timing constraints in order for this off chip memory to be used reliably. Since the memory access is a relatively standard procedure with limited variation it is possible to describe design modules which preserve the correct timing and can be integrated in user designs. The RAM access in both control and agent FPGAs is identical apart from the pin numbers. Since the flow of data is bidirectional, a read-write cycle involves not only time points for address change, data clocking, and SRAM write enable but also for the FPGA and SRAM output enables. Using a single symmetric clock, it has proved possible to implement such a cycle reliably at 8 MHz on NGEN. More sophisticated multi-clock cycles have also been developed.

5.4 Evolutionary Designs

One basic problem in the parallel simulation of interacting populations is to create a distributed mixing system with ergodic trajectories to allow a random pattern of inter-string collisions. The implementation of a dataflow switching network implementing a diffusion process is one way to achieve this, with the advantage of allowing spatially resolved studies of evolutionary reaction-diffusion systems[10]. A distributed random number generator has been developed, which uses irreducible polynomials in the signals from cell neighbours[11]. A simple example of an evolutionary process utilizing this mixing mechanism involves a genetic switch combining both reaction and diffusion. The switch has four states and determines whether two incoming strings are swapped or not (according to the local random numbers) or whether one of them is copied (overwriting the other). As a simple example, the string corresponding to the highest binary number may be copied. Of course, more sophisticated models are being studied with the hardware[12]. Since the simulation is at the level of individuals, no additional overhead is required when all sequences in the population are different and the full sequence dependence of every interaction must be processed. According to the formula $G = 2*F*N_p/N_s/L_s$, the number of generations/sec depends on the number of processors N_p, number of strings N_s, length of string in bits L_s, and clock speed F. For example, NGEN can handle populations of millions of strings of length 100. With populations of the oder of 10^4, a million generations per second are possible.

6. Performance and Conclusions

This contribution describes the hardware of the computer NGEN. Since so much of the computer is configurable, it must be regarded as a documentation of potential constraints on total flexibility, rather than as a prescription of how the computer operates. Examples of control FPGA and agent FPGA designs which can be loaded by software are available [13] [14]. While floating point applications may be developed, the main benefits lie in NGEN's flexibility with regard to integer and logical processing

NGEN performance is of course user dependent, in that the quality of user designs limits the density and speed of logic. A summary of basic hardware statistics of NGEN is given in Table 2. It is possible for example to configure the machine to perform 5760 simultaneous 16-bit add/subtract operations at 50 MHz placing it in the quarter tera-operation per second range. Applications to image processing may profit directly from such a capability. With 2592 processors (8 bit, 16 CLB) at 16 Mhz, 41 472 MIPS and with 11664 processors (2 bit, 4CLB) at 16 Mhz, 186 624 MIPS can be achieved. Much more important for an effective evaluation however, is the communication bandwidth, which enables problems to be scaled up to the size of the available SRAM without compromising efficiency.

Parallel computing with FPGAs took form in the early nineties[15]. Perhaps the most similar FPGA based computers to NGEN are the basically one dimensional configurable computer which has been built primarily for dedicated sequence analysis, e.g. SPLASH[16]. The designers of SPLASH 2 opted for a central memory store to ease communication with the host machine. The NGEN capabilities of transparent

Table 2. NGEN Hardware Statistics

Processors	**93312** max, user defined, typically 10^3-10^4
Architecture	**1D, 2D, 3D, Torus, Ncube** and novel
FPGAs Chips	**162** Xilinx XC4008
Config. Logic Blocks	**324**(18*18)/FPGA
FPGA IO-Pins	**144** User Pins/Chip
Distr. Mem Chips	**SRAM** (15ns 8*32k)
Distributed Memory	**42 MB** (=1296 SRAMs)
Bus	**VME** 32 (6U) 21 Slot
Boards	**18** Processor, 1 of 2 Frontplanes, Clock Card
Board Interconnect	**288** (=32*9) lines/board
Clock	**2KHz-100MHz** user ,typically **8MHz**
Power Consumption	<= **1800W**, 11 Amp/card, air cooled
NGEN Runtime Env.	**C++** Shell and library, with graphical interface
Host	**HP743i** Workstn Card, 100MHz, 128 MB
Host Operating System	**Unix** HPUX 9,10

problem size extension beyond the limit set by the number of simultaneously realisable processors are not pronounced on this machine. Other groups are using small numbers of configurable chips as a configurable co-processor to speed serial calculations[17] on conventional workstations. NGEN on the other hand is dedicated to massively parallel tasks. Logic simulators with multiple FPGAs do not have the distributed memory resources of NGEN.

Recently, DNA computing has become a popular topic in the computer science community[18], holding out promises of massively parallel computation up to the 10^{20} level and the possiblity of harnessing genetic processing for digital computation have loomed closer in the imagination. NGEN is not designed for the hardware simulation of DNA computing of this type, but rather primarily with processes of self-organizing computation occuring in autonomously evolving populations of catalysts and DNA and RNA.

This work was initiated to help bridge the gap between conventional and biological computation. Although a first approximation, NGEN has already demonstrated the efficacy of configurable hardware within the modern palette of computers. The authors hope to be able to improve both hardware and software in the coming years. In particular, the self-organizing computer will require hardware microconfigurability i.e. the ability to rapidly reconfigure small parts of a design. In future versions of the hardware, software configurable global routing will also be approached. Work in this direction is in progress.

References

[1] McCaskill J.S. "Architektur und Verfahren zum Konfigurieren eines Parallelrechners" International PCT Patent Application P4302297.9 28 Jan. 1993.

[2] McCaskill, J.S. "Polymer Chemistry on Tape: A Computational Model for Emergent Genetics", Interner Bericht des Max-Planck-Instituts für Biophys. Chemie Göttingen 1988.

[3] Eigen M. and Schuster P. "The Hypercycle: A principle of natural self-organization" Naturwissenschaften (1977) **64** 541-565 *ibid* (1978) **64** 7-41 *ibid* (1978) **65** 341-369

[4] Boerlijst M.C. and Hogeweg P. "Spiral-wave structure in prebiotic evolution: Hypercycles stable against parasites" (1991) Physica D **48** 17-28

[5] Thürk, M. "Ein Modell zur Selbstorganisation von Automatenalgorithmen zum Studium molekularer Evolution" Doct. Friedrich-Schiller-Univ. Jena (Informatik) 1993.

[6] Fontana, W. "Algorithmic Chemistry" (1990) Technical Report LA-UR 90-1959, Los Alamos Natl. Lab.

[7] The XC4000 Data Book, Xilinx Inc., San Jose, California, USA 1992.

[8] Conway, J., 1970, unpublished; see Gardner, M. "Mathematical Games" Sci. Amer. **226** Jan. 104.

[9] Tangen U., Gemm U., McCaskill J.S."Schematic entry with Viewlogic (and Control FPGA Design Decription)", Technical Report Institute for Molecular Biotechnology, Jena 1994.

[10] McCaskill, J.S. "Efficient dataflow architectures for diffusion in chemistry and population dynamics", in preparation, 1996.

[11] Breyer J., Ackermann J., Böddeker B., Tangen U., McCaskill J.S. "A minimal logic parallel random number generator for configurable hardware" in preparation, 1996.

[12] Breyer J., "A flow processor for massively parallel simulation of evolution in biopolymer reaction-amplification systems: PAD", in preparation, 1996.

[13] U.Tangen, "Schematic Entry with Viewlogic", Technical Report IMB, Jena 1994.

[14] J.S.McCaskill Programming NGEN Hardware with C(++) and the XACT environment. " Technical Report IMB, Jena 1994.

[15] Linde, A., Nordström, T. and Taveniku, M. "Using FPGAs to Implement a Reconfigurable Highly Parallel Computer.", 1992, in Lect. Notes in Computer Science **705,** H. GRÜNBACHER AND R.W. HARTENSTEIN EDS. 199-210.

[16] Arnold, J.M., Buell, D.A. und Davis, E.G. "SPLASH 2" in Proc. 4th Annual ACM Symp. on Parallel Algorithms and Architectures 1992, 316-322.

[17] Gray, J.P. and Kean, T.A. "Configurable Hardware: A New Paradigm for Computation" in "Advanced Research in VLSI" Proceedings of the 1989 Decennial Caltech Conf. Ed. C.L.Seitz (MIT Press, Cambridge,Mass.).

[18] Lipton, R. J. "DNA Solution of Hard Computational Problems" Science **268** 542-545.

Architecture of Cell Array Neuro-Processor

Takayuki Morishita and Iwao Teramoto

Okayama Prefectural University, Soja, Okayama 719-11, Japan

Abstract. We propose Cell Array(CA) to develop a PCA(Processing Cell Array) architecture as advocated in Reference [1] to improve the free degree of the composition of the instruction control unit and the memory. Also, we propose a connection method among cells to establish the architecture of the entire processor element. Our goal is to check the architecture, control sequence and it's performance. We took the back-propagation learning law as an example, because it is the typical model of the neural network.

1 Introduction

Optimizing of algorithm is indispensable in executing calculation at high speed. The optimizing of the whole arithmetic unit which includes software and hardware is necessary. A conventional processor must optimize software while considering an agreement with the hardware because the free degree of the hardware is low. Therefore, the optimal algorithm in the software causes such problems as taking processing time on the hardware. In an effort to optimize the program, agreement with the structure and the individual hardware must be sufficiently considered.

We believe the hardware should change structure more flexibly according to the software as the software changes structure according to the algorithm. Not only the software, but also the hardware should be optimized. If the hardware can be faithfully reconfigured according to the software, the whole system should be able to be optimized by optimizing only the software. We aim for such reconfigurable hardware.

In the case of neural network calculation, various models have been proposed, but we do not have an optimal model as of yet. Therefore, the processor must do high-speed calculations after each technique. When we apply an idea like the one mentioned above to the calculation of the neural network, it thought of improving arithmetic unit itself flexibility first. Filed Programmable Gate Arrays (FPGAs) is excellent technology for implementing flexible neural network hardware [2]. However, archtecture and circuit designs take much time and need high technology. We propose prcessing cell array architecture to obtain a flexible hardware. It can freely construct the basic arithmetic units such as the multiplier and the divider. In this paper, we examine the architecture of the whole processor element to get the reconfiguration of the hardware. Furthermore, we use the well-known back-propagation learning law [6] of the neural network model as an example.

2   Conceptual Architecture of Processing Cell Array

We proposed Dynamically Reconfigurable Pipeline Architecture to develop the
hardware which computes a neural network model at high speed [3]. This was
architecture which can change the composition of Pipeline by choosing the connection route of the arithmetic units such as the multiplier and the subtracter.
Especially, in the backpropagation algorithm, seven kinds of pipeline to use for
the computation could be executed by the composition which is the optimal
for each computation. As a result, high calculation speed, 18 MCUPS (Mega
Connection Updated Per Second) per a processor element, was obtained in 8-bit
precision neuron. However, in this architecture, the composition of the arithmetic
unit is limited to the preset one. Moreover, because the calculation precision of
the arithmetic unit is fixed, the high-precision circuit must be used when processing low-precision problem. In the neural network, the necessary precision
changes with the problem fairly [4]. There is an example which needs equal to
or more than 16-bit precision while there is an example in the 8-bit precision.
Moreover, the number of the models of the neural network increases more in
recent years. A computation order and a necessary arithmetic unit are fairly di-
versified. Therefore, we proposed the 2-dimension processing cell array structure
which was arranged matrix in the calculation unit for the optional calculation
precision to be realized.

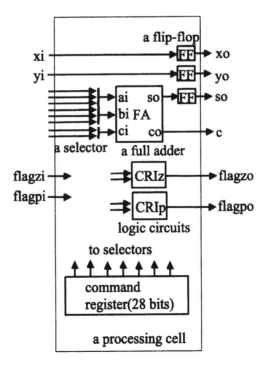

Fig. 1. A Processing Cell

A processing cell was composed of a full-adder, some flip-flops, some selectors and some logic circuits as shown in Fig.1. The data in a command register controls these selectors and they change connection signals. The command register is directly accessible from the host MPU. Two inputs to a full-adder, ai and bi, are changed by these selectors. The input of carry, ci, is chosen from three values of zero, one and a carry output from upper cell, co. Zero and one is used to take two pieces of complement and co is used to obtain a addition or a substraction. flagp is a flag that shows which we add or substract. flagz is a flag that shows which a coefficient is zero or not. xi and yi are latched by flip-flops and send to following neighbour cell. CRIz and CRIp are logic circuits. In case of multiplication, coefficients of booth algorithm are calculated in these circuits. These flag signals are transferred to all cells on a line as flagzi and flagpi. All values which must send to next cells are transferred from latch to latch between a clock and it gurarantees operation of the processor.

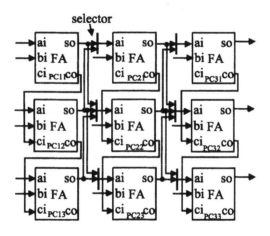

Fig. 2. Processing Cell Array

Fig.2, in which a processing cell is shown in a full adder(FA), is a processing cell array. We connect each calculation unit with upper, neighbour and lower units on the right of the unit. 1-bit right or left shift of data was made of this forwarding. As a simple example, we consider a multiplication with three bits precision, x*y. x and y are input to xi and yi of PC11, PC12 and PC13. ai and bi are connected to xi and yi in cells on the first column, PC11, PC12 and PC13, respectively. bi is connected to yi in cells on the second and last column. The upper co is directly connected to lower ci in same column. A selector chooses so of lower cell. At the first column, CRI circuits in PC11 determine adding x or not. At the second and last columns, these are calculated in PC22 and PC33. Final result is output from the so on the last column. The number of the cell to use decides calculation precision. We showed that multiplying by algorithm of Booth and dividing by the restoring method could be executed in Reference [1].

This processing cell array structure seems to be similar to the parallel computer of systolic array [5] but the fundamental thought is different. In systolic array, cell has only a single function, and each cell works independently. On the other hand, cells in the specified range of processing cell array cooperate as a single arithmetic unit. That is, the cells form a multiplier or a divider which are used in the conventional computer. The free degree becomes low for the function of cell to be specified. On the other hand, there is an advantage that the conventional program can form urgent hardware easily.

The element which is necessary to execute pipeline is memory which supplies arithmetic units with data and preserves data of the calculation result, counter which supplies memory with address and counter which controls timing of whole pipeline. In processing cell array, the kind of the model which can correspond receives an agreement by the limitation of the free degree of the memory and the counter part and the following problem occurs.

(1) memory can not be arranged in the necessary number with the optional precision.

(2) The number of independent memory which can supply an arithmetic unit with data at the same time is lacking and pipeline stops.

(3) The necessary number of counter can not be composed.

Therefore, in the following chapter, we make whole processor, which includes memory and counter part, be cell array structure. We change for the necessary number of control unit and memory part to be supplied with the necessary calculation precision. The multiplied inner bus which connects between the mutual connection of the arithmetic unit which composes pipeline and memory, and the arithmetic unit, address and the counter output of memory is explained.

3 Cell Array Structure of a processor

This neuro-processor speeds up calculation by arranging more than one PE like a matrix. A PE becomes the basic unit of the calculation and executes the calculation of the neural network. The maximum feature lies in not fixing hardware inside PE. The hardware is able to reconfigure in the form optimal for the program. Therefore, it is possible to form more than one pipeline at the same time inside a PE and to make more than one multiplication in a pipeline. In this processor, the agreement of the hardware can not make a pipeline length long faithfully to the program. Because the software receives an agreement from the hardware hardly, the optimizing of the software agrees with the optimizing of algorithm.

The processing element is composed of a sub-control, a Random Access Memory and Cell Arrays (Fig.3). Cell Arrays are composed of CCA (Counter Cell Array), MCA (Memory Cell Array) and PCA (Processing Cell Array). Each array is arranged like a matrix. Each Cell can fix a cell's function by preserving connection information at the control latch inside a cell. Counter cell array composes the counter part which supplies an address to a memory and computes a looping index of a for-sentence. Memory cell array supplies and saves data

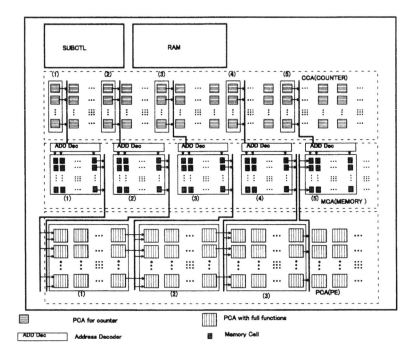

Fig. 3. Diagram of Cell Array

to and from the processing cell. It is used for the preservation of data and initial value. The processing cell array executes a basic numerical calculation like adding, subtracting, multiplying or dividing.

Fig.3 is a diagram of a processing element. In this example, three multipliers, (1)-(3), are composed by processing cell array. From four memory cells, (1)-(4), data is supplied to these multipliers. The result is stored in the fifth memory. The address of each memory is supplied from these five counters, (1)-(5).

If making the composition of each Cell a one bit unit, free degree of the number of bits becomes maximum but redundant circuits increase. It decreases gate density.

CC(Counter Cell) can realize an up-down counter with an optional precision. As shown in Fig.4, one Cell has a J-K Flip Flop, a selector and an and-gate. The and-gate checks which count-up is finished or not. Output from the counter is connected to a bus line and it is supplied to the memory cell address. Looping index of a for-sentence is formed by this counter cell array. Also, timing signal of pipelines is generated in this counter. However, when index of the for-loop needs an addition or a subtraction, the counter can be made in processing cell array.

If kind, calculation precision, and connection order of an arithmetic unit in a pipeline are decided, timing signal, which controls of the pipeline, is decided. A counter for this purpose is made in this CCA.

MC (Memory Cell) is composed of one bit memory as shown in Fig.3. Address

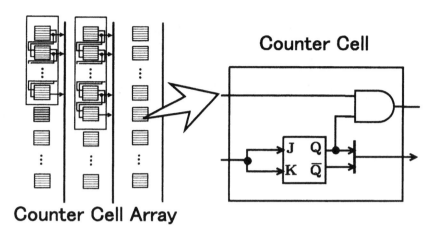

Fig. 4. Counter Cell

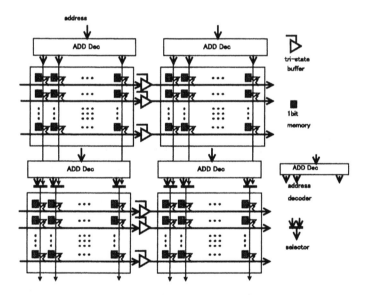

Fig. 5. Memory Cell

from counter cell array is decoded at the decoder. A unit of memory cells is constructed with 4 times 4 cells. If we need bigger memory bits than the unit, we connect a lower unit cell. We pause a lower address decoder and connect lower unit cell word lines to them. If we need longer memory length than the units, we connect neighboring unit cell. We connect two buses by switching on a tri-state buffer. Configuration information is saved to a control latch in the memory cell array.

These counters, memories and calculation parts are mutually connected through

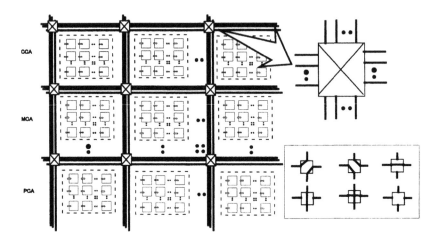

Fig. 6. Internal Bus Structure in Cell Array

a multiplicated bus (Fig.6). In this figure, upper, middle and lower sections show CCA, MCA and PCA, respectively. These bus-lines are connected or disconnected at the intersection of the bus. The type of connection is shown at the lower left side of the figure. Along these lines, output from a counter is sent to the memory cell array address. Output from the memory cell array is supplied to processing cell array. This routing information is stored to control latches at the intersection of the bus. The host MPU can directly access these control latches. At the same time, this bus is improving faulttrerance, by being multiplicated.

This Cell Array Neuro-Processor can change a connection among the arithmetic units according to the connection information. A program is made not by composing an order of calculation by microcode like a convensional processor. It is made by the structure of these counters, memory, an arithmetic unit and information (i.e. "Connection Code") from the mutual connection. Because the arithmetic unit can be freely composed, it composes more than one pipeline at the same time in one PE and arranges more than one multiplier in one pipeline. Therefore, the high-speed calculation, which is impossible in convensional microprocessor and DSP, can be expected. When computing the neural network which follows back-propagation (BP) learning law in parallel, a synaptic weight is generally divided into more than one processing element. When computing output of the neuron, it takes a vector product between the synaptic weight and the input vector to the neuron and the summation of these partial products. Therefore, the addition on a line becomes necessary. The bus that disperses and computes to more than one PE is simply described. We propose a high-speed bus to arrange more than one PE in the 2-dimension figure of arrangement and to connect it in Reference [1]. This bus is composed of a global bus and local buses as shown in Fig.7. The global bus supplies control data simultaneously to all PEs. It reads and saves data from each PE to host MPU. The local buses exchange data among PEs which belong to the same line. Because an inverse matrix becomes neces-

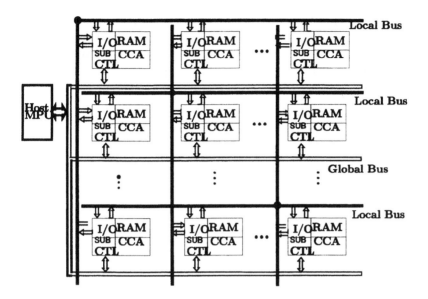

Fig. 7. Multi-Local Bus Structure

sary in BP learning computation, the operation such as transverse of a whole matrix becomes necessary. Using these buses, it is possible to execute transverse operation of matrix and transfer data between PEs on the same bus line. One of PEs on the same bus line becomes a master controller of data transfer and others become support devices like memories.

4 The operation in computation of the exercise

Although a compiler is not developed, a program is described by c-language to explain a method of a program. We extend c language to a parafor-sentence and a parallel sentence for parallel computation. In the case of calculation, these are six agreements:

(1) It divides a loop, which is specified by a parafor-sentence, into processing elements as many as a loop number.

(2) It assumes that there is not relation among variables in a parafor-sentence and we do parallel processing inside the same processing element.

Also, we assume that variables at the right and left hand side of an equation are treated as variables. It is automatically unified in one variable when parafor-sentence is finished.

(3) In the case of parallel computation with several sentences, parallel sentence is used. In this case, each variable breaks synchronization. To maintain the causality of each variable, we should use parafor-sentence.

(4) A loop inside a parafor-sentence makes a pipeline.

(5) All memories inside a parafor-sentence are maintained inside a processing element and they are not shared among other processing elements.

(6) When other processing elements quotate variables, we transfer the variables by using multi-local bus.

Table 1. Tipical Instruction Set

Command	Function	PE NO.	index	param.1	param.2	param.3
WRITE	write data from ext.	PENO	index	memory NO.	address	
READ	read data to ext.	PENO	index	memory NO.	address	
CSET	counter setting	PENO	"cx"	counter NO.	cn	
MSET	memory setting	PENO	"mx"	memory NO.	n	m
PSET	processing unit setting	PENO	"mltx"	kind of processing	bit	option
CINIT	c-init setting	PENO	index	counter NO.	value	
MINIT	m-init setting	PENO	index	memory NO.	value array	...
CONNECT	connection	PENO	index	c or m NO.	pin NO.	
DISCONNECT	connection	PENO	index	c or m NO.	pin NO.	
DELETE	delete cell	PENO	index			
START	start calculation	PENO				
BREAK	break calculation	PENO				
INIT	reset	PENO				

A typical instruction set is shown in table.1 as an outline. Data is read from memories and stored to memories by the first two commands. The processing element number is the address of each cell array. When this value is a PEALL, it can access all cell arrays at the same time. Counter, memory and processing cells are set up by three set commands, CSET, PSET and MSET, . It composes PSET instruction in the calculation precision to have specified four kinds of calculation devices such as a multiplier and a divider. CSET instruction composes the counter with optional precision. MSET instruction composes a memory with m depth and n bit length. Connect command specifies a connection among the counter, memory and processing unit. After the start command, it executes calculation until a ready signal is detected.

The learning phase according to the back-propagation learning law of the neural network is shown. L and W show the number of layers and a synaptic weight, respectively. The synaptic weights are stored at the $M[l]*M[l+1]$ distributed memories at the cell arrays. Eq.(1) is computation of delta in an output layer, Eq.(2) is preparation for delta computation of a middle layer, Eq.(3) is correction of a synaptic weight and Eq.(5) is computation of delta in a middle layer. Teaching signal t, output of each layer, o, and differential function of a sigmoid function, g, are divided into $M[l]$.

$l = L - 1;$
$parafor(m2 = 0; m2 < M[l]; m2 + +)\{$
$\quad for(n2 = 0; n2 < N[l]; n2 + +)\{$
$\quad\quad \delta[m2][n2] = (t[m2][n2] - o[l][m2][n2]) \times g(o[l][m2][n2]); \qquad (1)$
$\quad \}$
$\}$
$for(l = L - 2; l > 0; l - -)\{$
$\quad parafor(m1 = 0; m1 < M[l]; m1 + +)\{$
$\quad\quad for(n1 = 0; n1 < N[l]; n1 + +)\{$
$\quad\quad\quad parafor(m2 = 0; m2 < M[l + 1]; m2 + +)\{$
$\quad\quad\quad\quad s[m2][m1][n1] = 0;$
$\quad\quad\quad\quad for(n2 = 0; n2 < N[l + 1]; n2 + +)\{$
$\quad\quad\quad\quad\quad s[m2][m1][n1] = s[m2][m1][n1]$
$\quad\quad\quad\quad\quad\quad + \delta[m2][n2] \times w[l][m2][n2][m1][n1]; \qquad (2)$
$\quad\quad\quad\quad\quad w[l][m2][n2][m1][n1] = w[l][m2][n2][m1][n1]$
$\quad\quad\quad\quad\quad\quad + \eta \times \delta[m2][n2] \times o[l][m1][n1]; \qquad (3)$
$\quad\quad\quad\quad \}$
$\quad\quad\quad \}$
$\quad\quad\quad sum[m1][n1] = 0;$
$\quad\quad\quad for(m2 = 0; m2 < M[l]; m2 + +)\{$
$\quad\quad\quad\quad sum[m1][n1] = sum[m1][n1] + s[m2][m1][n1]; \qquad (4)$
$\quad\quad\quad \}$
$\quad\quad\quad \delta[m1][n1] = sum[m1][n1] \times \delta[m1][n1]; \qquad (5)$
$\quad\quad \}$
$\quad \}$
$\}$

In Eq.(1), because a single loop of parafor-sentence is used, the for-sentence is computed at the same time in M[l] PEs. One subtractor and one multiplier are composed. Next, we compute Eq.(2) and Eq.(3) in M[l]*N[l+1] PEs with 2-dimensional arrangement because there are double loops of parafor-sentence. Also, two equations exist in a for-sentense but they are calculated at the same time as two pipelines. It can be computed independently because there is not causality among those variables. Finally, it computes delta in Eq.(5).

To show the effectiveness of cell array, it is compared with the conventional pipeline. The processing of Eq.(2) and Eq.(3) shows the most characteristic function of this processor. They include triple multiplications and two pipelines. According to agreement (1), variable at the right hand side of an equation uses a value before being renewed and it is automatically allocated to another variable at the left hand side of the equation. Because there is no causality between the two sentences, these sentences can simultaneously be calculated in this cell array. These parallel computations are done in one PE (Fig.8). Three times the speed of CPU which has only a multiplier is shown in this timing chart. A similar performance might be possible if three CPU were connected. However, the programs become more complicated than this cell array technique.

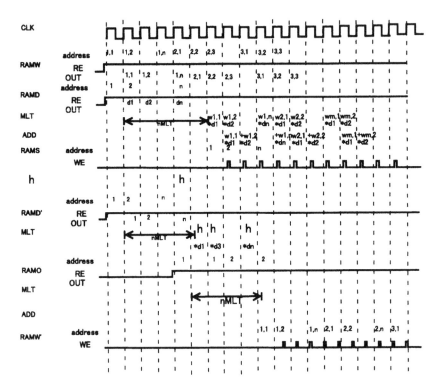

Fig. 8. Example of Timing Chart

5 Conclusion

We proposed and examined the neuro-processor of cell array architecture which can set the arithmetic unit by arbitary precision. It composes the arbitary structure of pipeline. We took the back-propagation learning law as an example and we evaluated the program method and its performance. We confirmed the effectiveness of this processor. The problem left is to confirm application of the cell array structure to other models of the neural network and to design a circuit to prove operation efficiency.

Because the calculation of the neural network is often described by the product summation calculation which doesn't contain exception processing, this architecture doesn't have a branch instruction and a loop direction. Therefore, it is difficult to apply this architecture to the general parallel computation. To investigate the extension of architecture which can correspond to the component of the classy languages such as C language is also future problem.

References

1. Morishita, T., and Teramoto, I.: Processing Cell Array Neural Network Multiprocessors proc. of IEEE International Conference on Neural Networks (1995) 1470–1473
2. Eldredge, J.G. and Hitchings, B.L.: RRANN: A Hardware Implementation of the Backpropagation Algorithm Using Reconfigur-able FPGAs. proc. of IEEE International Conference on Neural Networks (1994) 2097–2102
3. Morishita, T., Tamura, Y., Satonaka, T., Inoue, A., Katsu, S., and Otsuki, T.: A Digital Neural Network Coprocessor with a Dynamically Reconfigurable Pipeline Architecture IEICE Trans. Electron., Vol.E76-C, No.7 (1993) 1191-1196.
4. Mauduit, N., Duranton, M., and Gobert, J.:"Lneuro 1.0: A Piece of Hardware LEGO for Building Neural Network Systems", IEEE Trans. Neural Network, Vol.3, No.3 (1992) 414–422
5. Fortes, J.A.B., and Wah, W. B.:"Systolic Arrays- From Concept to Implementation", IEEE Computer, 20, 7, (1987) 12–17.
6. Rumelhart, D.E., McClelland, J.L., and the PDP Research Group: Parallel Distributed Processing. MIT Press, Cambridge, 1986.

Special-Purpose Brainware Architecture for Data Processing

Tadashi Ae, Hikaru Fukumoto, Saku Hiwatashi

Electrical Engineering, Faculty of Engineering, Hiroshima University,
1-4-1 Kagamiyama, Higashi-Hiroshima, 739 Japan.
Fax: +81-824-22-7195
Email: ae@aial.hiroshima-u.ac.jp

Abstract. A new architecture SBA is proposed for real-time data processing. The SBA is derived from systematic realization of human brain, two important features (adaptiveness and stability) of which are carefully introduced. The scheme of dataflow on SBA seems to be similar to that of a conventional parallel machine, but the architecture is different in realizing directly two types of memory, i.e., STM (Short Term Memory) and LTM (Long Term Memory). Learning is performed on these two memories, since SBA supports both types of learning; one for STM and the other for LTM.
As a result , the SBA may become a new type of neural AI (Artificial Intelligence) computer.

1 Introduction

In this paper, a new architecture SBA, which is derived from human brain model (see Fig.1), is proposed for real-time data processing based on neural AI (artificial intelligence). When we focus on systematic behavior in the human brain, three important operations as a system are given as follows [1];

i) Adaptiveness ——— Cerebellum.
ii) Stability ———— Basal Ganglia.
iii) Purposefulness —— Limbic System.

(The left-hand sides and the right-hand sides are terminologies in system and in human brain, respectively.)

Though the third feature of brain (purposefulness) is disregarded, the SBA is derived by realizing two features (adaptiveness and stability). The scheme of dataflow in SBA seems to be similar to that of a conventional parallel machine, but the architecture is different in realizing directly two types of memory, i.e., STM (Short Term Memory) and LTM (Long Term Memory). The STM and the LTM in SBA work mainly for adaptiveness and stability, respectively.

Learning is performed on these two memories, since the SBA supports both types of learning; one for STM and the other for LTM (with assistance of STM), and therefore, the SBA is naturally a kind of memory-based architectures.

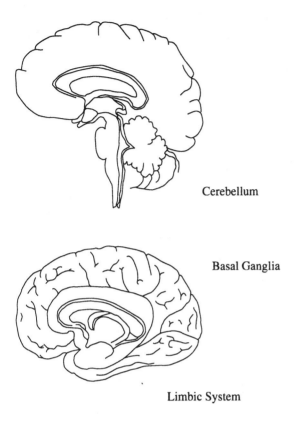

Cerebellum

Basal Ganglia

Limbic System

Fig. 1. Brain

The terminology "Brainware" is proposed by G.Matsumoto[2], and we show its concept as in Fig.2. The computer includes both software and hardware, but the brainware is derived from human brain model.

In this sense , the SBA can be regarded as an extension from mixed systems of neural networks, AI systems and computers. Moreover, the device technology is also important for fabrication, and nanometer IC technology will be expected. The SBA is a general system, but we focus on real-time data processing in this paper.

2   Special-Purpose Brainware Architecture

The global dataflow of the special-purpose brainware architecture (in short, SBA) is shown as in Fig.3, where FFN and FBN are the feed-forward network and the feedback network, respectively. The dataflow shows only a schematic representation of data transmission, but the FBN is important in the SBA, because the system with *passive* FBN coincides with a neural network without feedback (e.g., MLP: Multilayer Perceptron net), where *passive* means no function or identity function. By contrast, the system with passive FFN is the same

Artificial Intelligence

Computer

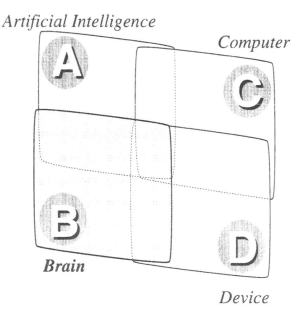

Brain

Device

Fig. 2. Brainware; Brain-computing

as the recurrent network. The system with both (*active* FFN and *active* FBN) is essentially important in the SBA.The SBA is shown as in Table1, when the systems are classified by "activity" of networks (FFN and FBN).

Table 1. SBA (Special-purpose Brainware Architecture) on Classification depending on Activity of Network.

FBN \ FFN	Passive	Active
Passive	——	MLP, Kohonen net, etc.
Active	Recurrent network	**SBA**

Notice 1: Passive means no function or identity function.
Notice 2: Either FFN or FBN becomes passive for the special case of SBA.

The configuration in Fig.3 seems to be the same as the dataflow in a conventional parallel machine, but the essential difference is that the SBA works by *not programming but learning*. The learning-based system is obtained by a memory-based architecture. To make it clear, we show a memory-based architecture as in Fig.4, where FFN and FBN include the memory each. The memory in FFN and the memory in FBN are the STM (short-term memory) and the LTM (long-term memory), respectively.

In general, the LTM and the STM are given as follows;

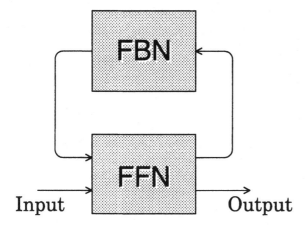

FBN: FeedBack Network
FFN: Feed-Forward Network

Fig. 3. Global Dataflow of SBA(Special-purpose Brainware Architecture)

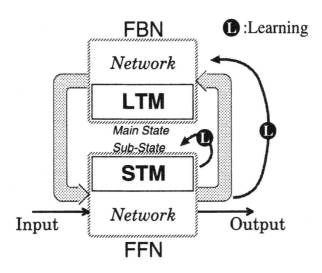

STM: Short-Term Memory
LTM: Long-Term Memory

Fig. 4. Memory-Based Architecture

LTM: Active Ensemble Memory + Passive Ensemble Memory.
STM: Active Ensemble Memory + Passive Ensemble Memory.

If *active* is more powerful than *passive*, it is called *active*, and vice versa. "Ensemble" means a kind of addressable-in-parallel memory. (About details of definitions see the appendix.)

Learning in STM is made within FFN, but learning in LTM is made within the loop including both FFN and FBN. Both learning are inductive, and learning in STM is the same as the conventional learning in the neural network. We do not restrict the type of neural networks in FFN, but focus mainly on MLP and Kohonen net, because these two are popular and features each are also well known. As learning, the BP (error-back-propagation) algorithm and the LVQ (leaning vector quantization) algorithm are supposed to be used for MLP and Kohonen net, respectively.

The type of FBN is not so easily determined, because inductive leaning in the loop structure provides a difficult problem. One candidate is the recurrent network with (an extended) BP algorithm, but the application is restricted because it is the case with passive FFN. Then, we show a type of FBN including the active FFN.

In the memory-based architecture in Fig.4, the memory in FFN (i.e., STM) represents the sub-state, while the memory in FBN (i.e., LTM) does the main state. This is the systematic reason why we call the former the short-term memory and do the latter the long-term memory.

The inductive learning for LTM should be a constructive inductive learning, and the structure must be obtained by learning procedure. In other words, the structure of automatic truth maintenance or the structure of minimal systematic representation must be constructed by learning. The rule-based notation or the logical (predicate or propositional) notation is used for the structure construction of automatic truth maintenance. The machine (or automaton) notation is used for the structure construction of minimal systematic representation. It should be noted that this learning procedure must be applied not only for symbols but also for numerical values. The notation depends on the application, but each notation must be extended to the numerical case from the symbolic case.

In the SBA adaptiveness (corresponding to Cerebellum in brain) is understood to be a parameter tuning, i.e., a small derivation in learning for a single data. On the other hand, stability (corresponding to Basal Ganglia in brain) is understood to be a convergence, i.e., not diverse in learning for a set of data (a large number of data). At a tuning for a single data, the loop for learning is easily tunable, and therefore, this process is realized mainly by STM. For a large number of data, however, we need a delicate loop-control for convergence, and realize it by LTM and also STM (as assistance).

The inductive learning on SBA is summarized as follows;

begin

    LF: Logical Formula is given. { on LTM }
    Local Induction Step.   { Neural Net Learning on STM }

Global Induction Step:
Consistency Check of LF
Simplification of LF   { Most Probable LF }
if Result is OK *then* terminate
else continue.
end.

3   Brainware Architecture for Data Processing

A brainware architecture for data processing (i.e., one of the SBA) is shown as
in Fig.5. The data separated from the input is represented as DB (Database) ,
and is used for learning.

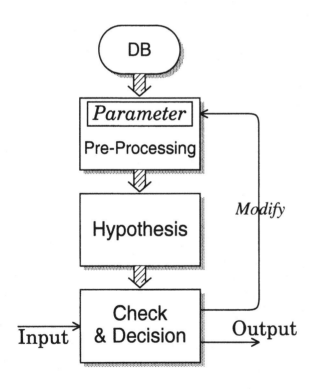

Fig. 5. Brainware Architecture for Data Processing

First, the DB is given to Pre-processing part, and next, to Hypothesis part.
The output yields a set of hypothesis, and is compared with the input. The
comparison (i.e., verification) decides whether or not this procedure continues.

While it continues, the results of verification modify the parameters of Pre-processing part, and it makes a loop. Otherwise, the results are given to the output.

In this paper we discuss two cases using this architecture as follows;

Case 1. Data Mining.

We divide data mining into two cases.

One case is the case where no input is given (or it is given only for synchronization). The essential procedures are in the loop, and the total complexity (of Pre-processing part, Hypothesis part and the number of loops) becomes large. Its complexity is the same as in data mining in the conventional database[3].

The other case is the case where the input is given while data mining is proceeding. In this case, the input is given on the way of learning process, and therefore, the results in the output may not be correct. In this case the input can be also regarded as the data for learning, and corresponds to a real-time system. The input is given by environment outside the system, or it may be given by a user. The database plays a main role and the input is subsidiary.

Both cases are treated in Fig.5, but the latter case is systematically similar to the following case.

Case 2. Real-Time Imaging.

This is the case where the input plays a main role and the database is subsidiary. An example is the real-time imaging. We use MLP net or Kohonen net as the FFN(Kohonen net in [4]), but must specify an adequate one depending on the purpose as the FBN.

4   Data Mining Using SBA

For this case we do not need essentially the input as in Fig.5, when we focus on the case where no input is given. For data mining using the SBA, we need to decide the neural networks for FFN and FBN.

FFN: Any neural network seems not to be appropriate for data mining, because learning is strongly connected with FBN. The learning-less part needs not to be realized by neural networks, and therefore, FFN should be realized by a logic-based system. As a brainware architecture, FFN is disregarded for data mining, and the part is included in FBN.

FBN: The feedback type of neural networks can be a candidate. For the recurrent network, however, we know only several BP-oriented learning algorithms. Unfortunately, these are not appropriate for data mining. We focus on two types: Hopfield net[5] and Chaotic neural net[6].

i) Hopfield net.

The Hopfield net provides an approximate algorithm for a class of combinatorial problems. The data mining procedure includes also a combinatorial problem, and the Hopfield net can be applied for Hypothesis part in Fig.5.

ii) Chaotic neural net.

The Hopfield net is not perfect for solving the combinatorial problems, since it may enter the local minima. To avoid this a type of chaotic neural nets seems to be applicable[6]. For this application, however, such a research is in the primary stage, and for our purpose we should expect to use it in future.

In the remainder of this section, we will state a realization of Hypothesis part in Fig.5 using Hopfield net. Suppose that the database be given by a table form as in Fig.6, where each column consists of numerical values of an attribute and each row is a vector form of attribute values.

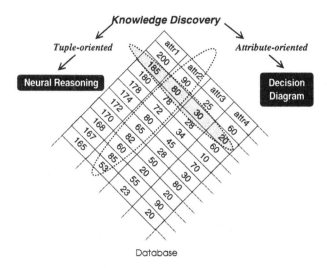

Fig. 6. Knowledge Discovery from Database: Two Approaches

We have two approaches for data mining;

a) Row Decomposition.

This is a typical type of classifications (e.g., Binary Decision Diagram, in short, BDD), and the problem to find the simplest representation is known to be NP-hard. For practice, however, several techniques (e.g., Bayesian method, ID3,..etc.) are developed, but parallel processing is indispensable for real-time processing. Therefore, the parallel algorithms (e.g., an Ordered BDD, in short, OBDD) are discussed, but such an algorithm provides an approximated method. We are also trying to develop an approximated method to solve OBDD using Hopfield net[7], and a method using neural networks is appropriate for SBA. An experimental result for solving of OBDD problem is shown as in Fig.7, which means that the required time for solving can be greatly reduced for n, a large number of attributes.

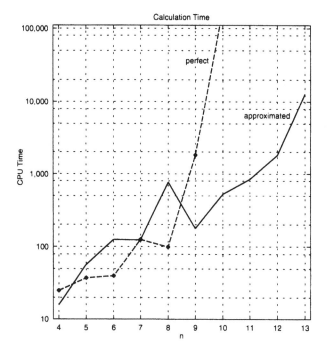

Fig. 7. Approximated Method using Hopfield Net ; An Example(n: number of attributes)

b) Column Decomposition.

This case is developed for extraction for a subset of rules desired by a user's view[8] such that the set of rules are required to derive a result in a designated attribute. For instance, it is a case to seek a set of rules such that "Value of an attribute is more than 0.6". A trivial answer is to show a local condition, but may not be correct because the consistency does not hold. For instance, let $\{x_i \rightarrow \bar{x}_j , x_j \rightarrow \bar{x}_i\}$ (where x_i and x_j are of attributes with logical value)be a local condition. In this case we have no solution such that both x_i and x_j become "true". This means that the consistency on the total database decides also the local condition.

For the general database we have more relaxed conditions, but we need to develop how to obtain the solution. The problem is also NP-hard if we need the best solution[8], and therefore, approximated solving ways must be developed. We proposed also the way of using Hopfield net for this problem, and show a good result[8].

a) is the case to seek the total knowledge in a compact form (e.g., a decision tree), and b) is the case to seek the local knowledge under holding consistency of the total database. b) is derived from a) and vice versa. If a) or b) is the solution that a user desires, a) or b) should be chosen, respectively. If the solution is intermediate, either will be chosen. For either solution the SBA plays an important role to support implementation.

5 Real-Time Imaging Using SBA

The real-time imaging scheme is just as in Fig.5, where the input plays a main role. A simple example is one of real-time pattern recognition, but the general case may be beyond real-time processing. Therefore, we introduce the recognition of a hand-written phrase.

Phrase Recognition.
Let a sample phrase be *Brainware Architecture.*

First: Character Recognition.
For *Brainware*, each character, that is, '*B*', '*r*', '*a*', '*i*', '*n*', '*w*', '*a*', '*r*' or '*e*' must be recognized, and for *Architecture* the similar process is required. For such a process we can use any neural network which is fit in pattern recognition. Considering hardware realization, however, the Kohonen net is one of the most appropriate neural networks, and a chip realization is already performed[9][10] and its system is already evaluated[4]. These case is restricted to digit symbols due to chip size, and therefore, an extension is now being prepared. It should be noted that this character recognition is performed only in FFN.

Second: Word Recognition.
In this stage, leaning of grammar or machine is required, but a simple grammar or a deterministic finite automaton is enough powerful for word recognition.

Third: Sentence Recognition.
The general case becomes complicated as in the recognition of a natural language. In our case, however, the third stage is included in the second stage because we focus on a simple case such as a phrase (e.g., *Brainware Architecture*).

The hand-written phrase such as a signature does not have the complicated structure. Instead, the character is not so easy to be recognized since the sequence of characters is not easily separable into characters. This is a reason why we use the neural network for its recognition. Moreover, the real-time processing is indispensable for recognition of a large number of signatures.

For real-time processing we are also preparing hardware realization using special-purpose chips for FFN. A prototype chip consisting of 18K-gates has been fabricated[10], and the next version is being prepared. (In Fig.8, we show an example of real-time learning using the prototype chip, although it is available only for digit characters.)

6 Conclusions

The reason why we are developing the brainware architecture is to realize a human-like brain which seems to be one of the best realization of a brain of robot.

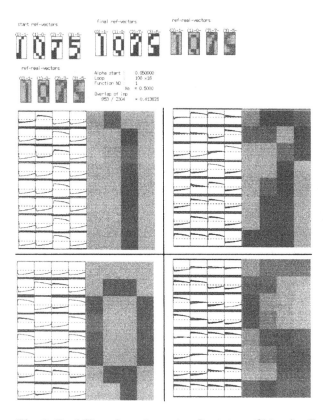

Fig. 8. Real-Time Learning using Prototype Chip; An Example

The brain of robot must satisfy the following;

1) *Real-time data processing* is performed.
2) *Small-size and low-power* are realized.

We also expect *real-time learning* as well as real-time input-output response (i.e., *recall* in the neural network). To realize such an artificial brain, we have introduced the following;

a) Memory-Based Architecture consisting of STM and LTM.
b) Chip realization with (possible) advanced technology.

Several types of memory-based architectures are already proposed, but we focus on the memory-based AI architecture[11][12]. In this paper, we specify more clearly the memory-based AI architecture with STM and LTM. As hardware realization we have several candidates on this architecture, and it depends on the application. We will state each performance in future.

About chip technology we are now using the conventional one (e.g., $0.5\mu m$ design-rule in [10]). In future, however, an artificial brain for robot will be realizable using more advanced nanometer device technology.

Appendix. Memories for Memory-Based Architecture.

Definition 1. Ensemble Memory.
An ensemble memory $M(t)$ with n elements is given as

$$M(t) = \{m(1,t), \ldots\ldots, m(n,t)\} \text{ at time } t,$$

where $m(p,t)$ is a memory element with address pointer p.

Definition 2. Passive Ensemble Memory.
A passive ensemble memory PEM is given as

PEM; $t \longleftarrow 0$;
　　　$M(t) \longleftarrow x = (x_1, \ldots, x_n)$ {write x into $M(t)$}
　　　repeat
　　　　　$(r_1, \ldots, r_n) \longleftarrow M(t)$　　{ read memory,
　　　　　　　　　　　　　　　　　　　　　　where $(r_1, \ldots, r_n)$ is a read variable }
　　　　　$t \longleftarrow t + 1$　　　　　{ next time }
　　　　　until $t < d$　　　　　{ d: positive number }
　　　　　$M(t) \longleftarrow 0$　　　　　{ reset memory } .

　　In PEM, if d is bounded by a small constant, it is called *short-term*, and if d is unbounded, it is called *long-term*.

Definition 3. Active Ensemble Memory.
An active ensemble memory AEM is given as

AEM; $t \longleftarrow 0$;
　　　$M(t) \longleftarrow x = (x_1, \ldots, x_n)$ {write x into $M(t)$}
　　　repeat
　　　　　$M(t+1) \longleftarrow M(t)$　　{ change memory contents from t to $t+1$}
　　　　　$(r_1, \ldots, r_n) \leftarrow M(t+1)$ { read memory,
　　　　　　　　　　　　　　　　　　　　　where $(r_1, \ldots, r_n)$ is a read variable }
　　　　　$t \longleftarrow t + 1$　　　　　{ next time }
　　　　　until $t < d$　　　　　{ d: positive number }
　　　　　$M(t) \longleftarrow 0$　　　　　{ reset memory } .

　　In AEM, if d is bounded by a small constant, it is called *short-term*, and if d is unbounded, it is called *long-term*.

Note1: In STM or LTM, PEM and AEM each are used depending on d is bounded or unbounded.

Note2: In AEM,

$$M(t+1) \longleftarrow M(t) \quad \{\text{change memory contents from } t \text{ to } t+1\}$$

is important. In LTM we use the neural network with feedback (e.g., Hopfield net) and this part is given by description of such a neural network. In STM we use the feed-forward neural network (e.g., MLP, Kohonen net) and this part is given by description of such a neural network.

References

1. M.Ito: Expecting for Brain Research in the 21st Century, Extended Abstract of Brainware Workshop, Tsukuba, FED-148, pp.3-5 (1996, in Japanese).
2. G.Matsumoto, Y.Shigematsu, M.Ichikawa: Brain-computing, i.b.i.d., pp.31-36 (1996, in Japanese).
3. R.Agrawal: Database Mining: A Performance Perspective, IEEE Transaction on Knowledge and Data Engineering, Vol.5, No.6, pp.914-925 (1993).
4. H.Araki, H.Fukumoto, T.Ae: Image Processing using Simplified Kohonen Network, Proceedings SPIE, San Jose, Vol.2661, pp.24-33 (1996).
5. J.J.Hopfield: Neurons with Graded Response Have Collective Computational Properties like Those of Two-State Neurons, Proceedings National Academy of Science, USA81, pp.3088-3092 (1984).
6. K.Aihara: Chaos and Brainware, Extended Abstract of Brainware Workshop, Tsukuba, FED-148, pp.8-12 (1996).
7. K.Murota, X. Wu, T.Ae: High-Speed Processing for Knowledge Representation from Decision Diagram, JSAI Technical Report, SIG-PPAI-9502-2 (1995, in Japanese).
8. H.Fukumoto, S.Hiwatashi, X.Wu, T.Ae: Knowledge Discovery by Neural Reasoning, JSAI Technical Report, SIG-PPAI-9503-6 (1996, in Japanese).
9. T.Ae, T.Toyosaki, H.Fukumoto, K.Sakai: ONBAM: An Objective-Neuron-Based Active Memory, Proceedings 1st ICA3PP, Brisbane, Vol.1, pp.231-234 (1995).
10. T.Toyosaki, T.Ae: A Neuro-Chip for Kohonen's LVQ Algorithm, Proceedings 6th ISIC, Singapore, pp.107-111 (1995).
11. T.Ae: New Functional Machine, (New) Information Processing Handbook (JIPS Ed.), 3.8.7., pp.455-459, Ohm Co. Ltd. (1995, in Japanese).
12. M.Kawada, X.Wu, T.Ae: A Construction of Neural-Net Based AI Systems, Proceedings 1st ICECCS, Ft. Lauderdale, pp.424-427 (1995).

Evolvable Systems

Evolvable Hardware: An Outlook

Bernard Manderick[1] and Tetsuya Higuchi[2]

[1] Computer Science Department. Free University Brussels
Pleinlaan 2. B-1050 Brussels, Belgium
[2] Computational Models Section, Electrotechnical Lab
1-1-4. Umezono. Tsukuba. Ibaraki 305, Japan

Abstract. In this paper. we explore the potential of Evolvable Hardware (EHW) for online adaptation in real-time applications. We follow a top–down approach here. We first review existing adaptation and learning techniques and take a look at their suitability for driving hardware evolution. Then we discuss some research problems whose solution will improve the performance of EHW.

1 Introduction

In recent decades many learning and adaptation techniques have been developped. Examples are more classical machine learning techniques like induction of decision trees and reinforcement learning [Narendra89], and techniques inspired by biological metaphors: Neural networks(NNs) [Rumelhart86] are inspired by the architecture and the working of the brain and Evolutionary algorithms(EAs), including Genetic algorithms(GAs) [Goldberg89], Evolution strategies(ESs), Evolutionary programming(EPs) [Fogel66] and Genetic programming(GPs) [Koza91], by the Neo-Darwian theory of natural evolution.

The current state of the art shows that these techniques are capable of solving difficult optimization problems like graph bipartitioning and traveling salesmen problems [Goldberg89], of evolving navigation strategies for robots [Harvey92], of learning classification tasks like handwritten character ecognition [Rumelhart86] and so on.

However, these techniques work either offline, i.e. the adaption phase proceeds the performance phase during which no adaptation is going on anymore. An example here is the evolution of robot navigation strategies [Harvey92]. Or if these techniques work online, i.e. adaptation and performance run in parallel, the adaptation is to slow to cope with all kind of changes in the environment. In other words, the current state of the art doesn't allow adaptation in real-time which is very important for real-world application like robotics and ATM–networks.

In the beginning of the nineties, at the Electrotechnical Lab in Japan, the idea grew to build on new technological developments concerning hardware-reconfigurable logic devices, e.g. Field Programmable Gate Arrays (FPGAs) [Lattice90, Xilinx90] to realize *adaptation in real-time*. This will be possible since adaptation now proceeds at the hardware level and the process will be speed up by many orders of magnitude. Moreover, EHW will be flexible and fault-tolerant since it can change its own structure in the case of environmental change or hardware error. Note that the notion of EHW used here is a special case of the general notion presented in [Yao96] which covers related work on EHW in other parts of Japan and the world.

The rest of the paper is organized as follows. First, we take a look at different adaptation techniques. Second, we discuss the conceptual architecture of EHW and evaluate the suitability of the above adaptation techniques for it. Next, we take a look at different existing hardware architectures used for EHW and point out related research problems whose solution will improve the performance. Finally, we take a look at two sets of applications which are here called prediction and control applications and associated problems which have to be solved for the successful application of EHW.

2 Adaptation Techniques

In this section, we briefly discuss neural networks, reinforcement learning and evolutionary algorithms. In the next section, we will evaluate their suitability for EHW.

In this paper we restrict ourselves to feedforward and recurrent networks which learn by backpropagation of errors since our main concern is adaptation. Such a network consists of a collection of units called neurons which are interconnected by weighted directed connections. Each unit has a state which is a continuous real value between zero and one and outputs this state to the units with which it is connected. The state of a unit is continuously updated as follows: The new state is a function of the weighted sum of its inputs and usually this function is a sigmoid one.

Some of the units are designated as input units, i.e. their state is set by the environment and others are designated as output units, i.e. their state is fed back into the environment. Learning proceeds by changing the weights of the connections and this change depends on the error, i.e. the difference between the actual and the desired output of the network, given the input. It is an example of supervised learning since a teacher is needed which provides the desired output.

In contrast, reinforcement learning is an example of unsupervised learning. The output of the network is evaluated as favorable or unfavorable by the enviroment and this evaluation is fed back as a positive or negative reinforcement signal to the network. There is no need for a teacher which provides the desired output. Moreover this feedback can be delayed, i.e. the reinforcement signal is provided after a number of outputs rather than after each output. Therefore, the applicability of reinforcement learning is much broader than supervised learning.

Actually reinforcement learning [Sutton92] is a set of techniques which can be implemented in different ways. Here, we will restrict ourselves to stochastic learning automata[Narendra89] because their implementation in hardware seems easiest.

A stochastic learning automata (SLA) has a number of states and can get an input from a set of possible inputs. Based on the current input and the current state, the SLA will generate an output and change its state. As opposed to deterministic automata, the output and the new state is not uniquely determined by the current input and state. For each combination of input and state there are a finite number of possible new states and outputs. Probabilities determine which ones will be selected.

Learning in SLAs proceeds by changing the above probabilities according to the received reinforcement. A number of different reinforcement schemes are available [Narendra89].

Hardware implementation of SLAs should be easy since they are basically Finite State Machines with some added randomness.

As opposed to neural networks and reinforcement learning which operate on one structure, evolutionary algorithms (EAs) manipulate a population of structures. Each of these structures is assigned a fitness which is a measure of how well it functions in its environment. Learning continuously updates this population by a combination of selection and recombination. First, structures are selected according to their fitness, the higher the fitness the higher the probability to be selected. These selected structures are then recombined by genetic operators like mutation and crossover. Mutation randomly changes each structure with a low probability and crossover exchanges parts of pairs of structures. The motivation behind mutation is that some of these changes will be improvements and behind crossover is that useful parts of a pair will be combined in a better overal structure. Of course, the resulting structure might be less fit but there will always remain enough high fit ones in the population and so the overall performance will not degrade. Instead, as long as better structures continue to be generated by selection and recombination the performance will improve.

Two variants of EAs which are immediately relevant for EHW are Genetic algorithms(GAs) [Goldberg89] and Genetic programming(GPs) [Koza91]. They differ in the kind of structures which are manipulated by the EA.

In GAs, these structure are usually bit strings which encode some phenotype. In the case of EHW, this will be architecture bits describing a combinatorial circuit.

In GP, these structures are parse trees of computer programs. Usually, these programs are written in LISP but any interpreted language where programs can be manipulated as data will do.

3   Conceptual Architecure for EHW

Recent technological developments allow changes at the hardware level, e.g. Field Programmable Gate Arrays (FPGAs) [Lattice90, Xilinx90], and show that these changes can be many orders of magnitude faster than similar changes at the software level. Moreover, this reconfigurable hardware can be produced at very low price in large quantities.

The question we will address here is how we can make these changes adaptive in order to arrive at online adaptation in real-time. Other desirable properties are flexibility and fault-tolerance, i.e. the hardware continues to function in case of environmental change and hardware error, respectively.

Hardware which meets these criteria will be called Evolvable hardware (EHW). This evolution opens the way to Evolvable hardware, a generally applicable adaptive device for real-time applications. Note that this notion of EHW is more restricted than the general notion presented in [Yao96].

We will answer this question here in a top-down way. In order to be fault-tolerant, EHW has to maintain a multitude of structures. If one starts to malfunction then one of the others can take over its role.

Since we have already a population of structures it is a natural step to apply an evolutionary algorithm to this population. Besides fault-tolerance we also get a form of flexibility. However, since adaptation provided by EAs is a rather slow process, the obtained flexibility might be to slow for real-time applications. Enhanced flexibility can be obtained by applying individual learning to each structure in the population.

So, EHW should combine a combination of evolutionary and individual adaptation. In Section 2, we have given a overview of existing adaptation techniques. The question now becomes: Which techniques are suitable for the different kinds of EHW.

In the next section, we introduce the different hardware architectures for EHW, determine the adaptation techniques suitable for them and pose some research questions to be solved in order to improve their performance.

4 Hardware Architectures

In EHW, adaptation takes the form of direct modification of the hardware structures according to rewards received from or fitness assigned by the environment. This results in a number of advantages. Adaptation in real-time is feasible due to a speed-up by many orders of magnitude. The long-term goal is to implement EHW on one chip so that it can be utilized as a generally applicable adaptive device for real-time applications.

There exist different kinds of EHW and they are discussed in [Yao96]. A first kind implements evolution at the gate level and is based on FPGA-technology, a second kind implements evolution at the function level and is now being developed at ETL.

The genetic algorithm is a good way to implement genetic learning for hardware evolution at the gate level while this is the case for genetic programming at the function level.

The stochastic learning automaton is suitable to implement individual adaptation. It is an example of unsupervised learning and a form of reinforcement which can implemented at the gate level.

5 Application Domain

The application domain of EHW can be divided in two large sets which will be called predicition and control applications here.

In a prediction application, EHW gets information from the environment in which it is embedded and tries to predict the next state of the environment. Evolving a model of a ATM-node and prediction of time series are examples. This kind of applications don't pose problems for EHW. First of all, finite state machines is often a good formalism for this kind of applications and are suitable for EHW. And second, since each individual in the population can later on compare its prediction with the future state of the environment, there is enough information avaible to determine the fitness of that individual or to provide it with a reinforcement signal.

In a control application, EHW not only gets information from the environment but also affects that environment. Robot navigation is an example here. Based on sensory information a robot can decide to move to the left or to the right. But different moves will generate different sensory information at the next time step and this information will be used to evaluate the fitness or to generate the reinforcement signal for the corresponding move.

Of course, only one of the possible moves can be executed and so a conflict resolution strategy has to be applied to select one of the possible moves. For this move, the fitness and the reinforcement signal can be determined but for the other ones not because their effects are not known.

This poses a major problem for EHW since at each generation only part of the population (the individuals will be called controllers in this kind of application) can be evaluated. Especially if several alternatives are available then only a small fraction of the controllers will be evaluated. Moreover, the population size is usually small in EHW.

Solutions have to be found if we want to apply EHW to control problems. We should try to evaluate each time step all controllers maintained by EHW. One possible solution is to evolve reliable models of the environment in which the controllers have to operate. This can be done using a second EHW-module. Evolving models doesn't pose problems since this is a prediction application, see above. Each controller can then simulate its proposed action in the best model of the environment available and get evaluated. Now, all controllers of the populations are assigned a fitness and or get a reinforcement signal.

6 Conclusions

In this paper, we have tried to answer the question: What should be the conceptual architecture of EHW in order to make changes as a result of hardware reconfigurations adaptive and possible in real-time. Such reconfigurations are now technically possible at high speed and large quantities of such hardware can be produced at low price. This evolution opens the way to Evolvable hardware, a generally applicable adaptive device for real-time applications.

We have argued that the adaptive mechanism driving such a device should be a combination of genetic and individual learning. Since genetic learning operates on a population of structures it makes EHW fault-tolerant. Many good structures are still available if one of them malfunctions. It also offers flexibility as a result of the adaptive nature of genetic learning. However, this adaptive process might be too slow for real-time applications. Fortunately, it can be supplemented by individual learning which improves the performances of individual structures using reinforcement signals from the environment.

The genetic algorithm is a good way to implement genetic learning for hardware evolution at the gate level while this is the case for genetic programming at the function level.

The stochastic learning automaton is suitable to implement individual learning. It is an example of unsupervised learning and a form of reinforcement which can implemented at the gate level.

References

[Armstrong91] Armstrong, W., Dwelly, A., Liang, J., Lin, D. and Reynolds, S., "Learning and generalization in adaptive logic networks in artificial neural networks", *Proc. of the 1991 International Conference on Artificial Neural Networks*, pp1173-1176, 1991.

[DeJong75] De Jong, K.A., *An Analysis of the behavior of a class of genetic adaptive systems*, Doctoral Dissertation, University of Michigan, 1975.

[Fogel66] Fogel, L.J., Owens, A.J. and Walsh, M.J., *Artificial Intelligence through Simulated Evolution*, Wiley, 1966.

[Goldberg89] Goldberg, D.E., *Genetic Algorithms in search, optimization and machine learning*, Addison Wesley, 1989.

[Harvey92] Harvey, I., "Evolutionary Robotics and SAGA: the Case for Hill Crawling and Tournament Selection", CSRP 222, the University of Sussex, 1992.

[Higuchi92] Higuchi, T., Niwa, T., Tanaka, T., Iba, H., de Garis, H. and Furuya, T., "Evolvable Hardware with genetic learning", in *Proc. of Simulated Adaptive Behavior*, MIT Press, 1993.

[Kitano91] Kitano, H., Hendler, J., Higuchi, T., Moldovan D. and Waltz D., "Massively Parallel Artificial Intelligence" *Proc. of IJCAI-91*, 1991.

[Koza91] Koza, J., "Evolution of Subsumption Using Genetic Programming", *Proc. of the first European Conf. on Artificial Life*, 1991.

[Kube93] Kube, C.R., Zhang, H. and Wang, X., "Controlling Collective Tasks With an ALN", *Proc. of IROS 93*, 1993.

[Lattice90] Lattice Semiconductor Corporation, "GAL Data Book", 1990.

[Mead89] Mead, C., "Adaptive Retina", in C. Mead and M. Ismail (eds.), *Analog VLSI Implementation of Neural Systems*, pp213-246, Kluwer, 1989.

[Narendra89] Narendra, K.S. and Tathachar, M.A.L., *Learning Automata: An Introduction*, Prentice Hall, 1989.

[Natarajan91] Natarajan, B.K., *Machine Learning: A Theoretical Approach*, Morgan Kaufmann, 1991.

[Rumelhart86] Rumelhart, D.E. and McClelland, J.L. (1986) *Parallel Distributed Processing: Explorations in the Microstructure of Cognition*, MIT Press, 1986.

[Spofford91] Spofford. J.J. and Hintz. K., "Evolving Sequential Machines in Morphous Neural Networks", in *Artificial Neural Network*, Elsevier. 1991.

[Sutton92] Sutton, R.S. (ed.).*Machine Learning, Vol. 8, Numbers 3/4: Special Issue on Reinforcement Learning*. Kluwer Academic Publishers. 1992.

[Xilinx90] Xilinx Semiconductor Corporation."LCA Data Book", 1990.

[Zhou86] Zhou. H. and Grefenstette. J.J., "Induction of finite automata by genetic algorithms". *Proc. of the 1986 IEEE International Conference Systems, Man and Cybernetics*. pp170-174. Atlanta. GA. 1986.

[Yao96] Yao. X. and Higuchi. T.. "Promises and Challenges of Evolvable Hardware". *This proceedings*.

Reuse, Parameterized Reuse, and Hierarchical Reuse of Substructures in Evolving Electrical Circuits Using Genetic Programming

John R.Koza[1]   Forrest H Bennett III[2]

David Andre[3]   Martin A. Keane[4]

1) Computer Science Department, Stanford University
Stanford, California 94305 USA
koza@cs.stanford.edu   http://www-cs-faculty.stanford.edu/~koza/

2) Visiting Scholar, Computer Science Department,
Stanford University

3) Computer Science Department, University of California, Berkeley, California

4) Econometrics Inc., 5733 West Grover
Chicago, IL 60630 USA

Abstract: Most practical electrical circuits contain modular substructures that are repeatedly used to create the overall circuit. Genetic programming with automatically defined functions and the recently developed architecture-altering operations provides a way to build complex structures with reused substructures. In this paper, we successfully evolved a design for a two-band crossover (woofer and tweeter) filter with a crossover frequency of 2,512 Hz. Both the topology and the sizing (numerical values) for each component of the circuit were evolved during the run. The evolved circuit contained three different noteworthy substructures. One substructure was invoked five times thereby illustrating reuse. A second substructure was invoked with different numerical arguments. This second substructure illustrates parameterized reuse because different numerical values were assigned to the components in the different instantiations of the substructure. A third substructure was invoked as part of a hierarchy, thereby illustrating hierarchical reuse.

1. Introduction

Computer programs are replete with modular substructures that are repeatedly used within the overall program. For example, segments of useful computer code are typically encapsulated as subroutines and then reused. Moreover, subroutines may be called with different instantiations of their formal parameters (dummy variables), thereby creating parameterized reuse. In addition, subroutines may be called hierarchically, thereby creating hierarchical reuse.

Practical electrical circuits are replete with modular substructures that are repeatedly used (sometimes with different numerical values) within the overall circuit. These observations suggest that a mechanism for the automated design of a complex structure such as an electrical circuit should incorporate some kind of mechanism to exploit such modularities by reuse, parameterized reuse, and hierarchical reuse.

2. Genetic Programming

Genetic programming is an extension of the genetic algorithm described in John Holland's pioneering *Adaptation in Natural and Artificial Systems* (1975).

The book *Genetic Programming: On the Programming of Computers by Means of Natural Selection* (Koza 1992) provides evidence that genetic programming can solve, or approximately solve, a variety of problems. Additional details are in Bennett, Koza, Andre, and Keane in this volume. See also Koza and Rice 1992.

The book *Genetic Programming II: Automatic Discovery of Reusable Programs* (Koza 1994a, 1994b) describes a way to evolve multi-part programs consisting of a main program and one or more reusable, parameterized, hierarchically-called subprograms. Specifically, an *automatically defined function (ADF)* is a function (i.e., subroutine, subprogram, DEFUN, procedure, module) that is dynamically evolved during a run of genetic programming and which may be called by a calling program (or subprogram) that is concurrently being evolved. When automatically defined functions are being used, a program in the population consists of a hierarchy of one (or more) *reusable* function-defining branches (i.e., automatically defined functions) along with a main result-producing branch. Typically, the automatically defined functions possess one or more dummy arguments (formal parameters) and are reused with different instantiations of these dummy arguments. During a run, genetic programming evolves different subprograms in the function-defining branches of the overall program, different main programs in the result-producing branch, different instantiations of the dummy arguments of the automatically defined functions in the function-defining branches, and different hierarchical references between the branches.

Genetic programming with automatically defined functions has been shown to be capable of solving numerous problems. More importantly, the evidence so far indicates that, for many problems, genetic programming requires less computational effort (i.e., fewer fitness evaluations to yield a solution with a satisfactorily high probability) with automatically defined functions than without them (provided the difficulty of the problem is above a certain relatively low break-even point). Also, genetic programming usually yields solutions with smaller average overall size with automatically defined functions than without them (provided, again, that the problem is not too simple). That is, both learning efficiency and parsimony appear to be properties of genetic programming with automatically defined functions.

Moreover, there is also evidence that genetic programming with automatically defined functions is scalable. For several problems for which a progression of scaled-up versions was studied, the computational effort increases as a function of problem size at a *slower rate* with automatically defined functions than without them. In addition, the average size of solutions similarly increases as a function of problem size at a *slower rate* with automatically defined functions than without them. This observed scalability results from the profitable reuse of hierarchically-callable, parameterized subprograms within the overall program.

When automatically defined functions are being evolved in a run of genetic programming, the architecture of the overall program must be determined in some way. The specification of the architecture consists of (a) the number of function-defining branches (i.e., automatically defined functions) in the overall program, (b) the number of arguments (if any) possessed by each function-defining branch, and (c) if there is more than one function-defining branch, the nature of the hierarchical references (if any) allowed between the function-defining branches (and between the function-defining branches and the result-producing branch(es).

The user may supply the specification of this architectural information as a preparatory step that occurs prior to executing the run of genetic programming. However, it is preferable, in many situations, to automate these architectural decisions so that the user is not required to prespecify the architecture. Recent work on genetic programming has demonstrated that it is possible to evolve the architecture of an overall program dynamically during a run of genetic programming using six recently developed architecture-altering operations, namely branch duplication, argument duplication, branch deletion, argument deletion, branch creation, and argument creation (Koza 1995). These architecture-altering operations provide an automated way to enable genetic programming to dynamically determine, during the run, whether or not to employ function-defining branches, how many function-defining branches to employ, and the number of arguments possessed by each function-defining branch. The architecture-altering operations and automatically defined functions together provide an automated way to decompose a problem into a non-prespecified number of subproblems of non-pre-specified dimensionality; to solve the subproblems; and to assemble the solutions of the subproblems into a solution of the overall problem.

The six architecture-altering operations are motivated by the naturally occurring mechanisms of gene duplication and gene deletion in chromosome strings as described in Susumu Ohno's book *Evolution by Gene Duplication* (1970). In that book, Ohno advanced the thesis that the creation of new proteins (and hence new structures and new behaviors in living things) begins with a gene duplication.

Recent work in the field of genetic programming is described in Kinnear 1994, Angeline and Kinnear 1996, Koza, Goldberg, Fogel, and Riolo 1996.

3 . The Problem of Circuit Synthesis

The problem of circuit synthesis involves designing an electrical circuit that satisfies user-specified design goals. A complete design of an electrical circuit includes both its topology and the sizing of all its components. The *topology* of a circuit consists of the number of components in the circuit, the type of each component, and a list of all the connections between the components. The *sizing* of a circuit consists of the component value(s) of each component.

Evolvable hardware is one approach to automated circuit synthesis. Early pioneering work in this field includes that of Higuchi, Niwa, Tanaka, Iba, de Garis, and Furuya (1993a, 1993b); Hemmi, Mizoguchi, and Shimohara (1994); Mizoguchi, Hemmi, and Shimohara (1994); and the work presented at the 1995 workshop on evolvable hardware in Lausanne (Sanchez and Tomassini 1996).

The design of analog circuits and mixed analog-digital circuits has not proved to be amenable to automation (Rutenbar 1993). CMOS operational amplifier (op amp) circuits have been designed using a modified version of the genetic algorithm

(Kruiskamp 1996; Kruiskamp and Leenaerts 1995); however, the topology of each op amp was one of 24 topologies based on the conventional human-designed stages of an op amp. Thompson (1996) used a genetic algorithm to evolve a frequency discriminator on a Xilinx 6216 reconfigurable processor. In Gruau's innovative cellular encoding technique (1996), genetic programming is used to evolve the architecture, weights, thresholds, and biases of neurons in a neural network.

4. Circuit Synthesis Using Evolution

Genetic programming can be applied to circuits if a mapping is established between the kind of rooted, point-labeled trees with ordered branches used in genetic programming and the line-labeled cyclic graphs encountered in the world of circuits. Developmental biology provides the motivation for this mapping. The starting point of the growth process used herein is a very simple embryonic electrical circuit. The embryonic circuit contains certain fixed parts appropriate to the problem at hand and certain wires that are capable of subsequent modification. An electrical circuit is progressively developed by applying the functions in a circuit-constructing program tree to the modifiable wires of the embryonic circuit (and, later, to both the modifiable wires and other components of the successor circuits).

These functions manipulate the embryonic circuit (and its successors) so as to produce valid electrical circuits at each step. The functions are divided into four categories: (1) connection-modifying functions that modify the topology of the circuit (starting with the embryonic circuit), and (2) component-creating functions that insert components into the topology of the circuit, (3) arithmetic-performing functions that appear in arithmetic-performing subtrees as argument(s) to the component-creating functions and specify the numerical value of the component, and (4) calls to automatically defined functions.

Each branch of the program tree is created in accordance with a constrained syntactic structure. Branches are composed from construction-continuing subtree(s) that continue the developmental process and arithmetic-performing subtree(s) that determine the numerical value of the component. Connection-modifying functions have one or more construction-continuing subtrees, but no arithmetic-performing subtrees. Component-creating functions have one construction-continuing subtree and typically have one arithmetic-performing subtree. Structure-preserving crossover with point typing (Koza 1994a). then preserves the constrained syntactic structure.

4.1. The Embryonic Circuit

The developmental process for converting a program tree into an electrical circuit begins with an embryonic circuit.

The bottom of figure 1 shows an embryonic circuit for a one-input, two-output circuit. The energy source is a 2 volt sinusoidal voltage source **USOURCE** whose negative (−) end is connected to node 0 (ground) and whose positive (+) end is connected to node 1. There is a source resistor **RSOURCE** between nodes 1 and 2. There is a modifiable wire (i.e., a wire with a writing head) **Z0** between nodes 2 and 3, a second modifiable wire **Z1** between nodes 2 and 6, and third modifiable wire **Z2** between nodes 3 and 6. There is an isolating wire **ZOUT1** between nodes 3 and 4, a voltage probe labeled **UOUT1** at node 4, and a fixed load resistor **RLOAD1** between nodes 4 and ground. Also, there is an isolating wire **ZOUT2** between nodes 6 and 5, a voltage probe labeled **UOUT2** at node 5, and a load resistor **RLOAD2** between nodes 5 and ground. All three resistors are 0.00794 Kilo Ohms.

All of the above elements of this embryonic circuit (except **Z0**, **Z1**, and **Z2**) are fixed forever; they are not subject to modification during the process of developing the circuit. Note that little domain knowledge went into this embryonic circuit. Specifically, (1) the embryonic circuit is a circuit, (2) this embryonic circuit has one input and two outputs, and (3) there are modifiable connections **Z0**, **Z1**, and **Z2** providing full point-to-point connectivity between the one input (node 2) and the two outputs **UOUT1** and **UOUT2** (nodes 4 and 5).

A circuit is developed by modifying the component to which a writing head is pointing in accordance with the associated function in the circuit-constructing program tree. The figure shows L, C, and C functions just below the LIST and three writing heads pointing to **Z0**, **Z1**, and **Z2**. The L, C, and C functions will cause **Z0**, **Z1**, and **Z2** to become an inductor and two capacitors, respectively.

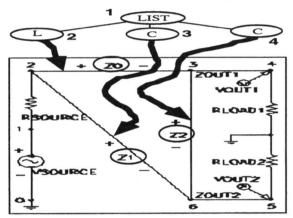

Figure 1   One-input, two-output embryonic electrical circuit.

4.2.   Component-Creating Functions

Each circuit-constructing program tree in the population contains component-creating functions and connection-modifying functions. Each component-creating function inserts a component into the developing circuit and assigns component value(s) to the component. We use the inductor-creating L function and the capacitor-creating C function in this paper. Space here does not permit a detailed description of these functions (or the functions in the next section). For details, see Bennett, Koza, Andre, and Keane (1996) in this volume; Koza, Andre, Bennett, and Keane (1996); and Koza, Bennett, Andre, and Keane (1996a, 1996b, 1996c, 1996d).

4.3.   Connection-Modifying Functions

Each connection-modifying function in a circuit-constructing program tree modifies the topology of the developing circuit.   The SERIES function creates a series composition; PSS and PSL a creates parallel composition; FLIP reverses polarity; NOP performs no operation; T_GND is a via-to-ground; PAIR_CONNECT connects points; SAFE_CUT cuts connections; and END ends growth.

5. Preparatory Steps

A *two-band crossover* (woofer and tweeter) filter is a one-input, two-output circuit that passes all frequencies below a certain specified frequency to its first output port (the woofer) and that passes all higher frequencies to a second output port (the tweeter). Our goal is to design a two-band crossover filter with a crossover frequency of 2,512 Hz.

Before applying genetic programming to circuit synthesis, the user must perform seven major preparatory steps, namely (1) identifying the embryonic circuit that is suitable for the problem, (2) determining the architecture of the overall circuit-constructing program trees, (3) identifying the terminals of the to-be-evolved programs, (4) identifying the primitive functions contained in the to-be-evolved programs, (5) creating the fitness measure, (6) choosing certain control parameters (notably population size and the maximum number of generations to be run), and (7) determining the termination criterion and method of result designation.

The one-input, two-output embryo of figure 1 is suitable for this problem.

Since the embryonic circuit has three writing heads – one associated with each of the result-producing branches – there are three result-producing branches (called RPB0, RPB1, and RPB2) in each program tree. The number of automatically defined functions, if any, will be determined by the evolutionary process using the architecture-altering operations. The automatically defined functions are called ADF0, ADF1, ... as they are created. Each program in the initial population of programs (generation 0) has a uniform architecture with three result-producing branches and no automatically defined functions.

The function sets are identical for all three result-producing branches of the program trees. The terminal sets are identical for all three result-producing branches.

For the three result-producing branches, the initial function set, $\mathcal{F}_{\text{ccs--rpb-initial}}$, for each construction-continuing subtree is

$\mathcal{F}_{\text{ccs--rpb-initial}} = \{$L, C, SERIES, PSS, PSL, FLIP, NOP, T_GND_0, T_GND_1, PAIR_CONNECT_0, PAIR_CONNECT_1$\}$.

For the three result-producing branches, the initial terminal set, $\mathcal{T}_{\text{ccs-initial}}$, for each construction-continuing subtree is

$\mathcal{T}_{\text{ccs-rpb-initial}} = \{$END, SAFE_CUT$\}$.

For the three result-producing branches, the function set, $\mathcal{F}_{\text{aps-rpb}}$, for each arithmetic-performing subtree is,

$\mathcal{F}_{\text{aps-rpb}} = \{+, -\}$.

For the three result-producing branches, the terminal set, $\mathcal{T}_{\text{aps-rpb}}$, for each arithmetic-performing subtree consists of

$\mathcal{T}_{\text{aps-rpb}} = \{\Re\}$,

where $\Re$ represents floating-point random constants from -1.0 to $+1.0$.

The architecture-altering operations create new function-defining branches (automatically defined functions). The set of potential new functions, $\mathcal{F}_{\text{potential}}$, is

$\mathcal{F}_{\text{potential}} = \{$ADF0, ADF1, . . . $\}$,

where ADF0, ADF1, . . . are automatically defined functions.

For this problem, the set of potential new terminals, $\mathcal{T}_{\text{potential}}$, is limited to

$\mathcal{T}_{potential} = \{ARG0\}$,

where ARG0 is a dummy variable (formal parameter) to an automatically defined function.

For each newly created function-defining branch, the function set, $\mathcal{F}_{aps\text{-}adf}$, for an arithmetic-performing subtree is,

$\mathcal{F}_{aps\text{-}adf} = \mathcal{F}_{aps\text{-}rpb} = \{+, -\}$.

For each newly created function-defining branch, the terminal set, $\mathcal{T}_{aps\text{-}adf}$, for an arithmetic-performing subtree is

$\mathcal{T}_{aps\text{-}adf} = \{\mathfrak{R}\} \cup \mathcal{T}_{potential} = \{\mathfrak{R}\} \cup \{ARG0\}$.

The architecture-altering operations progressively change the function set for the construction-continuing subtrees of the three result-producing branches. After ADF0 is created,

$\mathcal{F}_{ccs\text{-}rpb\text{-}0} = \mathcal{F}_{ccs\text{-}rpb\text{-}initial} \cup ADF0$.

After ADF1 is created,

$\mathcal{F}_{ccs\text{-}rpb\text{-}1} = \mathcal{F}_{ccs\text{-}rpb\text{-}initial} \cup ADF0 \cup ADF1$,

and so forth.

Since hierarchical references are to be permitted among the progressively created automatically defined function, the architecture-altering operations progressively change the function set for the construction-continuing subtrees of the function-defining branches. After ADF0 is created, the function set for the construction-continuing subtrees of ADF0 is

$\mathcal{F}_{ccs\text{-}adf\text{-}0} = \mathcal{F}_{ccs\text{-}rpb\text{-}initial}$.

However, after ADF1 is created, the function set for the construction-continuing subtrees of ADF1 is

$\mathcal{F}_{ccs\text{-}rpb\text{-}1} = \mathcal{F}_{ccs\text{-}rpb\text{-}initial} \cup ADF0$,

and so forth. Thus, ADF1 can potentially refer to ADF0 and ADF2 can potentially refer to both ADF1 and ADF0.

The evaluation of fitness for each individual circuit-constructing program tree in the population begins with its execution. This execution applies the functions in the program tree to the embryonic circuit, thereby developing the embryonic circuit into a fully developed circuit. A netlist describing the fully developed circuit is then created. The netlist identifies each component of the circuit, the nodes to which that component is connected, and the value of that component. Each circuit is then simulated to determine its behavior. The 217,000-line SPICE (Simulation Program with Integrated Circuit Emphasis) simulation program (Quarles et al. 1994) was modified to run as a submodule within the genetic programming system.

The SPICE simulator is requested to perform an AC small signal analysis and to report the circuit's behavior at the two probe points, **VOUT1** and **VOUT2**, for each of 101 frequency values chosen from the range between 10 Hz and 100,000 Hz. Each decade of frequency is divided into 25 parts (using a logarithmic scale), so there are 101 fitness cases for each probe point and a total of 202 fitness cases.

Fitness is the sum, over the 101 **VOUT1** frequency values, of the absolute weighted deviation between the actual value of voltage that is produced by the circuit at the first probe point **VOUT1** and the target value for voltage for that first probe point *plus* the sum, over the 101 **VOUT2** frequency values, of the absolute weighted deviation between the actual value of voltage that is produced by the circuit at the second probe point **UOUT2** and the target value for voltage for that second probe point. The smaller the value of fitness, the better. A fitness of zero represents an ideal filter. Specifically, the standardized fitness is

$$F(t) = \sum_{i=0}^{100} [W_1(d_1(f_i), f_i) d_1(f_i) + W_2(d_2(f_i), f_i) d_2(f_i)]$$

where $f(i)$ is the frequency (in Hertz) of fitness case i; $d_1(x)$ is the difference between the target and observed values at frequency x for probe point **UOUT1**; $d_2(x)$ is the difference between the target and observed values at frequency x for probe point **UOUT2**; $W_1(y,x)$ is the weighting for difference y at frequency x for probe point **UOUT1**; and $W_2(y,x)$ is the weighting for difference y at frequency x for **UOUT2**.

The fitness measure does not penalize ideal values; it slightly penalizes every acceptable deviation; and it heavily penalizes every unacceptable deviation.

Consider the woofer portion and **UOUT1** first. The procedure for each of the 58 points in the woofer passband interval from 10 Hz to 1,905 Hz is as follows: If the voltage equals the ideal value of 1.0 volts in this interval, the deviation is 0.0. If the voltage is between 970 millivolts and 1,000 millivolts, the absolute value of the deviation from 1,000 millivolts is weighted by a factor of 1.0. If the voltage is less than 970 millivolts, the absolute value of the deviation from 1,000 millivolts is weighted by a factor of 10.0. This arrangement reflects the fact that the ideal voltage in the passband is 1.0 volt, the fact that a 30 millivolt shortfall satisfies the design goals of the problem, and the fact that a voltage below 970 millivolts in the passband is not acceptable.

For the 38 fitness cases representing frequencies of 3,311 and higher for the woofer stopband, the procedure is as follows: If the voltage is between 0 millivolts and 1 millivolt, the absolute value of the deviation from 0 millivolts is weighted by a factor of 1.0. If the voltage is more than 1 millivolt, the absolute value of the deviation from 0 millivolts is weighted by a factor of 10.0. This arrangement reflects the fact that the ideal voltage in the stopband is 0.0 volts, the fact that a 1 millivolt ripple above 0 millivolts is acceptable, and the fact that a voltage above 1 millivolt in the stopband is not acceptable.

For the two fitness cases at 2,089 Hz and 2,291 Hz, the absolute value of the deviation from 1,000 millivolts is weighted by a factor of 1.0. For the fitness case at 2,512 Hz, the absolute value of the deviation from 500 millivolts is weighted by a factor of 1.0. For the two fitness cases at 2,754 Hz and 3,020 Hz, the absolute value of the deviation from 0 millivolts is weighted by a factor of 1.0.

The fitness measure for the tweeter portion is a mirror image (reflected around 2,512 Hz) of the arrangement for the woofer portion.

Many of the circuits that are randomly created in the initial random population and many that are created by the crossover and mutation operations are so bizarre that they cannot be simulated by SPICE. Such circuits are assigned a high penalty value of fitness (10^8).

The population size, M, was 640,000. The architecture-altering operations are used sparingly on each generation. The percentage of operations on each generation after generation 5 was 86.5% one-offspring crossovers; 10% reproductions; 1% mutations; 1% branch duplications; 0% argument duplications; 0.5% branch deletions; 0.0% argument deletions; 1% branch creations; and 0% argument creations. Since we do not want to waste large amounts of computer time in early generations where only a few programs have any automatically functions at all, the percentage of operations on each generation before generation 6 was 78.0% one-offspring crossovers; 10% reproductions; 1% mutations; 5.0% branch duplications; 0% argument duplications; 1% branch deletions; 0.0% argument deletions; 5.0% branch creations; and 0% argument creations. A maximum size of 200 points was established for each of the branches in each overall program. The other minor parameters were the default values in Koza 1994a (appendix D).

This problem was run on a medium-grained parallel Parsytec computer system consisting of 64 80 MHz Power PC 601 processors arranged in a toroidal mesh with a host PC Pentium type computer. The distributed genetic algorithm was used with a population size of $Q = 10,000$ at each of the $D = 64$ demes. On each generation, four boatloads of emigrants, each consisting of $B = 2\%$ (the migration rate) of the node's subpopulation (selected on the basis of fitness) were dispatched to each of the four toroidally adjacent processing nodes. See Andre and Koza 1996 for details.

6. Results

There are no automatically defined functions in any of the 640,000 individuals of the initial random generation. The best individual program tree of generation 0 has a fitness of 410.3 and scores 98 hits (out of 202). Its first result-producing branch has 15 points; its second result-producing branch has 103 points; and its third result-producing branch has 12 points. Figure 2 shows the behavior of this best-of-generation circuit from generation 0 in the frequency domain. As can be seen, the intended lowpass (woofer) output **UOUT1** has the desired value of 1 volt for low frequencies, but then drops off in a leisurely way and reverses and rises to around 1/2 volt for higher frequencies. The intended highpass (tweeter) output **UOUT2** has the desired value of 0 volts for low frequencies but then slowly rises to only about 1/2 volt.

Automatically defined functions are created starting in generation 1; however, the first pace-setting best-of-generation individual with an automatically defined function does not appear until generation 8. The circuit for the best-of-generation individual from generation 8 has a fitness of 108.1 and scores 91 hits). Its three result-producing branches have 187, 7, and 183 points, respectively. Its one automatically defined function, ADF0, has 17 points. Figure 3 shows the behavior of the best-of-generation circuit from generation 8 in the frequency domain. Although the general shape of the two curves now resembles that of a crossover filter, the rise and fall of the two curves is far too leisurely.

The best-of-generation circuit from generation 158 has a fitness of 0.107 and scores 200 hits (out of 202). This fitness compares favorably with the fitness of 0.7807 and 192 hits achieved in a previously reported run of this problem without the architecture-altering operations (Koza, Bennett, Andre, and Keane, 1996b). Figure 4 shows the behavior of this circuit in the frequency domain.

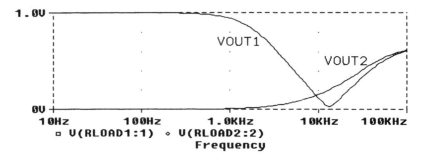

Figure 2 Frequency domain behavior of best circuit of generation 0.

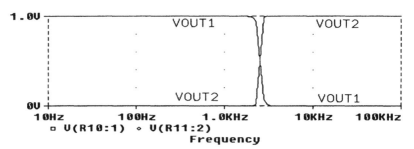

Figure 3 Frequency domain behavior of best circuit of generation 8.

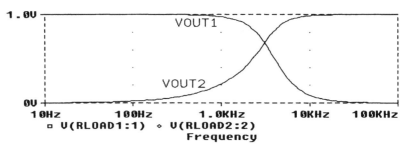

Figure 4 Frequency domain behavior of best circuit of generation
158.

Figure 5 shows the best-of-generation circuit from generation 158. Its three result-producing branches have 69, 158, and 127 points, respectively. This circuit has five automatically defined functions with 6, 24, 101, 185, and 196 points, respectively. Boxes indicate the use of ADF2 , ADF3, and ADF4.

There is an intricate structure of reuse and hierarchical reuse of structures in this evolved circuit. Result-producing branch RPB0 calls ADF3 once; RPB1 calls ADF3 once; and RPB2 calls ADF4 twice. ADF0 and ADF1 are not called at all. ADF2 is hierarchically called once by both ADF3 and ADF4. Note that ADF2 is called a total of five times – one time by RPB2 directly, two times by ADF3 (which is called once by RPB0 and RPB1), and two times by ADF4 (called twice by RFP2).

ADF2 has two ports and supplies one unparameterized 259 µH inductor **L147**. Its dummy variable, ARG0, plays no role. ADF0 and ADF1 are not used.

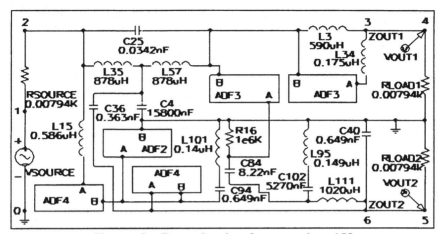

Figure 5   Best circuit of generation 158.

Figure 6 shows ADF3 of the best-of-generation circuit from generation 158. ADF3 has two ports.   It supplies one unparameterized 5,130 uF capacitor **C112**. ADF3 is interesting in two ways. First, it has one parameterized capacitor **C39** whose value is determined by ADF3's dummy variable, ARG0. Second, it has one hierarchical reference to ADF2 (which, in turn, supplies one unparameterized 259 μH inductor).   Thus, the combined effect of ADF3 is to supply two capacitors (one of which is parameterized) and one inductor.

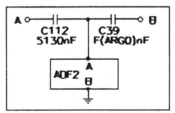

Figure 6   Automatically defined function ADF3.

Figure 7 shows ADF4 of the best-of-generation circuit from generation 158. ADF4 has three ports. It supplies one unparameterized 3,900 uF capacitor **C137** and one unparameterized 5,010 uF capacitor **C149**. ADF4 has one hierarchical reference to ADF2 (which, in turn, supplies one unparameterized 259 μH inductor). Thus, the combined effect of ADF4 is to supply two capacitors and one inductor.

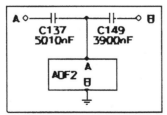

Figure 7   Three-ported automatically defined function ADF4.

An electrical engineer knows that one conventional way to realize a crossover filter is to connect a lowpass filter between the input and the first output port and to connect a highpass filter between the input and the second output port. In this neat decomposition, the only point of contact between the woofer part of the circuit feeding **VOUT1** and the tweeter part feeding **VOUT2** is the node that provides the incoming signal from **VSOURCE** and **RSOURCE**. There is no fitness incentive in a run of genetic programming to evolve a circuit that employs a neat decomposition of the problem into two disjoint parts. Figure 8 shows the best-of-generation circuit from generation 158 after all components have been substituted in lieu of the automatically defined functions. As can be seen, the evolved circuit is holistic in the sense that there are numerous interconnections between the parts feeding **VOUT1** and **VOUT2**.

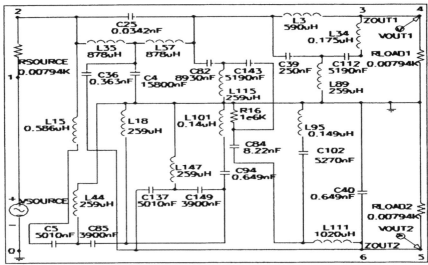

Figure 8   Best circuit of generation 158 after substitution.

7.   Another Example of Reuse of Evolved Substructures Using Automatically Defined Functions

The usefulness of automatically defined functions has been demonstrated in other examples of the evolutionary design of electrical circuits. For example, in Koza, Andre, Bennett, and Keane 1996, both the topology and sizing of a fifth order elliptic (Cauer) lowpass filter was evolved.

In one run, the best circuit from generation 0 consisted of one inductor and one capacitor in the topology of one rung of the classical ladder. For a lowpass filter, this classical topology (used in Butterworth or Chebychev filters) consists of repeated instances of series inductors and vertical shunt capacitors (Van Valkenburg 1982).

When electrical engineers design Butterworth or Chebychev filters, additional rungs on the ladder (in conjunction with properly chosen numerical values for the components) generally improve the level of performance of the filter (at the expense of additional power consumption, space, and cost). As the run continued from generation to generation, the best circuit from generation 9 had the classical two-rung ladder topology. The two rungs were produced by a twice-called automatically defined function that supplied the equivalent of a 154,400 μH inductor as a two-ported substructure. The improved behavior of this circuit was a consequence of the two rungs of the ladder.

The best circuit from generation 16 consists a three-rung ladder topology and was better than its predecessors. A thrice-called automatically defined function providing a suitable inductance created the three rungs. The best circuit from generation 20 consisted of a four-rung ladder. An automatically defined function providing a suitable inductance created the four rungs of the ladder and was responsible for the improved performance.

The best circuit in generation 31 satisfied all the design requirements of the problem. An automatically defined function constructed a three-ported substructure that was used five times. This genetically evolved 100% compliant circuit is especially interesting because it had the topology of an elliptic (Cauer) filter. The circuit had the equivalent of six inductors horizontally across the top of the circuit and five vertical shunts. Each shunt consisted of an inductor and a capacitor. At the time of its invention, the Cauer filter was a significant advance (both theoretically and commercially) over the Butterworth and Chebychev filters (Van Valkenburg 1982). For example, for one illustrative set of specifications, a fifth-order elliptic filter can equal the performance of an eighth order Chebychev filter. The benefit is that the fifth order elliptic filter has one few component than the eighth order Chebychev filter.

The best circuit in generation 35 has a fitness that is about an order of magnitude better than that of the best-of-generation individual from generation 31. This 100% compliant circuit exhibits two-fold symmetry involving the repetition of four modular substructures. The symmetry of this circuit is a consequence of its quadruply-called three-ported automatically defined function. Two inductors and one capacitor form a triangle with the substructure produced by the automatically defined function. There is an additional induction element branching away from the triangle at one node.

8. Conclusion

Genetic programming with automatically defined functions and architecture-altering operations successfully evolved a design for a two-band crossover (woofer and tweeter) filter with a crossover frequency of 2,512 Hz. Both the topology and the sizing (numerical values) for each component of the circuit were evolved during the run. The evolved circuit contained three different noteworthy substructures. One substructure was invoked five times thereby illustrating reuse. A second substructure was invoked with different numerical arguments. This second substructure illustrates parameterized reuse because different numerical values were assigned to the components in the different instantiations of the substructure. A third substructure was invoked as part of a hierarchy, thereby illustrating hierarchical reuse.

Acknowledgments

Jason Lohn, Simon Handley, and Scott Brave made helpful comments on various drafts of this paper.

Related Paper in this Volume

See also Bennett, Koza, Andre, and Keane (1996) in this volume.

References

Andre, David and Koza, John R. 1996. Parallel genetic programming: A scalable implementation using the transputer architecture. In Angeline, P. J. and Kinnear, K. E. Jr. (editors). *Advances in Genetic Programming 2*. Cambridge: MIT Press.

Angeline, Peter J. and Kinnear, Kenneth E. Jr. (editors). 1996. *Advances in Genetic Programming 2*. Cambridge, MA: The MIT Press.

Bennett III, Forrest H, Koza, John R., Andre, David, and Keane, Martin A. 1996. Evolution of a 60 decibel op amp using genetic programming. In this volume.

Gruau, Frederic. 1996. Artificial cellular development in optimization and compilation. In Sanchez, Eduardo and Tomassini, Marco (editors). 1996. *Towards Evolvable Hardware*. Lecture Notes in Computer Science, Volume 1062. Berlin: Springer-Verlag. Pages 48 – 75.

Hemmi, Hitoshi, Mizoguchi, Jun'ichi, and Shimohara, Katsunori. 1994. Development and evolution of hardware behaviors. In Brooks, R. and Maes, P. (editors). *Artificial Life IV* Cambridge, MA: MIT Press. Pages 371–376.

Higuchi, Tetsuya, Niwa, Tatsuya, Tanaka, Toshio, Iba, Hitoshi, de Garis, Hugo, and Furuya, Tatsumi. 1993a. In Meyer, Jean-Arcady, Roitblat, Herbert L. and Wilson, Stewart W. (editors). *From Animals to Animats 2: Proceedings of the Second International Conference on Simulation of Adaptive Behavior*. Cambridge, MA: The MIT Press. 1993. Pages 417 – 424.

Higuchi, Tetsuya, Niwa, Tatsuya, Tanaka, Toshio, Iba, Hitoshi, de Garis, Hugo, and Furuya, Tatsumi. 1993b. *Evolvable Hardware – Genetic-Based Generation of Electric Circuitry at Gate and Hardware Description Language (HDL) Levels*. Electrotechnical Laboratory technical report 93-4. Tsukuba, Japan: Electrotechnical Laboratory.

Holland, John H. 1975. *Adaptation in Natural and Artificial Systems*. Ann Arbor, MI: University of Michigan Press.

Kinnear, Kenneth E. Jr. (editor). 1994. *Advances in Genetic Programming*. Cambridge, MA: The MIT Press.

Koza, John R. 1992. *Genetic Programming: On the Programming of Computers by Means of Natural Selection*. Cambridge, MA: MIT Press.

Koza, John R. 1994a. *Genetic Programming II: Automatic Discovery of Reusable Programs*. Cambridge, MA: MIT Press.

Koza, John R. 1994b. *Genetic Programming II Videotape: The Next Generation*. Cambridge, MA: MIT Press.

Koza, John R. 1995a. Evolving the architecture of a multi-part program in genetic programming using architecture-altering operations. In McDonnell, John R., Reynolds, Robert G., and Fogel, David B. (editors). 1995. *Evolutionary Programming IV: Proceedings of the Fourth Annual Conference on Evolutionary Programming*. Cambridge, MA: The MIT Press. 695–717.

Koza, John R., Andre, David, Bennett III, Forrest H, and Keane, Martin A. 1996. Use of automatically defined functions and architecture-altering operations in automated circuit synthesis using genetic programming. In Koza, John R., Goldberg, David E., Fogel, David B., and Riolo, Rick L. (editors). 1996. *Genetic Programming 1996: Proceedings of the First Annual Conference, July 28-31, 1996, Stanford University*. Cambridge, MA: The MIT Press.

Koza, John R., Bennett III, Forrest H, Andre, David, and Keane, Martin A. 1996a. Toward evolution of electronic animals using genetic programming. *Artificial*

Life V: Proceedings of the Fifth International Workshop on the Synthesis and Simulation of Living Systems. Cambridge, MA: The MIT Press.

Koza, John R., Bennett III, Forrest H, Andre, David, and Keane, Martin A. 1996b. Four problems for which a computer program evolved by genetic programming is competitive with human performance. *Proceedings of the 1996 IEEE International Conference on Evolutionary Computation.* IEEE Press. Pages 1–10.

Koza, John R., Bennett III, Forrest H, Andre, David, and Keane, Martin A. 1996c. Automated design of both the topology and sizing of analog electrical circuits using genetic programming. In Gero, John S. and Sudweeks, Fay (editors). *Artificial Intelligence in Design '96.* Dordrecht: Kluwer. 151-170.

Koza, John R., Bennett III, Forrest H, Andre, David, and Keane, Martin A. 1996d. Automated WYWIWYG design of both the topology and component values of analog electrical circuits using genetic programming. In Koza, John R., Goldberg, David E., Fogel, David B., and Riolo, Rick L. (editors). 1996. *Genetic Programming 1996: Proceedings of the First Annual Conference, July 28-31, 1996, Stanford University.* Cambridge, MA: The MIT Press.

Koza, John R., Goldberg, David E., Fogel, David B., and Riolo, Rick L. (editors). 1996. *Genetic Programming 1996: Proceedings of the First Annual Conference, July 28-31, 1996, Stanford University.* Cambridge, MA: The MIT Press.

Koza, John R., and Rice, James P. 1992. *Genetic Programming: The Movie.* Cambridge, MA: MIT Press.

Kruiskamp, Marinum Wilhelmus. 1996. *Analog design automation using genetic algorithms and polytopes.* Eindhoven, The Netherlands: Data Library Technische Universiteit Eindhoven.

Kruiskamp Marinum Wilhelmus and Leenaerts, Domine. 1995. DARWIN: CMOS opamp synthesis by means of a genetic algorithm. *Proceedings of the 32nd Design Automation Conference.* New York, NY: Association for Computing Machinery. Pages 433–438.

Mizoguchi, Junichi, Hemmi, Hitoshi, and Shimohara, Katsunori. 1994. Production genetic algorithms for automated hardware design through an evolutionary process. *Proceedings of the First IEEE Conference on Evolutionary Computation.* IEEE Press. Vol. I. 661-664.

Ohno, Susumu. 1970. *Evolution by Gene Duplication.* New York: Springer-Verlag.

Quarles, Thomas, Newton, A. R., Pederson, D. O., and Sangiovanni-Vincentelli, A. 1994. *SPICE 3 Version 3F5 User's Manual.* Department of Electrical Engineering and Computer Science, University of California, Berkeley, CA. March 1994.

Rutenbar, R. A. 1993. Analog design automation: Where are we? Where are we going? *Proceedings of the 15th IEEE CICC.* New York: IEEE. 13.1.1-13.1.8.

Sanchez, Eduardo and Tomassini, Marco (editors).1996.*Towards Evolvable Hardware.* Lecture Notes in Computer Science, Vol. 1062. Berlin: Springer-Verlag.

Thompson, Adrian. 1996. Silicon evolution. In Koza, John R., Goldberg, David E., Fogel, David B., and Riolo, Rick L. (editors). 1996. *Genetic Programming 1996: Proceedings of the First Annual Conference, July 28-31, 1996, Stanford University.* Cambridge, MA: MIT Press.

Van Valkenburg, M. E. 1982. *Analog Filter Design.* Fort Worth, TX: Harcourt Brace Jovanovich.

Machine Learning Approach to Gate-Level Evolvable Hardware

Hitoshi Iba   Masaya Iwata   Tetsuya Higuchi

1-1-4,Umezono,Tsukuba,Ibaraki,305,Japan
Electrotechnical Laboratory

Abstract   Evolvable Hardware (EHW) is a hardware which modifies its own hardware structure according to the environmental changes. EHW is implemented on a programmable logic device (PLD), whose architecture can be altered by downloading a binary bit string, i.e. *architecture bits*. The architecture bits are adaptively acquired by genetic algorithms (GA). The target task of EHW is "Boolean concept formation", which has been intensively studied in machine learning literatures. Although many evolutionary or adaptive techniques were proposed to solve this class of problems, there have been very few comparative studies from the viewpoint of computational learning theory. This paper describes machine learning approach to the gate-level EHW, i.e. 1) MDL-based improvement of fitness evaluation, and 2) comparative studies of efficiency by PAC criterion. We also discuss the current extension of EHW and related works.

1. Introduction

Evolvable Hardware (EHW) is a hardware built on a software-reconfigurable logic device, such as PLD (Programmable Logic Device) and FPGAs (Field Programmable Gate Arrays). EHW architecture can be reconfigured through the evolutionary method so as to adapt to the new environment. If hardware errors occur or a new hardware functionality is required, EHW can alter its own hardware structure in order to accommodate such changes.

The target task of the gate-level EHW is a Boolean concept formation. An n-variable Boolean function is defined as a function whose ranges and domain are constrained to have 0 (false) or 1 (true) values, i.e.

$$y = f(x_1, x_2, \cdots, x_n) = \begin{cases} 0, & \text{False value} \\ 1, & \text{True value} \end{cases} \qquad (1.1)$$

where

$$x_1 \in \{0,1\} \ \wedge \ x_2 \in \{0,1\} \ \wedge \ \cdots \ \wedge \ x_n \in \{0,1\}. \qquad (1.2)$$

The goal of Boolean concept formation is to identify an unknown Boolean function, from a given set of observable input and output pairs, i.e.,

$$\{(x_{i1}, x_{i2}, \cdots, x_{in}, y_i) \in \{0,1\}^{n+1} \mid i = 1, \cdots, N\}, \qquad (1.3)$$

where N is the number of observations. In general, N is less than the maximum possible number of distinct n-variable Boolean functions (2^{2^n}). Since the ratio of the size of the observable data to the size of the total search space i.e. ($\frac{N}{2^{2^n}}$) drastically decreases with n, effective generalizing (or inductive) ability is required for Boolean concept learning.

Although earlier algorithmic approaches to Boolean concept learning such as decision trees or enumeration [Anthony92] proved to be sound and complete, they suffered from computational complexity. Alternatively several stochastic or evolutionary methods have been proposed, which aim at improving efficiency by using probabilistic search at the expense of completeness. However there have been few comparative studies between their performances from the viewpoint of computational learning theory [Anthony92].

This paper describes machine learning approach to EHW, i.e., 1) MDL-based improvement for fitness evaluation of EHW, and 2) the comparative experiments in Boolean concept learning based on PAC (Probably Approximately Correctly) learnability theory.

The rest of this paper is structured as follows. Section 2 explains theoretical backgrounds of our approach, i.e. MDL and PAC. Section 3 describes the fundamental principle of EHW and its improvement by means of MDL-based evaluation. Section 4 shows the PAC-based comparative experiments in learning Boolean functions and discusses the results. Section 4 discusses the current extension of EHW and related works, followed by some conclusions.

2. Machine Learning Backgrounds

2.1 Complexity-based Fitness Evaluation

Complexity-based fitness is grounded on a *"simplicity criterion"*, which is defined as a limitation on the complexity of the model class that may be instantiated when estimating a particular function. For example, when one is performing a polynomial fit, it seems fairly apparent that the degree of the polynomial must be less than the number of data points. Simplicity criteria have been studied by statisticians for many years.

The complexity of an algorithm can be measured by the length of its minimal description in some language. The old but vague intuition of Occam's razor can be formulated as the *minimum description length criterion*, i.e., given some data, the most probable model is the model that minimizes the sum [Weigend94]:–

$$\text{MDL}(model) = \text{Desc\_L}(data\ given\ model) + \text{Desc\_L}(model) \longrightarrow min. \quad (2.1)$$

Desc_L(*data given model*) is the code length of the data when encoded using the model as a predictor for the data. The sum MDL(*model*) represents the tradeoff between residual error (i.e., the first term) and model complexity (i.e., the second term) including a structure estimation term for the final model. The final model (with the minimal MDL) is optimum in the sense of being a consistent estimate of the number of parameters while achieving the minimum error [Tenorio90].

More formally, suppose that z_i is a sequence of observations from the random variable Z, which is characterized by probability function $p_Z(\theta)$. The dominant form of the MDL is,

$$\text{MDL}(k) = -\log_2 p(z \mid \hat{\theta}) + \frac{k}{2} \log_2 N, \tag{2.2}$$

where $\hat{\theta}$ is the maximum likelihood estimate of θ, $p(z \mid \hat{\theta})$ is the likelihood of the estimated density function of $p_Z(\theta)$, k is the number of parameters in the model and N is the number of observations. The first term is the self-information of the model, which can be interpreted as the number of bits necessary to encode the observations. The second term can be also interpreted as the number of bits needed to encode the parameters of the model. Hence, the model which achieves the minimum of MDL is the most efficient model to encode the observations [Tenorio90,p.103].

Another criterion is the *Akaike information criterion* (AIC) [Akaike77]. The essential idea here is to figure out how many parameters, k, to include in a model. Minimizing AIC means minimizing k minus the log-likelihood function for the model, based on some assumed variance, $\hat{\sigma}^2$. In particular, if k is allowed to get too large, it does not matter that the likelihood of the data given a k-parameter model is very great; one will not achieve a minimal AIC. Unfortunately, the log-likelihood function cannot be calculated without an assumed family of distributions and a reasonable estimate of $\hat{\sigma}^2$. Nevertheless, the AIC has an important structural feature and that is the existence of a penalty term for the model complexity [Seshu94,p.220]. AIC is an approximation of the idealized Kullback-Leibler distance between the "true" data generating distribution and the model, which involves the expectation operation.

Assuming the above condition of $\{z_1, \cdots, z_N\}$, the AIC estimator is given as follows:-

$$AIC(k) = -2\ln p(z \mid \hat{\theta}) + 2k \tag{2.3}$$

By comparison with the MDL(k) criterion, we see that the difference is the crucial second term, k (AIC) versus $\frac{k}{2} \log_2 N$ (MDL). Therefore, the MDL(k) criterion penalizes the number of parameters asymptotically much more severely [Rissanen89,p.94]. Moreover, under some conditions, it is assumed that learning generally converges much faster for MDL than for AIC (see [Yamanishi92] for details).

Wallace proposed a similar measure called MML (i.e. Minimum Message Length). The coded form has two parts. The first states the inferred estimates of the unknown parameters in the model, and the second states the data using an optimal code based on the data probability distribution implied by those parameter estimates [Wallace87]. The total length might be interpreted as minus the log joint probability of estimate and data, and minimizing the length is therefore closely similar to maximizing the posterior probability of the estimate. MML is almost identical MDL. However, they differ in the implementations and philosophical views as to prior probabilities. The details of those differences can be found in [Wallace87] and its discussions.

The other criteria proposed are the *cross-validation* measure and the Maximum Entropy principle. It is shown that qualitatively and asymptotically the cross-validation criterion is equivalent to AIC. We may consider that the Maximum Entropy principle as a special case of the MDL principle, namely one where the model class is restricted to be of the special form. Within the statistical community, there is a considerable debate about both the proper viewpoint and the nature of the penalty term [Seshu94].

The goal shared by these complexity-based principles is to obtain accurate and parsimonious estimates of the probability distribution. The idea is to estimate the simplest density that has high likelihood by minimizing the total length of the description of the data. Barron introduced the index of resolvability, which may be interpreted as the minimum description-length principle applied on the average. It is has been shown that the rate of convergence of minimum complexity estimators is bounded by the index of resolvability [Barron91].

Another useful criterion proposed is PLS (i.e., Predictive Least Square) or PSE (i.e., Predicted Square Error) [Rissanen89,p.122]. This is mostly aimed at solving the "selection-of-variables" problem for linear regressions. The problem is solved by using the stochastic complexity and the sum of the prediction errors as a criterion, the latter either considered as an approximation of the former or as providing an independent extension of the LS principle. Rissanen described how to achieve the PLS solution to the posed regression problem, and revealed that the PLS criterion is a special case of the MDL principle. The detailed discussions are given in [Rissanen89,Ch.5]

The complexity-based fitness evaluation can be introduced in order to control genetic algorithms (GA) search strategies. For instance, when applying GAs to the classification of genetic sequences, [Konagaya93] employed the MDL principle for GA fitness in order to avoid overlearning caused by the statistical fluctuations. They presented a GA-based methodology for learning stochastic motifs from given genetic sequences. A stochastic motif is a probabilistic mapping from a genetic sequence (which has been drawn from a finite alphabet) to a number of categories (cytochrome, globin, trypsin, etc). They employed Rissanen's MDL principle in selecting an optimal hypothesis [Yamanishi91].

When applying the MDL principle to genetic programming, redundant structures should be pruned as much as possible, but at the same time premature convergence (i.e., premature loss of genotypic diversity) should be avoided [Zhang95]. Zhang proposed a dynamic control to fix the error factor at each generation and change the complexity factor adaptively with respect to the error. Let $E_i(g)$ and $C_i(g)$ denote the error and complexity of ith individual at generation g. Assuming that $0 \le E_i(g) \le 1$ and $C_i(g) \ge 0$, they defined the fitness of an individual i at generation g as follows:-

$$F_i(g) = E_i(g) + \alpha(g)C_i(g), \tag{2.4}$$

$\alpha(g)$ is called the adaptive Occam factor and expressed as

$$\alpha(g) = \begin{cases} \frac{1}{N^2} \frac{E_{best}(g-1)}{\check{C}_{best}(g)}, & \text{if } E_{best}(g-1) > \epsilon \\ \frac{1}{N^2} \frac{1}{E_{best}(g-1) \cdot \check{C}_{best}(g)}, & \text{otherwise} \end{cases} \tag{2.5}$$

where N is the size of training set. E_{best} is the error value of the program which has the smallest (best) fitness value at generation $g-1$. $\check{C}_{best}(g)$ is the size of the best program at generation g estimated at generation $g-1$, which is used for the normalization of the complexity factor. The user-defined constant ϵ specifies the maximum training error allowed for the final solution. They have shown the experimental results in the genetic programming of sigma-pi neural networks. There results were satisfactory.

In [Iba94,96], MDL-based fitness functions were applied successfully to system identification problems by using the implemented system STRO-GANOFF. The results showed that MDL-based fitness evaluation works well for tree structures in STROGANOFF, which controls GP-based tree search.

2.2 PAC-based Comparison Criterion

In order to evaluate the performance of EHW, we conduct the comparative experiments in Boolean concept learning based on PAC (Probably Approximately Correctly) learnability theory.

The theoretical background for our experiments is as follows. Let N be the number of attributes and K the number of literals needed to write down the smallest DNF (Disjunctive Normal Form) description of the target concept. Let ϵ be the maximum percentage error that can be tolerated during the testing task. The number of learning examples we used is given by the following formula:-

$$\frac{K \times log_2 N}{\epsilon}. \tag{2.6}$$

Qualitatively the formula indicates that we require more training examples as the complexity of the concept increases or the error decreases [Pagallo90].

In our experiments we set $\epsilon = 10\%$ and used 2000 examples to test classification performance. Thus 90% (2000 × 0.9 = 1800 examples) correctness of testing data is the expected learning success rate.

Table 2.1. Test Functions.

Name	description	attributes	#terms	#literals(K)	#training data
dnf3	random DNF	32	6	33	1650
def.	$x_1x_2x_6x_8x_{25}x_{28}\overline{x_{29}} \lor x_2x_9x_{14}\overline{x_{16}}\ \overline{x_{22}}\ \overline{x_{25}} \lor x_1\overline{x_4}\ \overline{x_{19}}\ \overline{x_{22}}x_{27}x_{28}$ $\lor \overline{x_2}x_{10}x_{14}\overline{x_{21}}\ \overline{x_{24}} \lor x_{11}x_{17}x_{19}x_{21}\overline{x_{25}} \lor \overline{x_1}\ \overline{x_4}x_{13}\overline{x_{25}}$				
mx6	6-multiplexor	16	4	12	720[1]
def.	$x_{13}x_{16}x_1 \lor \overline{x_{13}}x_{16}x_7 \lor x_{13}\overline{x_{16}}x_4 \lor \overline{x_{13}}\ \overline{x_{16}}x_{10}$				
par4	4-parity	16	8	32	1280
def.	$x_1 \oplus x_5 \oplus x_9 \oplus x_{13}$ (where $\oplus$ is the XOR operator)				

We used 3 problems (target concepts) shown in Table 2.1 [Pagallo90]; dnf3 (randomly generated DNF, 32 attributes, 6 terms), mx6 (6-multiplexor, 16 attributes with 10 irrelevant attributes), and par4 (4-parity, 16 attributes with 12 irrelevant attributes).

The number of training data is derived from equation (2.6); i.e. 1280 ($= \frac{32 \times log_2 16}{0.1}$) training data are given for par4[1].

3. Evolving Hardware by GA

There is a clear distinction between a conventional hardware (CHW) and EHW. A designer can begin to design a CHW only after its detailed specification is given. In this sense, CHW is a top-down approach. However, EHW is applicable even when no hardware specification is known beforehand. EHW implementation is determined through a genetic learning in a bottom-up way. Thus, EHW will be applied totally differently from CHW. EHW is suitable for problem domains where both on-line adaptation and real-time response are required.

The basic idea of EHW is to regard the architecture bits of a PLD as a chromosome for GA (see Fig. 3.1). The hardware structure is adaptively searched by GA. These architecture bits, i.e. the GA chromosome, are downloaded onto a PLD, on and after the genetic learning. Therefore, EHW can be considered as an on-line adaptive hardware.

3.1 Fundamental Principle of Gate-Level EHW

This section describes the genetic learning method for evolving a hardware. We call our evolutionary method as a *gate-level EHW*, because PLD gates used are fairly primitive ones such as AND gates.

---

[1] The correct number for mx6 is 480 according to the equation (2.6). However, we used the number given in the original table [Pagallo90,p.91].

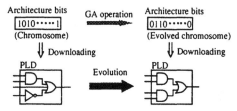

Figure 3.1. Evolvable Hardware (EHW)

3.1.1 Programmable Logic Device (PLD). A PLD consists of *logic cells* and a *fuse array* (see Fig. 3.2). In addition, architecture bits determine the architecture of the PLD. These bits are assumed to be stored in an architecture bit register (ABR). Each link of the fuse array corresponds to a bit in the ABR.

The fuse array determines the interconnection between the device inputs and the logic cell. It also specifies the logic cell's AND-term inputs. If a link on a particular row of the fuse array is switched on, which is indicated by a black dot in Fig. 3.2, then the corresponding input signal is connected to the row. In the architecture bits, these black and white dots are represented by 1 and 0 respectively.

Consider the example PLD shown in Fig. 3.2. The first row indicates that I_0 and $\overline{I_2}$ are connected by an AND-term, which generates $I_0\overline{I_2}$. Similarly, the second row generates I_1. These AND-terms are connected by an OR gate. Thus, the resultant output is $O_0 = I_0\overline{I_2} + I_1$.

As mentioned above, both of the fuse array and the functionality of the logic cell are represented in a binary string. The key idea of EHW is to regard this binary bit string as a chromosome for the sake of GA-based adaptive search.

The hardware structure we actually use is a FPLA device, which is a commercial PLD (Fig. 3.3). This architecture mainly consists of an AND and OR arrays. A vertical line of the OR array corresponds to a logic cell in Fig. 3.2.

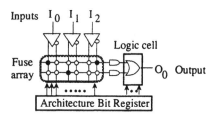

Figure 3.2. A simplified PLD (Programmable Logic Device) Structure

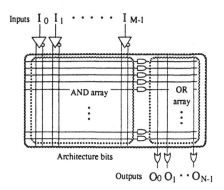

Figure 3.3. A FPLA Architecture for EHW

3.1.2 Variable length chromosome GA (VGA). In our earlier works, the architecture bits were regarded as the GA chromosome and the chromosome length was fixed. In spite of this simple representation, the hardware evolution was successful for combinatorial logic circuits (e.g. 6-multiplexer) and sequential logic circuits (e.g. 4-state machine, 3-bit counter [Higuchi94]).

However, this straightforward representation had a serious limitation in the hardware evolution. All the fuse array bits should have been included in the genotype, even when effective bits in the fuse array were only a few. This made the chromosome too long to be effectively searched by evolution.

Therefore, we have introduced a variable length chromosome called *VGA*, i.e. Variable length chromosome GA [Kajitani95]. VGA is expected to evolve a large circuit more quickly. The chromosome length of VGA is smaller than the previous GA, especially when evolving a circuit with large inputs. This is because VGA can deal with a part of architecture bits, which effectively determine the hardware structure. Because of this short chromosome, VGA can increase the maximum circuit size and establish an efficient adaptive search.

The coding method of VGA is as follows. An allele in a chromosome consists of a location and a connection type. The location is the position of the allele in the fuse array. For example, the fuse array in Fig. 3.4 (a) has 14 locations as shown in Fig. 3.4 (b). Therefore, the locations of the connected points in Fig. 3.4 (a) are denoted as 0, 4, 8, 9, 13 and 14. The connection type defines the input to be either positive or negative. For example, the connection type at location 0 is 1, i.e. the positive input. Thus, the chromosome for Fig. 3.4 (a) is (0,1) (4,1) (8,2) (9,1) (13,1) (14,1).

We use the roulette wheel selection strategy. Recombination operators are *cut* and *splice*, which are used in the messy GA [Goldberg93]. A mutation operator is applied so as to change the values of the location and the connection type randomly. Splice operator concatenates two chromosomes.

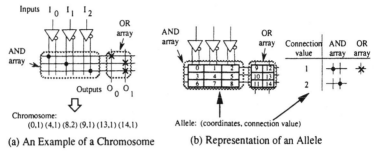

(a) An Example of a Chromosome

(b) Representation of an Allele

Figure 3.4. Chromosome Representation of Variable Length Chromosome GA

3.2 MDL-based Improvement

The fitness of GA is basically evaluated in terms of the output correctness for the training data. In addition, we introduce an MDL (Minimum Description Length) based fitness [Rissanen89] for evolving EHW. As can seen in the application of EHW to pattern recognitions [Iwata96], the robustness of the generated hardware is achieved as a result of this improvement.

3.2.1 MDL-based Fitness for EHW. We have introduced the above MDL criterion into the GA fitness evaluation. The purpose is to establish a robust learning method for EHW. In general, the greater the number of "don't care" [2] inputs, the more robust (i.e. noise-insensitive) the evolved hardware. Thus, we regard the number of "don't care" inputs as an index of MDL.

More formally, the MDL value for our EHW is written as follows:

$$\text{MDL} = A_c \log(C + 1) + (1 - A_c) \log(E + 1), \tag{3.1}$$

where C denotes the complexity of the EHW. E is the error rate of the EHW's output. Usually, A_c is increased with generations (see Table 4.4).

The C value (i.e. the complexity of the EHW) determines the performance of the MDL. We introduce three types of C definitions as follows:-

$$C_1 = \sum_i |AND_{Oi}|, \tag{3.2}$$

$$C_2 = |AND| \times |OR|, \tag{3.3}$$

$$C_3 = \sum_i |AND_{Oi}| \times |OR_{Oi}|. \tag{3.4}$$

---

[2] We call an input "don't care" if it is not included in the output expression. For instance, if $O = I_1 + I_2$ in case of a PLD shown in Fig. 3.2, then I_0 is a "don't care" input.

Table 4.1. Parameters for Classifier Systems.

Population Size	400
Crossover Rate	12%
Crossover TYPE	One-Point
Mutation Rate	0.1%
Payoff Quantity (R)	1000
Decay by Error (e)	80%
Bias for # (G)	4.0
Reference	**Boole** [Wilson87]

Table 4.2. Parameters for ALN.

Initial Nodes	29999
Node Types	AND, OR, LEFT, RIGHT
Reference	[Armstrong79]

Where $| AND_O |$ and $| OR_O |$ are the numbers of ANDs and ORs connected to the output O. $| AND |$ ($| OR |$) is the number of ANDs (ORs) on the AND (OR) array. Consider Fig. 3.4(a) for instance. ANDs and ORs are represented as black dots and $\times$ marks in the figure. The values of C_1, C_2 and C_3 are 3 ($= 1+2$), 9 ($= 3 \times 3$) and 5 ($= 1 \times 1 + 2 \times 2$) respectively, because $| AND_{O0} |$, $| OR_{O0} |$, $| AND_{O1} |$, $| OR_{O1} |$, $| AND |$, and $| OR |$ are 1, 1, 2, 2, 3 and 3.

The definition of C_1 is not very precise because it does not include the information of OR gates. On the other hand, C_2 and C_3 are expected to give more exact MDL values. We will show the comparative experiments by using these different MDL definitions in section 2.2.

The above MDL value is normalized so that it satisfies $0 \leq \text{MDL} \leq 1$. That is, the GA fitness is defined to be:

$$\text{Fitness} = 1 - \text{MDL}. \tag{3.5}$$

4. Experimental Results

In order to evaluate the performance of EHW, this section describes the comparative experiments in Boolean concept learning based on PAC (Probably Approximately Correctly) learnability theory. We compare the performance of EHW with those by neural networks (NN) [Rumelhart86], classifier systems (CS) [Wilson87] and adaptive logic networks (ALN) [Armstrong79].

All methods, i.e. CS, ALN, NN and EHW, were run according to standard operational criteria. The parameters shown in Tables 4.1, 4.2, 4.3 and 4.4 were used for each method.

Table 4.3. Parameters for Neural Networks.

Learning Rate	0.01
Momentum	0.5
# of Hidden Layers	1
# of Hidden Nodes	4 (3 for dnf3)
Reference	[Rumelhart86]

Table 4.4. Parameters for EHW.

Population Size		100
Maximum Generation		2000
Initial Chromosome Length		100
Cut and Splice Rate		0.01
Mutation Rate		0.01
A_C Coefficient	Gen:0–1000	0.0
	Gen:1001–1500	0.025
	Gen:1501–2000	0.05
Reference		[Higuchi94]

These parameters were chosen to obtain the most effective learning results after several experimental runs. Learning was terminated after convergence is attained. Thus the numbers of iterations needed in training phases differ for the 4 methods; i.e. $O(10000)$ for NN, $O(1000)$ for CS, $O(100)$ for ALN and $O(1000)$ for EHW. However, this number did not necessarily reflect the computational complexity, because each iteration included qualitatively different computations. We executed several independent runs for each test function.

We conducted experiments with the learning of both noiseless and noisy Boolean concepts. In noisy environments, learning attribute values are inverted from 1 to 0 or from 0 to 1 (with a probability less than 5%).

Tables 4.5 and 4.6 show the averages and the standard deviations of correctness for training and testing data for ten runs. Note that following equation (2.6)the success rate for this Boolean learning is expected to be above 90%. Although we cannot make any concluding remarks as to which is the method, the following points should be emphasized:

1. **Neural Networks (NN)**

   NN copes with noise relatively successfully; i.e. so called "graceful degradation" was observed. However, in noiseless cases (i.e. 0% noise), NN dose not always succeed in learning the training data. NN shows poor results for mx6 or par4. Although it is widely believed that NN performs boolean concept learning well, no significant superiority of NN was observed. This is because the distributed representations prevent NN from distinguishing between relevant and irrelevant attributes for mx6 and par4. Dnf3 is a hard problem for NN.

Table 4.5. Learning Performances (CS, ALN, and NN).

Noise	Func.	CS Train		Test		ALN Train		Test	
	mx6	100.0	0.0	100.0	0.0	100.0	0.0	98.9	0.7
0%	par4	100.0	0.0	100.0	0.0	100.0	0.0	98.6	1.3
	dnf3	90.0	1.8	87.8	3.1	100.0	0.0	87.6	1.6
	mx6	100.0	0.0	100.0	0.0	96.4	0.6	95.5	3.5
2%	par4	98.2	2.3	97.1	3.3	92.6	0.8	99.9	0.3
	dnf3	71.0	27.7	66.2	31.9	96.4	0.6	86.4	1.2
	mx6	98.3	2.5	98.4	2.4	90.9	1.2	99.8	0.4
5%	par4	44.8	2.2	36.6	1.8	74.4	6.7	71.5	23.7
	dnf3	27.0	29.0	20.3	32.2	90.1	0.8	89.9	2.5

Noise	Func.	NN Train		Test	
	mx6	99.0	1.2	98.7	1.4
0%	par4	89.1	12.7	85.9	18.3
	dnf3	96.7	0.8	92.7	3.0
	mx6	95.7	1.0	95.6	1.0
2%	par4	84.8	11.4	81.9	17.2
	dnf3	94.5	1.4	92.4	1.7
	mx6	90.7	1.1	90.1	0.5
5%	par4	76.8	8.8	74.7	13.1
	dnf3	92.0	0.8	90.7	0.8

Table 4.6. Learning Performances (EHW w and w/o MDL).

Noise	Func.	EHW(w/o MDL) Train		Test		EHW(w MDL) Train		Test	
	mx6	95.90	6.96	94.75	8.75	98.61	2.16	97.95	2.98
0%	par4	70.26	7.35	62.95	11.05	68.74	4.25	62.11	6.31
	dnf3	91.57	2.44	91.05	2.65	90.28	3.42	89.72	3.20
	mx6	94.11	5.16	96.45	3.89	94.68	4.29	96.17	6.29
2%	par4	68.74	6.48	70.91	11.18	74.09	6.75	71.50	10.64
	dnf3	88.56	2.11	88.19	2.93	87.10	2.67	85.28	3.64
	mx6	88.17	5.02	96.03	4.93	86.33	8.91	92.20	5.43
5%	par4	63.92	1.88	54.05	4.66	64.22	3.47	60.62	6.58
	dnf3	86.75	1.39	84.93	2.29	88.43	1.84	87.90	3.51

2. Classifier Systems (CS)

CS is superior to the other methods for mx6 and par4. CS can cope with the irrelevant attributes. Actually CS is successful in acquiring a perfect set of rules for mx6. For instance, the acquired rules are as follows:-

Condition ($x_1\ x_2\ x_3\ \cdots\ x_{16}$)	Action	Strength
# # # 0 # # # # # # # # 1 # # 0	0	5620
# # # # # # 0 # # # # # 0 # # 1	0	5526
# # # # # # # # # 1 # # 0 # # 0	1	5512
# # # # # # 1 # # # # # 0 # # 1	1	5503
1 # # # # # # # # # # # 1 # # 1	1	4222
0 # # # # # # # # # # # 1 # # 1	0	4090
# # # 1 # # # # # # # # 1 # # 0	1	3633
# # # # # # # # # 0 # # 0 # # 0	0	3060

Notice that these rules express the concept of mx6 by using significant bits $(x_1, x_4, x_7, x_{10}, x_{13}, x_{16})$ and ignoring the irrelevant attributes. On the other hand, CS fails to solve dnf3. This is because it is difficult for CS to represent the concept of dnf3 in the form of classifier rules. So many classifier rules are required to express 0-valued actions for dnf3 whereas # (wild-card) works very well for par4 and mx6. Therefore the rule size is an important factor for CS. For mx6 and par4, 400 rules were enough. On the other hand, $O(1000)$ rules were necessary for dnf3.

CS has poor records abruptly when the noise level exceeds 2%. Considering their high deviations, the performance of CS is not stable; that is, results of CS are likely to be influenced by noise.

3. **Adaptive Logic Networks (ALN)**

ALN performs better for all 3 tests in noiseless cases. However, considering that the average performance is below 90% for dnf3, ALN was not successful in generalizing the training data. This results from the fact that ALN simply memorizes part of the training data, and lacks the ability to generalize. For these reasons, ALN, in general, requires a large number of initial nodes (for instance, [Armstrong79] used $O(60000)$). Although the final node size might well be reasonable $(O(100))$, a small number of initial nodes results in failure. It should be noted that ALN's performance is heavily dependent upon the problem size. For example, ALN failed to solve par5 (5-bit parity problem with 27 irrelevant bits) Besides, as can be seen in mx6 (2% and 5% noise) and par4 (0% and 2%), overfitting phenomena were observed for ALN.

4. **EHW without MDL** (Table 4.6)

EHW gave satisfactory results for mx6, whereas it failed in learning par4 and dnf3. This is because the representation used by EHW is not suitable for expressing a long disjunctive clause. For instance, in one run the best expression acquired at the 1991th generation was as follows:-

$$x_0\overline{x_1}\overline{x_4}x_5x_6\overline{x_8}x_9\overline{x_{10}}\overline{x_{15}} \vee \overline{x_0}\overline{x_4}\overline{x_8}x_{12} \vee x_2\overline{x_3}x_4x_5\overline{x_{10}}x_{12}x_{13}\overline{x_{15}} \vee$$
$$\overline{x_0}\overline{x_4}x_7x_8\overline{x_{12}} \vee x_0x_1x_2\overline{x_3}x_4x_6\overline{x_9}x_{10}x_{11}\overline{x_{12}}x_{13}\overline{x_{15}} \vee$$
$$x_0x_2\overline{x_4}x_7\overline{x_8}\overline{x_{12}} \vee x_0\overline{x_4}x_8\overline{x_{10}}x_{12} \vee x_0x_4\overline{x_8}x_{12} \vee x_0x_4x_8\overline{x_{12}} \vee$$
$$\overline{x_0}x_4x_8x_{12} \vee x_0\overline{x_4}x_6\overline{x_8}x_9x_{11}\overline{x_{12}} \vee \overline{x_2}\overline{x_5}x_6x_8\overline{x_{10}}x_{11}x_{12}. \qquad (4.1)$$

The correctness of this expression was 84.70%. The chromosome length is 200. Considering that the correct par4 has 8 terms in a principle disjunctive canonical form and that each term consists of 4 variables, EHW seems to have fallen in a local extreme in this case.

In noisy environments, EHW performed slightly better than CS. However, the results were not necessarily satisfactory.

5. **EHW with MDL** (Table 4.6)

The performance of MDL-based EHW was better than non-MDL EHW in almost all cases. The generalization effect, i.e. the avoidance of overfitting, can be seen in mx6. For example, the typical expression acquired by non-MDL EHW was as follows:-

$$x_{10}\,\overline{x_{13}}\,\overline{x_{16}} \ \lor\ \overline{x_1}\,\overline{x_2}\,\overline{x_3}\,x_4\,\overline{x_8}\,\overline{x_{12}}\,x_{13}\,x_{14}\,\overline{x_{16}} \ \lor$$
$$x_1\,\overline{x_2}\,x_3\,\overline{x_5}\,x_6\,x_7\,\overline{x_9}\,\overline{x_{13}}\,x_{15} \ \lor\ x_1\,x_4\,\overline{x_5}\,\overline{x_{12}}\,x_{13}\,\overline{x_{14}} \ \lor$$
$$x_3\,x_5\,x_6\,x_7\,\overline{x_9}\,\overline{x_{10}}\,\overline{x_{12}}\,x_{16} \ \lor\ \overline{x_2}\,\overline{x_4}\,x_7\,x_8\,\overline{x_{11}}\,x_{14}\,\overline{x_{15}}\,x_{16} \ \lor$$
$$x_1\,x_{13}\,x_{16} \ \lor\ x_7\,x_8\,\overline{x_9}\,\overline{x_{12}}\,\overline{x_{13}}\,\overline{x_{14}} \ \lor\ \overline{x_2}\,\overline{x_3}\,x_4\,\overline{x_5}\,\overline{x_6}\,x_7\,x_9\,x_{13}\,x_{14}\,x_{16} \ \lor$$
$$x_4\,x_{13}\,\overline{x_{14}} \ \lor\ x_7\,\overline{x_{13}}\,x_{16} \ \lor\ \overline{x_2}\,\overline{x_3}\,\overline{x_4}\,x_5\,x_6\,\overline{x_7}\,\overline{x_{10}}\,\overline{x_{12}}\,\overline{x_{14}}\,\overline{x_{15}}, \qquad (4.2)$$

whereas MDL-based EHW resulted in the following expression,

$$x_{13}\,x_{16}\,x_1 \ \lor\ \overline{x_{13}}\,x_{16}\,x_7 \ \lor\ x_{13}\,\overline{x_{16}}\,x_4 \ \lor\ \overline{x_{13}}\,\overline{x_{16}}\,x_{10}. \qquad (4.3)$$

Although both expressions gave 100% correct outputs for the training data, yet non-MDL EHW did not succeed in learning the target expression, i.e. the correctness of the equation (4.2) was 99.4% for the testing data. Note that the equation (4.3) is equivalent to the definition of mx6 (see Table 2.1). This shows the more compact expression has been acquired by means of MDL-based fitness, which clearly shows the success of complexity-based evaluation of EHW.

In summary, we cannot conclude that EHW is superior to the other learning methods. The methods we compared can be classified roughly into analog approaches (NN) and into digital approaches (ALN, CS). As we have observed, both approaches have their own merits and demerits. In order to improve the EHW learning ability, we are currently working on the extension of EHW, which integrates analog and digital approaches.

5. Discussion and Conclusion

EHW has been applied to high-speed pattern recognition in order to establish a robust system in noisy environments [Iwata96]. This ability, i.e. robustness, seems to be the main feature of ANN. ANN is mostly run in a software-based way, i.e. executed by a workstation. Thus, current ANN may have difficulty with real-time processing because of the speed limit of the software-based execution.

Another desirable feature of EHW is its readability. The learned result by EHW is expressed as a Boolean function, whereas ANN represents it as thresholds and weights. Thus, the acquired result of EHW is more easily understood than that of ANN. We believe that this understandable feature leads to wider usage of EHW in industrial applications.

For the sake of achieving flexible recognition capability, it is necessary to cope with a pattern which is classifiable not by a linear function, but by a non-linear function. We have conducted an experiment in learning the exclusive-OR problem in order to check the above capability. From the simulation result, we confirmed that EHW can learn non-linear functions successfully [Murakawa96]. In other words, EHW is supposed to fulfill the minimum requirement towards the robust pattern recognition.

The EHW described so far is based on the hardware evolution at *gate-level*, in the sense that each gene of a chromosome corresponds to a primitive gate such as an AND gate or an OR gates. However, because of the limitation of the GA execution time, the size of a circuit allowable for this gate-level evolution may not be large enough for practical applications. If hardware is genetically synthesized from higher-level functions (e.g. adder, subtracter, sine generator, etc.), more practical facilities can be provided by EHWs.

Recently ETL initiated a function-level evolution approach, in which each gene corresponds to a real function such as a floating multiplication and a sine function. The function-level evolution using the FPGA model is described in [Murakawa96]. The FPGA model consists of 20 columns, each containing seven hardware functions; an adder, a subtracter, an if-then, a sine generator, a cosine generator, a multiplier, and a divider. Columns are interconnected by crossbar switches. The crossbars designate inputs to columns. In addition to these columns, a state register maintains a past output for dealing with temporal data. There are two inputs and one output. Data handled with this FPGA are floating point numbers.

Function-level evolution is expected to attain the satisfactory performance comparable to the neural networks. We have confirmed the validity of this model by experimenting with various problems, such as the classification of iris data and distinguishing two interwined spirals (for details of these experiments, refer to [Murakawa96]).

This paper presented a machine learning approach to gate-level EHW, i.e.,

1. MDL-based improvement of fitness evaluation has been introduced for the sake of the robustness of EHW.
2. Comparative studies have been constructed from a viewpoint of PAC learning theory, in order to evaluate the performance of several adaptive methods.

We believe this is a step toward the integration of the practical field, i.e. EHW, and the theoretical field, i.e. machine learning.

References

[Akaike77] Akaike, H., On the Entropy Maximization Principle, in *Applications of Statistics*, Krishnaiah,P.R. (ed.), North-Holland, 1977

[Anthony92] Anthony, M. and Biggs, N., Computational Learning Theory, Cambridge Tracts in Theoretical Computer Science 30, Cambridge, 1992

[Armstrong79] Armstrong, W.W. and Gecsei, J., Adaptation Algorithms for Binary Tree Networks, *IEEE TR. SMC*, SMC-9, No.5, 1979

[Barron91] Barron,A.R. Minimum Complexity Density Estimation, in *IEEE Tr. on Information Theory*, vol.37, no.4, pp.1034–1054, 1991

[Goldberg93] Goldberg, D. et al., Rapid Accurate Optimization of Difficult Problems using Fast Messy Genetic Algorithms, Proc. 5th Int. Joint Conf. on Genetic Algorithms (ICGA93), 1993

[Higuchi94] Higuchi, T. et al., Evolvable Hardware with Genetic Learning, in Massively Parallel Artificial Intelligence (eds. H. Kitano), MIT Press, pp 398–421, 1994.

[Iba94] Iba, H., deGaris, H. and Sato, T., Genetic Programming using a Minimum Description Length Principle, in *Advances in Genetic Programming*, (ed. Kenneth E. Kinnear, Jr.), MIT Press, pp.265–284, 1994

[Iba96] Iba, H., deGaris, H. and Sato, T., Numerical Approach to Genetic Programming for System Identification, in *Evolutionary Computation*, vol.3, no.4, 1996

[Iwata96] Iwata, M., Kajitani, I., Yamada, H., Iba, H. and Higuchi, T., A Pattern Recognition System Using Evolvable Hardware, in *Proc. of Parallel Problem Solving from Nature IV* (Berlin, Germany: Springer-Verlag), 1996

[Kajitani95] Kajitani, I. et al., Variable Length Chromosome GA for Evolvable Hardware in *Proc. of 3rd Int. Conf. on Evolutionary Computation (ICEC96)*, 1996 .

[Konagaya93] Konagaya, A. and Kondo, H., Stochastic Motif Extraction using a Genetic Algorithm with the MDL Principle, in *Hawaii Int. Conf. of Computer Systems*, 1993

[Murakawa96] Murakawa, M., Yoshizawa, S., Kajitani, I., Furuya, T., Iwata, M. and Higuchi, T., Hardware Evolution at Function Levels, *Proc. of Parallel Problem Solving from Nature IV* (Berlin, Germany: Springer-Verlag), 1996

[Pagallo90] Pagallo, G. and Hausslear, D. Boolean Feature Discovery in Empirical Learning, *Machine Learning*, vol.5, 1990

[Rissanen 89] Rissanen, J., Stochastic Complexity in Statistical Inquiry, World Scientific, 1989

[Rumelhart86] Rumelhart, D.E. and McClelland, J.L. Parallel Distributed Processing, MIT Press, 1986

[Seshu94] Seshu,R., Binary Decision Trees and an "Average-case" Model for Concept Learning: Implications for Feature Construction and the Study of Bias, in *Computational Learning Theory and Natural Learning Systems, Volume I: Constraints and Prospects*, Hanson, J. et al. (eds.), MIT Press, pp.213–248, 1994

[Tenorio *et al.*90] Tenorio, M.F. and Lee, W. Self-organizing Network for Optimum Supervised Learning, *IEEE Tr. Neural Networks*, vol.1, no.1, pp.100–109, 1990

[Wallace*et al.*87] Wallace,C.S. and Freeman,P.R. Estimation and Inference by Compact Coding, in *Journal of Royal Statistical Society*, Ser.B, vol.49, no.3, pp.240–265, 1987

[Weigend94] Weigend, A.S. and Rumelhart,D.E., Weight Elimination and Effective Network Size, in *Computational Learning Theory and Natural Learning Systems, Volume I: Constraints and Prospects*, Hanson, J. et al. (eds.), MIT Press, pp.457–476, 1994

[Wilson87] Wilson, S.W., Classifier Systems and the Animat Problem, *Machine Learning*,vol.2, no.3, 1987

[Yamanishi*et al.*91] Yamanishi, K., and Konagaya, A., "Learning Stochastic Motifs from Genetic Sequences", in *Proc. of 8th International Workshop on Machine Learning*, pp.467–471, 1991

[Yamanishi92] Yamanishi, K., A Learning Criterion for Stochastic Rules, in *Machine Learning*, vol.9, pp.165–204, 1992

[Zhang95] Zhang, B.-T. and Mühlenbein, H., Balancing Accuracy and Parsimony in Genetic Programming, in *Evolutionary Computation*, vol.3, no.1, pp.17–38, 1995

Evolvable Systems in Hardware Design: Taxonomy, Survey and Applications

Ricardo S. Zebulum †   Marco Aurélio Pacheco †§   Marley Vellasco †§

† ICA: Núcleo de Pesquisa em
Inteligência Computacional Aplicada
Departamento de Engenharia Elétrica,
PUC-Rio
Rua Marquês de S. Vicente, 225
CEP 22453-900, Rio de Janeiro - Brasil
E-mail: ICA@ele.puc-rio.br

§ Departamento de Sistemas de Computação
Universidade do Estado do Rio de Janeiro
Rua S. Francisco Xavier, 524/5º andar

CEP 20550-013, Rio de Janeiro - Brasil

ABSTRACT

This article proposes a taxonomy, presents a survey and describes a set of applications on Evolvable Hardware Systems (EHW). The taxonomy is based on the following properties: ***Hardware Evaluation Process, Evolutionary Approach, Target Application Area*** *and* ***Evolving Platform.*** *Recent reported applications on EHW are also reviewed, according to the proposed taxonomy. Additionally, a set of digital design applications, developed by the authors are presented. The applications consist in evolving basic digital devices, and the main objective is to evaluate the performance of an EHW system in terms of chromosome representation and evaluation.*

1 - INTRODUCTION

The evolutionary engineering approach to electronic circuits design has been intensively investigated recently [2, 5, 6, 10]. The idea of encoding elements and connections of programmable integrated circuits [7] into the chromosomes of a Genetic Algorithm (GA), has created a promising research area called Evolvable Hardware [12]. Evolvable Hardware (EHW) extends the concepts of genetic algorithms to the evolution of electronic circuits, capable of producing a correct (or approximately correct) output. Each possible electronic circuit can be represented as an individual or a chromosome of an evolutionary process, which performs standard genetic operations over the circuits, such as *selection*, *crossover* and *mutation*.

Hardware evolution and self-reproducing hardware is an alternative form of system conception that dispenses with the conventional hardware design methodology in solving complex problems in a variety of applications areas. The current belief is that the discovery of suitable evolutionary hardware design methodologies will lead to novel evolvable hardware systems with the ability to deliver powerful solutions in areas that challenges conventional design approaches. Although most of the work in this area has been on small-scale problems, evolvable hardware promises to be extensible to real world timed and sized problems in hardware design.

This article serves a threefold purpose. First, we propose a taxonomy of Evolvable Hardware Systems in section 2. In spite of being a new research area, Evolvable Hardware already incorporates a large set of applications, ranging from pattern recognition [12] to robotics [5]. In section 2, we classify these applications according to four important properties. Secondly, in section 3, we review some of the most significant projects supporting evolutionary hardware, which are presented according to the categories defined in section 2. Finally, in section 4, we describe a set of experiments being carried out at PUC-Rio, Brazil, to investigate the potential of hardware evolution. These experiments consist of genetically evolving basic digital devices, such as decoders, multiplexers, adders and an arbitrary string generator.

Section 5 presents the conclusion of the work and discusses some ideas related to possible future applications of Evolvable Hardware.

2 - TAXONOMY OF EVOLVABLE HARDWARE

Many different classes of evolvable hardware systems have been proposed [2]. Table 4 in section 3 presents some of the most notable EHW systems, together with their properties.

According to the current work in this area we can characterise EHW systems by four key properties:

1. *Hardware Evaluation Process;*
2. *Evolutionary Programming Approach;*
3. *Target Application Area;*
4. *Evolving Platform.*

In order to better visualise the proposed taxonomy, Figure 1 presents the complete taxonomy structure. The following sections discuss each of the properties mentioned above.

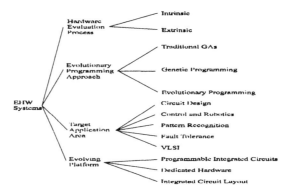

Figure 1 - Tree representation of the EHW Systems Taxonomy

2.1 - Hardware Evaluation Process

A possible way to define Evolvable Hardware is as being an integration of reconfigurable devices with genetic learning [12]. The Hardware Evaluation Process property deals with the kind of environment where chromosomes are evaluated during the genetic learning. Currently, there is a standard classification adopted by most authors [2, 10, 11, 12]:

- *Intrinsic EHW*
- *Extrinsic EHW*

In the Intrinsic EHW approach, evolution takes place in the reconfigurable device. Chromosomes or genotypes are downloaded into the reconfigurable device and are evaluated usually through hardware mechanisms. There are two major advantages in using this approach [2, 5, 6]: the speed of the evolution process and, more importantly, the fact that no simulation models are needed to evaluate the circuit. Indeed, in this case, all the solutions will behave exactly as they would do in a real time application [2].

The Extrinsic Approach consists of using software simulation models to evaluate evolving circuits. At the end of the evolutionary process, the final solution (best individual) is downloaded into the reconfigurable device. The Extrinsic Approach is simpler and, in many cases, the only practicable way to evaluate the fitness of a large number of individuals. Additionally, this approach helps to draw theoretical results into the Evolvable Hardware area.

2.2 - Evolutionary Approach

Evolvable Hardware uses artificial evolution to design circuits, but the way which researchers implement their evolutionary algorithm may vary. There are basically three different approaches.

1. *Traditional Genetic Algorithms;*
2. *Genetic Programming;*
3. *Evolutionary Programming.*

Traditional Genetic Algorithms (GAs), as they were originally proposed by John Holland [17], have the following basic features:
• Binary encoding to form chromosomes;
• One-point crossover and mutation as genetic operators;
• Generational replacement of population with elitism.

The main advantage of the traditional GAs is that they provide maximum number of schemas (building blocks) per unity of information. However, the binary representation can lead to large chromosomes in some applications.

In the genetic programming method (GP) [8], chromosomes are represented as trees with ordered branches, in which the internal nodes are functions and the leaves are terminals (variables and operands). In this case, crossover consists of exchanging sub-trees between chromosomes [8].

The Evolutionary Programming approach (EP) differs from Genetic Algorithms mainly in the way chromosomes are encoded and in the genetic operators used. Instead of using the binary representation, the EP approach usually adopts integers or real numbers to form the chromosomes. Additionally, EP uses special genetic operators which incorporate problem specific knowledge [24]. This approach may lead to shorter chromosomes and higher performance evolutionary systems due to the special genetic operators.

Besides these three basic methods, there are other alternative evolutionary programming approaches, not so commonly used, such as VGAs, *Variable Length Chromosomes Genetic Algorithms* [12], and PGAs, *Production Genetic Algorithms* [20].

2.3 - Target Application Area

As electronics itself involves many areas, the concepts of evolution can also be applied to a broad class of Evolvable Hardware applications. Evolvable Hardware comprises a variety of applications that can be classified according to the following categories:

1. *Circuit Design (Digital and Analog)*
2. *Control and Robotics*
3. *Pattern Recognition*
4. *Fault Tolerance*
5. *VLSI*

This classification is based on the implementations currently reported in the literature. In the following years, this number is likely to increase. In the next sub-sections we describe the main features of each category.

2.3.1 - Circuit Design

Electronic circuit design, either analog or digital, is the prime application of evolutionary electronics. The availability of a large range of circuit simulators for extrinsic experiments, as well as programmable devices, such as FPGAs, used in intrinsic EHW systems, has allowed researchers to study EHW systems.

2.3.2 - Control and Robotics

This is possibly one of the most promising fields of EHW, if we consider the number of sub-systems in control applications that can be evolved, such as Finite State Machines [5] and other basic models [12]. The particular area of *Evolutionary Robotics* [21], in which evolutionary techniques are used to develop control systems of robots [15], is also being intensively studied.

2.3.3 - Pattern Recognition

Artificial Neural Networks have been increasingly used for pattern recognition applications [18, 25]. Higuchi and fellows [12] have proposed the use of EHW in this area too, pointing the basic advantages of EHW in relation to Neural Networks. Table 1 summarises the comparison between EHW and Neural Networks made by Higuchi.

	EHW	**ANN**
Speed	Faster	Slower
Adaptiveness	On-Line	Generally Off-Line
Designer Knowledge	Little required	Little required
Hardware Conversion	Direct	More Complex

Table 1 - EHW x Neural Networks

As remarked by Higuchi's work, Evolvable Hardware solutions are potentially faster than Neural Networks because the results of adaptations, due to environmental changes or hardware failures, can be performed on-line, and the result of the adaptation is the new hardware itself [12]. We recall that the Neural Network adaptation or training is, up to now, realised mainly by software.

Additionally, Higuchi's work remarks that, as the result of the learning process in EHW is visible in terms of Boolean functions, it can be easily converted to a hardware structure. Therefore, maintenance of EHW systems would be easier than Neural Network based systems, which are more difficult to understand. Furthermore, they also argue that, as happens with Neural Networks, EHW usually requires much less knowledge of the particular process than in the case of conventional systems.

2.3.4 - Fault Tolerant Systems

The most common way to build fault tolerant hardware systems is through duplication of critical components, which is a non-economical and sometimes not feasible solution. We can think of a far more advantageous approach to the fault tolerance problem using the concept of adaptive machines [12].

Ideally, adaptive machines can change their hardware structure in response to environmental changes, such as malfunctions. For instance, an ideal adaptive computer should have the property of changing, if necessary, its architecture in each clock transition [1].

EHW is a suitable approach for fault tolerant systems, since the reconfigurable devices can change themselves through genetic learning. In this case, the process of genetic learning consists of seeking for the best hardware configuration given a particular environmental condition. As mentioned by Higuchi [12], the learned result can be directly translated into a new hardware system, making it suitable for real time applications.

2.3.5 - VLSI

Although at this point there are few works [22] reporting EHW application in VLSI design, this is an unquestionable area of future application of evolutionary systems. In VLSI design, the following problems are potential candidates to the EHW approach:

1. Circuit Layout (Analog and Digital)
2. Placement
3. Routing

In circuit layout, the evolving structures may be transistors, transmission gates, layers or even cells of a library.

Placement and routing are basically combinatorial optimisation problems that can largely benefit from evolutionary approaches.

2.4 - Evolving Platform

Evolving platform refers to the hardware platform over which the evolution of the circuit takes place. We can devise three main classes of Evolving Platforms:

1. *Programmable Integrated Circuits;*
2. *Dedicated Hardware;*
3. *Integrated Circuit Layout.*

Each of these platforms is described in the following sections.

2.4.1 - Programmable Integrated Circuits

The standard platform used in EHW consist of the programmable integrated circuits. These circuits can be divided into three categories [7]: *Memories*, *Microprocessors* and *Logic Circuits*. EHW focus mainly on the Programmable Logic Circuits. Logic circuits can be divided into three sub-categories [7]: PLD (Programmable Logic Device), CPLD (Complex Programmable Logic Device) and FPGA (Field Programmable Gate Array). A graph showing this hierarchy is illustrated in Figure 2.

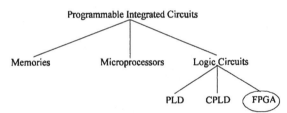

Figure 2 - Family of Reconfigurable Circuits

PLD circuits consist of an array of AND gates, which generates product terms from the system's inputs, and an array of OR gates, which generates the output of the system. According to their degree of programmability, PLDs can be classified into PROM (Programmable Read-Only Memory), PAL (Programmable Array Logic) or PLA (Programmable Logic Array) [7].

A CPLD can be seen as a combination of programmable cells, consisting usually of multiplexers or memories, and an interconnection network that selects the inputs of the programmable cells.

FPGAs are the most used reconfigurable devices in EHW [2,7]. They consist of an array of logic cells and I/O cells. Each logic cell consists of a universal function [7] (multiplexer, demultiplexer and memory), which can be programmed to realise a certain function.

The FPGAs of the XILINX [7] family has been intensively used by EHW researchers [2,7]. Each logic cell or functional block of a XILINX FPGA consists of a RAM or look-up table, in which one programs the truth table of the combinational logic circuit. The interconnections are controlled by the contents of a RAM, which means that this family of FPGA can be reprogrammed.

Instead of using FPGAs, there is also the possibility to use a memory as the evolving platform. In this case, the memory contents will be genetically programmed [5].

2.4.2 - Dedicated Hardware

The evolutionary process can be directed to achieve a dedicated architecture. A Neural Network dedicated architecture [25], for instance, can be evolved using *artificial neurons* as a basic element of the evolutionary process. *Fuzzy chips* [19] is another class of dedicated architecture that is a candidate to the Evolutionary Engineering design approach. Both the intrinsic and extrinsic approaches may be used.

2.4.3 - Integrated Circuit Layout
In this method, integrated circuit layers, such as metal, oxide and silicon, as well as complete devices, such as transistors or transmission gates, are handled by the evolutionary process to create a complete circuit layout. VLSI design is an example of this approach. In this case, the extrinsic approach is the only viable scheme to evaluate a large number of possible circuits. Areas that may benefit from the EHW approach in the future includes optical and Galium Arsenide technologies.

3- OVERVIEW OF EHW RESEARCH

In this section we briefly describe some of the most important works in the EHW area, according to the taxonomy presented in section 2.

3.1 - Hardware Evaluation Process
Thompson's work [2] presents a comparison study on the use of Extrinsic and Intrinsic EHW approaches in the design of digital circuits. In his work, a low frequency oscillator has been evolved using fast digital gates.
The evolving platform consisted of 100 gates, which can be configured to perform different boolean functions, such as NAND, AND, NOT etc. Both the Intrinsic and Extrinsic approaches have been applied, in order to identify behavioural differences between simulation and real FPGA circuits. According to Thompson, deviations occurred due to limitations in the simulation tools, such as:
- not adequately taking into account noise effects;
- considering digital gates ideal, when they are essentially high gain analog amplifiers.

3.2 - Evolutionary Approaches
Three evolutionary approaches for EHW applications have been devised in section 2.2. In this section, we present examples of these three different approaches.
Thompson [2] has employed *Conventional GAs* to evolve an *oscillator* and a *two level frequency detecting circuit*, obtaining good results in both cases. He has used a chromosome representation similar to the one used in *traditional GAs,* applying linear bit strings, each with 101 segments, to encode the circuits. Each segment corresponds to a particular circuit node. The segments encode the boolean function performed by the node and the input sources of the node's gate. The *Species Adaptation Genetic Algorithm (SAGA)* Theory [6] has been used to find the more adequate mutation rate. Table 2 shows Thompson's chromosome representation.

Bits	Meaning
0-4	Junk
5-7	Node Function
8-15	Pointer to First Input
16-23	Pointer to Second Input

Table 2 - Genotype segment for one node (Extracted from [2])

Genetic Programming (GP), the second approach, is an extension of John Holland's genetic algorithm, in which the population consists of computer programs of varying sizes and shapes and chromosomes are represented as trees [3].
John Koza and his fellows at Stanford [3] remarked on the difference between the trees used in GP and the graphs representing electric circuits:

1. **Genetic Programming chromosomes** - population of routed, point-labelled trees with ordered branches.

2. **Electric Circuits graphs** - cyclic graphs in which every line belongs to a cycle. All lines have two labels, indicating the component type (resistor, transistor, etc) and, if it is the case, its numerical value.

They proposed a *circuit constructing tree* as a way of making a link between the trees of GP and the cyclic graphs defined by electronic circuits. The individuals of circuit constructing programming trees can contain *component creating functions* and *connection-modifying functions*. The former inserts particular components such as capacitors and transistors into developing circuits; the latter modifies the developing circuit topology, for instance, creating series or parallel composition of a particular component. The starting circuit of the evolutionary process is called *embryonic electrical circuit*. Figure 3 shows the embryonic circuit used in this particular application, together with the circuit constructing tree. In this example, the function C will operate over the resistor R.

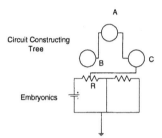

Figure 3 - Circuit constructing tree and the embryonic electrical circuit (extracted from [3])

Using the circuit constructing tree, a low pass analog filter has been evolved. Koza and his group have observed that the best individuals produced by the evolutionary process were similar to the Butterworth and Chebychev filter topologies.

Hemmi and fellows [16] have also used a tree-like chromosome representation. They have used production rules to translate Hardware Description Language (HDL) grammar into Hardware Description Language (HDL) programs, which are encoded into chromosomes. Five genetic operators have been used: *crossover*, *mutation*, *gene duplication*, *fusion* and *deletion*. Using this programming approach, they achieved good results evolving a sequential adder.

As an example of the third approach, *Evolutionary Programming*, Kitano [11] showed the effectiveness of the so-called *Grammar Encoding Method*. In this method, *rewriting rules* are used to develop graphs and are encoded through integer numbers into the chromosomes. Each chromosome allele value is an integer number ranging from 0 to 19. In the chromosome decoding process, matching rules are applied in order to construct a graph representing a circuit. Kitano has carried out two experiments using this approach: evolving Multiple XOR and six inputs multiplexer circuits.

3.3 - Target Area Applications

3.3.1 - Circuit Design
Although both analog and digital designs may be handled by EHW, a larger number of digital design applications have been reported. Table 3 presents a summary of the reported works applying EHW to general purpose circuit design:

Author	Application	Evolving Platform	Gates
Adrian Thompson[2]	Oscillator	Simulation and FPGA Xilinx	100 gates (NOT, AND, OR, etc)
Adrian Thompson[2]	Decoding tones	FPGA Xilinx	100 gates (NOT, AND, OR, etc)
Kitano [11]	Multiple XOR	Simulation	64 gates (NOT, AND, OR, etc)
Kitano [11]	6 input Multiplexer	Simulation	--------------------
Higuchi and others [12]	4-bit Comparator	CAD simulator and FPGA Xilinx	23 gates (ANDs, ORs and INVERTERS)
Hemmi and others [16]	Sequential Adder	adAM system	---------------

Table 3 - Digital EHW Applications

Table 3 shows that EHW has been able to evolve, though in some cases not optimally, basic digital components, such as multiplexers and comparators. The development of larger circuits is still an important challenge to EHW researchers. Hemmi and fellows [16] have investigated the potential of the EHW approach to evolve larger structures having some kind of functional regularity.

While EHW digital design applications use basic digital gates (i.e. NANDs, ORs, etc) as circuits building blocks, analog design applications use resistors, capacitors and transistors as the basic building blocks. As we have already mentioned, Koza and fellows [3] have used the genetic programming approach to evolve an analog low pass filter.

3.3.2 - Control and Robotics

There is an increasing number of EHW applications in this area. Thompson [5] has presented an application in which both a *1k x 8 bits RAM* contents and the *clock signal* were evolved in order to control a robot movement. He has attained a control system in which the original Finite State Machine (FSM), used to control the robot, was substituted by a *Dynamic State Machine*, in which synchronous and asynchronous variables are mixed. Additionally, the choice of variables to be used synchronously or asynchronously, has been determined genetically.

In another application, Mondada and Floreano [15] developed control systems for autonomous mobile robots using the evolutionary approach.

3.3.3 - Pattern Recognition

In their work, Higuchi and fellows [12] conducted an initial experiment to test EHW for pattern recognition, using the exclusive-OR problem. They have achieved promising results. In this particular case, EHW were able to classify patterns as efficiently as Neural Networks. They also related that they were developing a pattern recognition tool for more complicated problems.

3.3.4 - Fault Tolerant Systems

Higuchi and fellows [12] have implemented a fault tolerant EHW system applied to a robot controlling circuit. The system has the following features:

1. The EHW system functions as a backup for a *comparator logic circuit*.
2. While the main comparator works, the EHW system learns its behaviour.
3. Once the learning is over, if the comparator breaks down, the EHW system replaces the comparator circuit.

One important feature of this system is the concept of *real time genetic learning*; the backup circuit is evolved using the main circuit behaviour data.

3.3.5 - Summary

Table 4 presents a summary on the reported applications, according to the taxonomy of section 2. Besides these works, there are other studies on EHW: Gruau [9] used the *cell division theory* to generate parallel distributed computer networks; Hugo de Garis [10] conducts a project whose objective is the evolution of an *artificial brain*; Marchal and Mange [13, 14] have presented some applications using the *Embryonics concept* of very large scale integrated circuits project, endowed with living world properties.

Authors	Hardware Evaluation Process	Evolutionary Approach	Target Application Area	Evolving Platform	Genetic Operators	Conv. (Indiv.)
Thompson [2]	Intrinsic/ Extrinsic	Conventional GA	Digital Design (oscillator)	FPGA	1-Point Xover Mutation	1200
Thompson [2]	Intrinsic	Conventional GA	Digital Design (Decoding tones)	FPGA	1-Point Xover Mutation	3500
Thompson and fellows [6]	Intrinsic	Conventional GA	Robotics	Simulator/ "Mr.Chips" robot	Crossover Mutation	≅ 1000
Kitano[11]	Extrinsic	Evolutionary Programming	Digital Design (MXOR)	Logic Circuit Simulator	2-Point Xover Mutation	≅ 1000
Kitano[11]	Extrinsic	Evolutionary Programming	Digital Design (Multiplexer)	Logic Circuit Simulator	2-Point Xover Mutation	≅ 500
Higuchi and others[12]	Intrinsic/ Extrinsic	VGA	Fault Tolerance (Comparator)	CAD Simulator and FPGA	Cut, Splice	---------
Higuchi and others[12]	Intrinsic/ Extrinsic	VGA	Pattern Recognition (64pixels patterns)	Pattern Recognition Platform	Cut, Splice	----------
Hemmi and others [16]	-------------	Genetic Programming and PGAs	Digital Design (Sequential Adder)	adAM system (LSI Circuits - Computer Aided Design)	Crossover, Mutation, Gene Duplication, Fusion, Deletion	≅ 30000
Koza and others [3]	Extrinsic	Genetic Programming	Analog Design	SPICE Simulator	Crossover, Reproduction, Mutation	--------

Xover = Crossover; VGA - Variable Length Chromosome GA; PGA - Production GA
Table 4 - Summary of EHW applications summary

4 - CASE STUDIES

In this section we present a set of implementations on Extrinsic EHW developed at PUC-Rio. The objective was to evaluate the performance of an EHW system, when evolving a set of basic digital circuits. The following experiments have been performed:

1. Arbitrary digital string generator.
2. Multiplexer 4-1.
3. Decoder 3-8.
4. Sequential Adder.

To perform these experiments, we have designed an evolutionary programming system with a fix representation and standard genetic operators, described below.

4.1 - Representation

Representation is related to the way the circuits are encoded in chromosomes. Its specification is a critical step in evolutionary systems, first because invalid solutions must be avoided, and second because the possibility of deception [23].

In order to limit the number of possible solutions, the circuits are organised into levels of logic gates. The inputs of a particular level can only be chosen between the outputs of the predecessor level, limiting the number of possible connections.

In these experiments, instead of using the standard binary representation of traditional genetic algorithms, we adopted chromosomes formed by integer numbers. As stated in section 2.2, this representation leads to shorter chromosomes and may increase the efficiency of the evolutionary process. Each chromosome encodes the circuit connections and the boolean function performed by each gate. This structure is shown in Figure 4:

Figure 4 - Chromosome structure

As shown in Figure 4, each chromosome has N positions divided in two main parts: the first determines the connections of the circuit; and the second determines the gates that will form the circuit. The first part of the chromosome is divided in sub-sections, corresponding to the layers (or levels) of the circuit.

Figure 5 exemplifies the chromosome structure, showing a particular circuit together with the associated chromosome. In this particular example, we show a circuit with two logic levels, 8 inputs and one output, and 16 possible source inputs (I0, I1, ..., I15). The chromosome is logically divided in three sections, indicating respectively the first and second level connections and the type of gates. As it can also be seen, the chromosome evolves to a solution which uses signals 1 and 3 from the first level to generate the output.

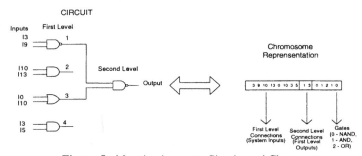

Figure 5 - Mapping between Circuits and Chromosomes

According to this representation, *the number of levels and the number of gates in each level is not genetically determined, but stated by the designer.*

4.2 - Genetic Operators

For this set of preliminary experiments, we opted for conventional genetic operators. The standard genetic operators *one-point crossover* and *mutation* have been used, with 70% and 1% probability rates, respectively. We shall note that by using the representation presented in section 4.1, these genetic operators will not produce invalid solutions.

4.3 - Digital String Generator

In this experiment, the hardware consisted of a digital logic to produce a 256-bits string output, previously chosen. There are 8 available inputs to the system; these inputs are square wave signals, with frequencies f, f/2, f/4, f/8, f/16, f/32, f/64 and f/128.

Each chromosome has been evaluated in a digital circuit simulator (Extrinsic EHW). The fitness evaluation function is realised through an accumulator that sums the hits in the output sequence, according to the following expression:

$$Acc = \sum_{i}^{n} t_i \qquad (1)$$

where: $t_i = 1$ if there is a hit and 0 otherwise. The sum is made over **n**, which is the string length (in the present case, n = 256).

Linear rank have been used to convert evaluation into fitness. Generational replacement with elitism have been used to change the population.

Tests have been performed using many different target strings and, as expected, the performance notably changes with the selected target sequence. We arrived at two basic (and somewhat expected) conclusions:

1. *For any sequence, the EHW system was able to evolve and converge to a particular solution.*
2. *For low frequency or smooth target sequences, the EHW system usually converges to a final solution, in which all the output sequence bits equals the target sequence. For "noisy" sequences, the EHW system usually does not converge to a solution with maximum fitness.*

For instance, we tried to generate the 256 bits sequence shown in Figure 6.

Figure 6 - Example of a "smooth" sequence

The sequence shown in Figure 6 is a signal with time varying behaviour. The first half of the sequence has frequency f1 while the second half has frequency f2, smaller than f1. Both f1 and f2 are of the same order of the input signal frequencies (i.e. f, f/2, f/4 etc).

For this particular test, we have chosen a circuit composed of three levels and 13 gates: 8 gates in the first level, 4 gates in the second level and 1 gate in the third level (see Figure 8). There are also 8 possible inputs to the first level, corresponding to the square wave input signals.

The graph of Figure 7 shows the average performance over 5 different executions. It can be seen that the best chromosome of the initial, randomly generated population, has, in average, a fitness value of 190 over 256. After 100 generations, the best individual has an average score of 230 over 256, with population of 300 individuals.

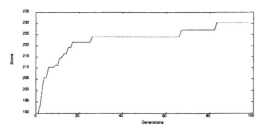

Figure 7 - Performance of the EHW Digital String Generator for the sequence in Fig. 6

The best average fitness achieved was 230, but in a particular execution the maximum fitness 256 has been achieved. The correspondent circuit is shown in Figure 8. As can be seen, the first circuit level is formed by NANDs and ORs gates and the other levels only by AND gates. Furthermore, some gates are not effectively used in the circuit.

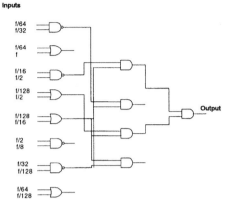

Figure 8 - Evolved Digital String Generator Circuit

We have implemented another test using a more complex target sequence. This sequence can be seen as a low frequency signal summed with a random noise. It has been verified that the EHW system converged to a circuit whose output sequence was the low frequency signal; the noisy component of the sequence could not be reproduced by the evolved solution in our experiments. Tests have been performed, both using more gate options, such as XORs, and increasing the number of circuit logic levels. However, the evolutionary program did not arrive to the desired circuit.

The graph of Figure 9 shows the average best chromosome score over 5 executions. The best solution was a circuit whose output was a sequence having 215 right bits over 256. The 40 wrong bits corresponded to the added noisy signal.

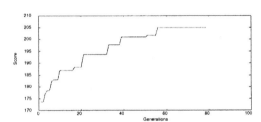

Figure 9 - EHW system performance for a noisy sequence

4.4 - Multiplexer

Using the same evolutionary programming approach, a 4-1 multiplexer has been evolved. A 19-position chromosome was used. The first 12 positions encodes the circuit connections and the last 7 positions encodes the Boolean Functions. The fitness evaluation method is similar to the one defined previously.

Figure 10 shows the evolved MUX circuit. It is a three-level circuit, consisting of four three-inputs NAND structure in the first level, two OR gates in the second level and an OR gate in the third level. As shown, the system have 12 inputs: 8 of them are genetically determined over the two multiplexer selection inputs and its inverted values (S1, S2, S1b and S2b); 4 are stated in the beginning of the evolutionary process as being the 4 multiplexer inputs (I1, I2, I3 and I4). The search space consists of approximately 10^{10} possible solutions.

Inputs

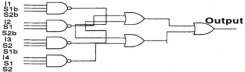
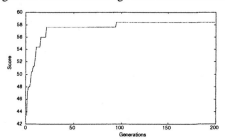

Figure 10 - 4-1 Multiplexer Evolved Circuit

The graph in Figure 11 shows the average best chromosome score over 5 executions.

Figure 11 - Average Performance for the MUX experiment

4.5 - Decoder

For this experiment we have introduced some changes in the basic representation defined above. The chromosome was organised in a different manner: eight genes have been defined, corresponding to eight gates; each gene encodes the Boolean function performed by the gate and its inputs. The idea was to keep connection and gate function closer, in order to promote short high performance building blocks. The gene representation is the same as the one defined in section 4.1.

Figure 12 shows an evolved 3x8 decoder using this approach. It can be seen that it corresponds to a one level circuit. The graph in Figure 13 shows the average best chromosome score over 5 executions. The system inputs are the three decoder selection signals and their inverted values (S1, S2, S3, S1B, S2B and S3B) and there are 8 outputs in this case (Y0, Y1, ..., Y7). The main task of the evolution process is to perform a parallel search to find 8 boolean terms corresponding to each decoder output. The search space consists of approximately 10^{22} possible solutions.

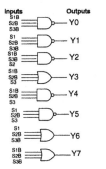

Figure 12 - 3x8 Decoder

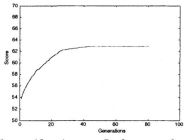

Figure 13 - Average Performance for the Decoder experiment

Both in the multiplexer and in the decoder, the best average fitness achieved was around 60, but in particular executions the maximum fitness 64 has been achieved.

4.6 - Sequential Adder

The sequential adder circuit has 3 inputs- the two summing bits and the *carry-in* bit, and two outputs, the result and the *carry-out* bit. We have separately evolved two circuits to produce the two outputs. As it is known, these circuits perform simple boolean functions, consisting of sum of products. Indeed, the convergence in this case is very fast. However, we remark that, even for simple boolean functions, a previous knowledge of the boolean expression complexity is necessary to design the EHW system.

5 - FUTURE WORK AND CONCLUSIONS

This work has presented a taxonomy of the EHW area, classifying the systems according to four basic properties: Hardware Evaluation Process, Evolutionary Approaches, Target Application Area and Evolving Platform. Furthermore, we have discussed some reported works using different approaches in these four properties.

We have also presented a set of experiments on Extrinsic EHW carried out at PUC-Rio. Basically, we have applied the Evolutionary Algorithm approach to the design of digital modules. The circuits are encoded into chromosomes using an integer number representation, corresponding to input and output circuit pins. We have limited the evolutionary algorithm search space by dividing the circuits into logic levels. We intend to expand our EHW research at PUC-Rio, particularly studying the VLSI and digital design areas.

As it has been verified through the presented EHW implementations, the artificial evolution approach can be successfully applied to the design of simple digital circuits. The authors devise future developments in the EHW area, such as:

1. *Project of larger and more complex digital circuits;*
2. *EHW VLSI design using circuit layers, transistors and transmission gates as the evolving components.*
3. *Using digital circuit simulating tools which take the limitations of digital gates more accurately into account.*

In our view, the main challenge in the design of efficient EHW systems consists in identifying adequate chromosomes representation and evaluation strategy for electronic circuits, which promote partially correct circuit parts, allowing building blocks evolution and assembly, as in conventional genetic algorithms.

References

[1] - Hugo de Garis, "Evolvable Hardware Workshop Resort", Technical Report, ATR Human Information Processing Research Laboratories, Evolutionary Systems Department, Kyoto, Japan, 1996.

[2] - Adrian Thompson, "Silicon Evolution", Proceedings of Genetic Programming 1996 (to be published), MIT Press.

[3] - John R. Koza, Forrest H. Bennet, III, David Andre, Martin A. Keane. "Towards Evolution of Electronic Animals Using Genetic Programming", Artificial Life V Conference in Nara, Japan, May 16-18, 1996.

[4] - Adrian Thompson, Inman Harvey, Philip Husbands, "The Natural Way to Evolve Hardware", Technical Report, School of Cognitive and Computing Sciences, University of Sussex, England.

[5] - Adrian Thompson, "Evolving Electronic Robot Controllers That Exploit Hardware Resources", Proceedings of The 3rd European Conference on Artificial Life, Springer Verlag, 1995.

[6] - Adrian Thompson, Inman Harvey, Philip Husbands, "Unconstrained Evolution and Hard Consequences", Lecture Notes in Computer Science - Towards Evolvable Hardware, Vol. 1062, pp136-165 Springer-Verlag, 1996.

[7] - E. Sanches, " Field Programmable Gate Array (FPGA) Circuits", Lecture Notes in Computer Science - Towards Evolvable Hardware , Vol. 1062, pp. 1- 18, Springer-Verlag, 1996.

[8] - M. Tomasini, "Evolutionary Algorithms", Lecture Notes in Computer Science - Towards Evolvable Hardware , Vol. 1062, pp. 19 - 47, Springer-Verlag, 1996.

[9] - F. Gruau, "Artificial Cellular Development in Optimization and Compilation", Lecture Notes in Computer Science - Towards Evolvable Hardware, Vol. 1062, pp. 48-75 Springer-Verlag, 1996.

[10] - H. de Garis, "CAM-BRAIN: The Evolutionary Engineering of a Billion Neuron Artificial Brain by 2001 Which Grows/Evolve at Electronic Speeds Inside a Cellular Automata Machine (CAM)", Lecture Notes in Computer Science - Towards Evolvable Hardware , Vol. 1062, pp. 76-98, Springer-Verlag, 1996.

[11] - H. Kitano, "Morphogenesis of Evolvable Systems", Lecture Notes in Computer Science - Towards Evolvable Hardware Vol. 1062, pp. 99-117, Springer-Verlag, 1996.

[12] - T. Higuchi, M. Iwata, I. Kajitani, H. Iba, Y. Hirao, T. Furuya, B. Manderick, "Evolvable Hardware and Its Applications to Pattern Recognition and Fault-Tolerant Systems", Lecture Notes in Computer Science - Towards Evolvable Hardware , Vol. 1062, pp. 118-135, Springer-Verlag, 1996.

[13] - P. Marchal, P. Nussbaum, C. Piguet, S. Durand, D. Mange, E. Sanches, A. Stauffer, G. Tempesti, "Embryonics: The birth of Synthetic Life", Lecture Notes in Computer Science - Towards Evolvable Hardware, Vol. 1062, pp. 166 - 196, Springer-Verlag, 1996.

[14] - D. Mange, M. Goeke, D. Madon, A. Stauffer, G. Tempesti, S. Durand, "Embryonics: A New Family of Coarse-Grained Field Programmable Gate Array with Self-Repair and Self-Reproducing Properties", Lecture Notes in Computer Science - Towards Evolvable Hardware , Vol. 1062, pp. 197-220, Springer-Verlag, 1996.

[15] - F. Mondada, D. Floreano, "Evolution and Mobile Autonomus Robotics", Lecture Notes in Computer Science - Towards Evolvable Hardware , Vol. 1062, pp. 221- 249, Springer-Verlag, 1996.

[16] - H. Hemmi, J. Mizoguchi, K. Shimonara, "Development and Evolution of Hardware Behaviors", Lecture Notes in Computer Science - Towards Evolvable Hardware , Vol. 1062, pp. 250 - 265, Springer-Verlag, 1996.

[17] - John H. Holland, "Adaptation in Natural and Artificial Systems", Ann Arbor: The University of Michingan Press, 1975.

[18] - G. Perelmuter, E. Carrera, M. Vellasco, M. Pacheco, M., "Recognition of Industrial Parts Using Artificial Neural Networks", Proceedings of EPMESC V-Education, Practice and Promotion of Computational Methods in Engineering Using Small Computers , pp. 481-486, Macao, August 1995.

[19] - A. P. Ungering, H. Bauer, K. Goser, "Architecture of a Fuzzy-Processor based on an 8-bit Microprocessor", IEEE, pp. 297-301, 0-7803-1869-X/94, 1994.

[20] - Jun'ichi Mizoguchi, Hitoshi Hemmi and Katsunori Shimohara, "Production Genetic Algorithms for Automated Hardware Design through an Evolutionary Process", IEEE Conference on Evolutionary Computation, 1994.

[21] - I. Harvey, P. Husbands and D. Cliff, "Genetic Convergence in a Species Evolved Robot Control Architectures", S. Forrest , editor, Proc. of 5th Int. Conf. on Genetic Algorithms, p636, Morgan Kaufmann, 1993.

[22] - Tobias Blickle, Jurgen Teich, Lothar Thiele, "System-Level Synthesis Using Evolutionary Algorithms", Computer Engineering and Communication Networks Lab (TIK), Swiss Federal Institute of Technology (ETH), TIK-Report, Nr. 16, April, 1996.

[23] - David E. Goldberg, "Genetic Algorithms in Search, Optimization & Machine Learning",Addison-Wesley Publishing Company, 1989.

[24] - Z. Michalewicz , "Genetic Algorithms + Data Structures = Evolution Programs", Springer-Verlag, 1994.

[25] -Treleaven P., Pacheco M.A., Vellasco M., "VLSI architectures for neural networks", IEEE micro, v.9,n.6, p.8-27,1989.

From some Tasks to Biology and then to Hardware

Jan Kazimierczak

Technical University of Wrocław
Wyb. Wyspiańskiego 27
50-370 Wrocław, Poland

Abstract. In this paper a class of some tasks which should be realized by the hardware is characterized. On the basis of that characterization the hypothesis of organization of such a part human brain that realizes these tasks is determined. Then on the basis of that hypothesis a formal model of hardware which models the mentioned part of brain is derived. The transformation of the formal model of such a hardware into its logical structure is shown.

Keywords: artificial intelligence, hardware design, transforming software into hardware, robotics.

1 Introduction

One of main goals of artificial intelligence is to construct such a computer hardware or robot hardware that should possess the same organization and behavior like human brain. This means that during realization of a given task the hardware should behave like human brain during realization of the same task. The achievement of the mentioned goal of AI is very hard, because we don't know in full the organization and behavior of human brain. However, solving a given formal task, we can imagine both the behavior our brain and its organization which are demanded to solve our task. Then on the basis of our knowledge, taken from books, about human brain and on the basis of our imagination we can assume a hypothesis of behavior and organization of such a part of human brain that participates in realization of the given task.

So designing a hardware which will be realized a certain task, it is necessary to assume a hypothesis of the part of human brain that realized the given task. Having that hypothesis we can design the hardware that will have the same organization like the part of brain appeared in hypothesis. The line of synthesis of the proposed hardware is shown in Fig. 1.

Notice that the given task is always associated with another tasks which form a class of similar tasks. Hence, the hypothesis mentioned above can be used to design hardware for all tasks contained in the given class. One of class of tasks which should be realized by hardware we characterize below. In the formal description of any task belonging to the mentioned class there are sequences of the same elements, but in each sequence the succession of elements is different. Each task of that class we can consider as a direct graph in which nodes denote states of realization of the given task. To each node there is attached a decision whose realization causes transition to another node. Edges of the graph are described by symbols of features of decision realization. There are a lot of nodes in the graph to which the same decision is attached.

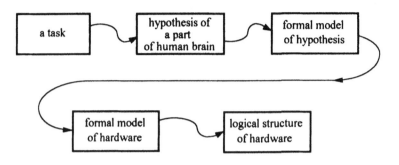

Fig. 1. The line of a hardware synthesis

For the purpose of design of such a hardware that will correspond to a hypothesis of a natural system the graph is divided into some subgraphs. Each subgraph must be connected and must contain only those nodes to which different symbols of decisions have been assigned. An auxiliary criterion is that any node belonging to a fixed subgraph cannot belong to any other subgraph of the given graph. After splitting the given graph into subgraphs, the overlaying one of the subgraphs with another one is performed. This overlaying is realized in such a way that nodes characterized by the same symbol of decision coincided at one node. Those operations result in a graph which is treated as the main part of the designed hardware.

2   Some Tasks for Realization by Hardware

One of tasks belonging to the considered class is transformation of a program written in an assembler language into hardware. This task refers to the problem of constructing some parts of operating system in the form of hardware [3]. Notice that a single program written in an assembler language is a sequence of instructions. Each instruction consists of two main components: the operation code (opcode) and the address of one or two operands. In the program there are many instructions with the same opcode. This means that the given opcode is stored in many locations of computer memory. Since the current operating system consists of a number of control programs, the number of memory location with the same of opcode is very large. Similar considerations apply to addresses of operands. The mentioned feature of the current OS we can treat as the disadvantage. Comparing that organization of computer memories with the organization of memories in human brain, it seems impossible that the same elementary unit of information appearing in descriptions of different events is stored in many cells of the human brain. Hence, in order to bring the organization of the considered hardware nearer to organization of memory in human brain, each opcode appearing in a considered program should be represented in the designed hardware only by one elementary memory element, i.e. by one flip-flop. The approach to formal synthesis of the mentioned hardware we describe below.

Let us assume that we have a program $\mathcal{A}_i$ of the operating system. This program will be transformed into the hardware. The program $\mathcal{A}_i$ is written down in an assembler language. The schematic diagram of the program $\mathcal{A}_i$ is shown in Fig. 2.

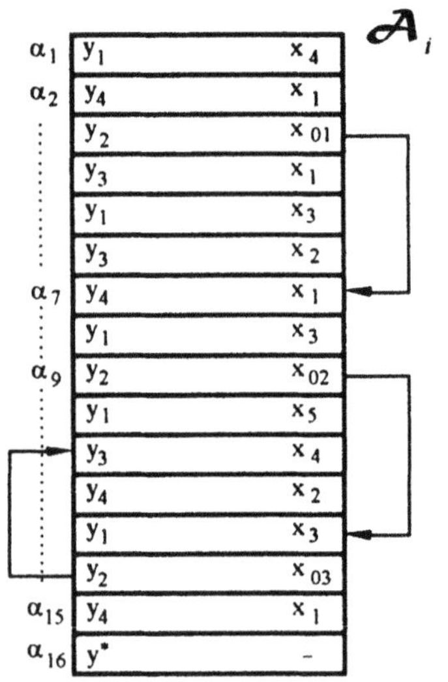

Fig. 2. The schematic diagram of a program $\mathcal{A}_i$

This program may be expressed in form a graph $\overline{G}_i$ treated as the transition state diagram of a finite automaton $\langle A_i \rangle$. The graph $\overline{G}_i$ is shown in Fig. 3. Nodes of the graph $\overline{G}_i$ designed by symbols $q_r \in Q_i$ represent internal states of the automaton $\langle A_i \rangle$. With respect to program $\mathcal{A}_i$ the given symbol q_r denotes the program state in which a computer instruction $\langle y, x \rangle$ is realized.

The symbol y denotes an opcode and the symbol x denotes the address of an operand or address of jump in the program. The edges of $\overline{G}_i$ are labeled by symbols $z_j \in Z_i$. Each symbol z_j ($j = 1, 2, 3$) either means a sign of the result of performing operation defined by opcode or only means that the given operation is finished (e.g. the symbol z_1).

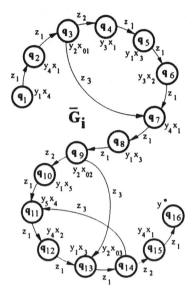

Fig. 3. The graph $\overline{G}_i$ of the program $\mathcal{A}_i$ treated as the state diagram of an automaton $< A_i >$

The graph $\overline{G}_i$ is split into two parts, i.e. into the operating part and into addressing part. Here we consider only the operating part which is shown in Fig. 4 as the graph G_i.

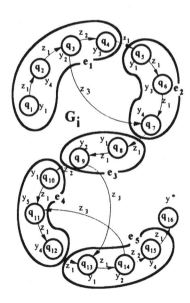

Fig. 4. The graph G_i of the operating part of the program $\mathcal{A}_i$ and its splitting into subgraphs

The graph G_i is split into such subgraphs that each from them includes nodes labeled by different opcodes $y_r \in Y$. Each subgraph of the graph G_i is labeled by symbol e_r $(r = 1,2,...)$ which is called a value of the internal parameter e of the hardware that will generate of opcodes of the program $\mathcal{A}_i$. In order to overlay the subgraphs of the graph G_i with each other the symbols q_r are replaced by symbols b_j $(j = 1,2,...)$ in such a way that all those symbols q_r to which the same symbol y_s is attached are coded by the symbol b_s. After the coding the subgraphs of the graph G_i are overlaid with each other in such a way that nodes characterized by identical symbols b_j coincide. It results in the graph G'_i shown in Fig. 5.

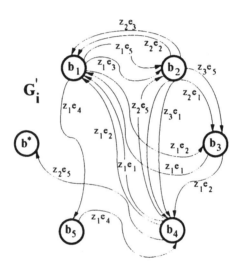

G'_i

Fig. 5. The graph G'_i of transitions between opcodes in a program $\mathcal{A}_i$

The graph G'_i is treated as the transition state diagram of the part $\langle B \rangle$ of hardware which generates opcodes $b_j = y_j$ under influence of parametric signal e_r given from a part $\langle E \rangle$ of the hardware. Notice that symbols e_r appear in description of edges of the graph G'_i from Fig. 3 and those symbols denote subgraphs of the graph G_i from Fig. 2. If we treat the subgraphs of the graph G_i as nodes e_r of another graph and those edges which lead from one to other subgraph we additionally denote by appropriate symbol b_j, then we shall obtain the graph $\check{G}_i$ being transition state diagram of the part $\langle E \rangle$ of the designed hardware. On the basis of the graphs G'_i and $\check{G}_i$ we can determine logical structure of the hardware which generates opcodes. This hardware consists of two levels: low level is represented by the part $\langle B \rangle$ of the hardware and higher level is represented by the part $\langle E \rangle$. Each active state on the higher level causes the transition of a configuration of states on the low level into active state.

The other task belonging to the considered class concerns with the moving of a mobile robot $\langle M \rangle$ which moves over a restricted plane R, discover the configuration of obstacles in the plane and return to the initial point along the shortest path [4]. We assume that the moving of the robot will be executed by a decision unit which should be realized in the form of a hardware. For formal description of the task, it is assumed that the plane R is partitioned into a finite number of identical squares. Within every square the central point is distinguished. While moving, the robot $\langle M \rangle$ applies appropriate moving strategies $y_i \in Y$, according to which it moves. It is assumed that the robot $\langle M \rangle$ moves along straight lines. The moving strategies and behavior of the robot $\langle M \rangle$ at meeting obstacle is shown in Fig. 6.

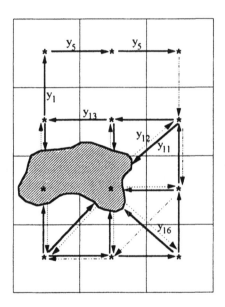

Fig. 6. Behaviour of a robot $\langle M \rangle$ at meeting obstacle

The moving strategies $y_i \in Y$ lead to definite results. It is assumed that the collection of features of the results consists of four elements: $Z = \{z_1, z_2, z_3, z_4\}$, where: z_1 – there are no obstacles when moving "forward"; z_2 – there is an obstacle when moving "forward"; z_3 – a lack of obstacles when moving "backward"; z_4 – there is an obstacle when moving "backward". With every square a single-bit memory is associated with an address $x_r \in X$. This address concerns to central point of the given square. Thus, value "1" stored at the memory element $x_r \in X$ means that the central point of the given square is inside of an obstacle.

The behavior of the robot during the moving over the plane we can expressed in the form of a graph G_j. Nodes of the graph G_j are labeled by symbols $q_i \in Q_j$ representing state of the task. With every q_i there is associated a couple (y_s, x_r). Edges of the graph G_j are labeled by symbols $z_i \in Z$ mentioned above. The graph G_j is split into subgraphs on the basis of the same principles as in the first task described above. After splitting, vertices of subgraphs are coded by symbols b_r and edges are additionally labeled by appropriate symbols e_i. Then the subgraphs are overlaid to cover nodes described by the same symbols b_r. In this way the transition state diagram G'_j of hardware part on the first level is obtained, then transition state diagram of hardware part on the second level is obtained.

The behavior of the designed hardware we can explain as follows. Assume that the hardware on the second level in state e_r. The state e_r distinguishes a configuration of hardware elements on the first level. In the graphical interpretation the symbol e_r distinguishes a subgraph of transitions between states b_j in the graph G'_j, that represents transition diagram. Those transitions depends on symbols z_j. Hence for the given state e_r and the given sequence of input symbols z_j on the output of hardware a sequence of moving strategies $y_r \in Y$ is generated.

The third type of task, belonging to the considered class, which should be realized by the hardware, is the retrieval of user programs from the knowledge base in an automatic programming system [5]. This system acts in such a way that on the basis of user's requirement expressed in natural language the computer performs the synthesis of a program in Pascal to solve user's task mentioned in the requirement. The knowledge base (KB) in this system consists of two parts. The first part contains the knowledge necessary for the natural language understanding. The second part of KB contains the knowledge needed to synthesize user programs, that is the knowledge needed for retrieving algorithm structures, declaration statements, sequences of opcodes, sequences of operands.

For security of the automatic programming (AP) system, made in the form of software, from undesirable copying, some components of the KB should be realized in the form of the hardware. We assume that the part of the KB containing knowledge needed for retrieving algorithm structures will be made in the form of hardware. This hardware has four levels. The highest level, i.e. fourth level, consists of one-bit memory elements. Each element represents symbol p_i of an algorithm structure. The third, second and first level of the hardware we denote by symbols e, f and b, respectively, where e means groups of structural components of algorithm structure, f means structural components and b means words in Pascal. The hierarchy of levels of the designed hardware is shown in Fig. 7. The behavior of the hardware we can easily explain applying graphical interpretation of the levels e, f

and **b**. Hence these levels we consider as graphs treated as transition state diagrams of these parts of the hardware.

Now assume that a user's requirement is entered into the computer. On the basis of knowledge contained in the first part of the KB, the computer found out that the user's requirement has its semantic counterpart with the symbolic name c_i in the KB. On the basis of the symbol c_i the symbolic name p_r of an algorithm structure is determined.

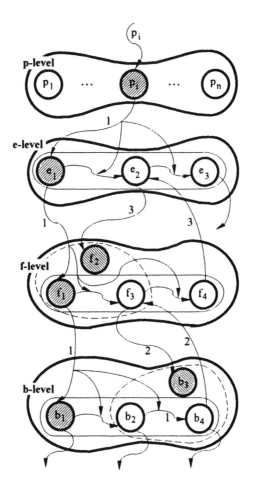

Fig. 7. Hierarchy of a hardware levels and transitions between states

The symbol p_r causes that one-bit memory element with name p_r on the level p becomes active. The signal "1" from that memory element is given to the level e, distinguishes a configuration of elements e_r $(r = 1, 2, \ldots)$ on this level and indicates on starting point in this configuration. In the other words, the symbol p_r given into e level causes that a configuration of elements on this level becomes excited and an element e_i representing initial state in this configuration becomes active. The signal from active element e_i is given to f-level, excites a configuration of elements f_n on this level and causes that initial element f_n in this configuration becomes active. Similarly as above, the signal from active element f_n is given to b-level, excites a configuration of elements b_j and causes that an element b_j representing initial state in this configuration becomes active (see Fig. 7). This configuration has the form of a sequence of states $b_j \in B$. To each element in this sequence a word in Pascal is assigned. When the given state b_j is active then a word in Pascal assigned to that state is generated on the output of the hardware. At beginning, the first state b_j of the given sequence is active and a word in Pascal is generated. After that the second state becomes active and the second word in Pascal is generated and so forth. When the last state b_j of the first sequence is active, the signal about this state is given on the f-level and causes that the second state f_n in the excited configuration becomes active. The signal about new active state on the f-level is given to b-level and excites new configuration of states b_j on this level. The above described process is repeated until the last state in the excited configuration becomes active, then signal about that state is given on the e-level and causes transition of next excited state into active form.

In general, one active state of e-level determines a sequence of several states of f-level, while one active state of f-level determines a sequence of several different states of b-level.

3   A Hypothesis about Logical Structure of Human Brain

In this paper we assume the following thesis: *if a designed hardware we assimilate to logical organization and behavior of a part of human brain, then we may suppose that this hardware can achieve such some intelligent features as possesses the human brain.* We want to prove this thesis, but first we have to determine whether our hardware is consistent with a hypothesis about logical structure of a part of human brain. Because we don't know such a hypothesis that would be corresponded to our hardware which we want to design, we must determine this hypothesis on the basis of our image about processing in human brain appearing during solving the given task. A hypothesis about logical structure of a part of human brain which would be corresponded to realize the tasks described in previous section we introduce below.

We assume that space of neurons in a part of human brain which realizes our tasks is split into a few levels. Not all levels can participate in performing a given task, this depend on the complexity of the task. At performing some tasks only two first levels can participate.

The first level is treated as lowermost one. Each neuron on the first level can represent elementary unit of information, e.g. word, symbol, etc. There are no two neurons which represent the same unit of information. There are also neurons which represent nil, but during knowledge acquisition associated with the given task, to some from them new units of information can be attached. Any neuron can be stayed in one of three states: passive state, excited state, active state. Neurons of the first level are connected by directed edges (axons in biological interpretation) which can also occur in one of three states, i.e. passive, excited, active. Some neurons of the first level are connected with some neurons of the second level by directed edges first-second level or second-first level. Under influence of a signal entered into the first level from an active neuron of the second level, on the first level is created a configuration of neurons. Then, under influence of some signals $z_j \in Z$, where any z_j ($j = 1, 2, \ldots$) represent either satisfied condition of transition between neurons (see the first task) or the result of executing the decision (see the second task) represented by the active neuron, in the excited configuration a sequence of neurons is distinguish. Units of information attached to those neurons which appear in the distinguished sequence can be generated outside. The signal from the last neuron of the distinguished sequence is given on the second level and causes the transition of another neuron of this level in the active state.

Each neuron on the second level represents more greater unit of information than neurons of the first level. Any active neuron of the second level excites a configuration of neurons on the first level. Under influence of signal from the active neuron of the second level one of neurons in the excited configuration becomes active. Because each neuron in the excited configuration represents an unit of information, the active neuron on the second level represents the unit of information being a composition of pieces of information represented by neurons of the excited configuration. Neurons of the second configuration are connected, similar like neurons of the first level. However, if in the realization of a task only the first and second levels of neurons participate, then connections between neurons of the second level are omitted. Below, for simplicity, we consider an example of the organization and behavior of the first and second levels only.

Each level of neurons can be viewed as weighted directed graph in which neurons are nodes and directed edges are connections between neuron outputs and neuron inputs. We shall only consider the first and second levels of neurons. For formal description of these levels we use some symbols as follows: neurons of the first levels we denote by symbols $b_j \in B = \{b_1, b_2, \ldots, b_j, \ldots, b_n\}$; neurons of the second level we denote by symbols $e_r \in E = \{e_1, e_2, \ldots, e_r, \ldots, e_m\}$; external signals or signal satisfying conditions of transition from one active neuron to the other neuron we denote by symbol $z_i \in Z = \{z_1, z_2, \ldots, z_i, \ldots, z_w\}$; units of information represented by neurons of the first level we denote by symbols $y_s \in Y = \{y_1, y_2, \ldots, y_s, \ldots, y_w\}$. As an example, we assume that on the second level there are three neurons e_1, e_2, e_3 and on the first level there are four neurons b_1, b_2, b_3, b_4.

We also assume that for realization of a task the active neuron e_1 excites a configuration of neurons of the first level. The picture of this configuration in the form of a graph is shown in Fig. 8. In this figure the obscured circle denotes the initial state (active neuron) of the configuration. The solid edges in the graph denote transitions between neurons, while the broken edges denotes signals given to some neurons of the second level. For example the transition from active neuron b_2 to next active neuron b_3 (see Fig. 8) causes that the signal, represented by couple z_1, b_2, is given to the neuron e_2 of the second level and causes that this neuron becomes active.

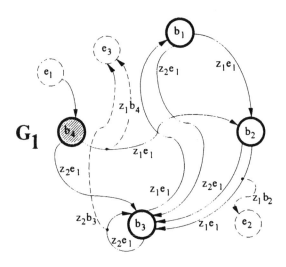

Fig. 8. The graph G_1 representing an excited configuration of neurons of the first level

For the construction of a formal model of the considered part of neurons space the graph G_1 from Fig. 8 is transformed into a symbolic expression as follows:

$$G_1^+ = {}^0(b_4{}^1(z_1e_1[e_3]b_2{}^2(z_2e_1b_3{}^3(z_1e_1b_1{}^4(z_1e_1b_2, z_2e_1b_3)^4, z_2e_1[e_3]b_3)^3,$$
$$z_1e_1[e_2]b_3)^2, z_2e_1b_3)^1)^0 \tag{1}$$

The expression G_1^+ we can interpret as follows: the symbols b_j appearing before open brackets denote nodes of the graph G_1; pairs z_je_r denoted edges; the symbol e_i in square bracket i.e. $[e_j]$ denotes a signal given to the neuron e_i. For example, the term $b_4{}^1(z_1e_1[e_3]b_2$ denotes the node b_4 from which the edge z_1e_1 comes out leading to the node b_2 and denotes that the signal represented by pair b_4z_1 is given to the neuron e_3 on the second level.

In the considered example the active neurons e_2 end e_3 of the second level excite two another configurations. The graphs G_2 and G_3 representing these configuration are shown in Fig. 9. Symbolic expressions describing these graphs are introduced below.

$$G_2^+ = {}^0(b_3{}^1(z_1e_2b_1{}^2(z_1e_2b_2{}^3(z_1e_2b_4{}^4(z_1e_2b_3,z_2e_2[e_3]b_2)^4,z_2e_2[e_3]b_1)^3, \\ z_2e_2b_2)^2,z_2e_2b_2)^1)^0$$

(2)

$$G_3^+ = {}^0(b_2{}^1(z_1e_3b_3,z_2e_3b_1{}^2(z_2e_3b_3{}^3(z_1e_3[e_1]b_1,z_2e_3b_3)^3,z_1e_3b_4{}^3(\\ z_1e_3[e_1]b_4,z_2e_3b_4)^3)^2)^1)^0$$

(3)

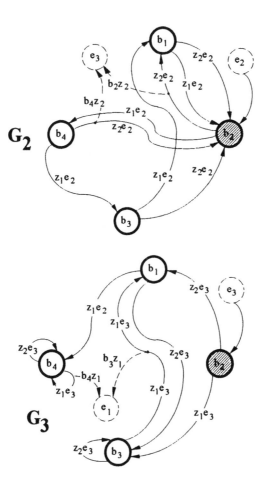

Fig. 9. The graphs G_2 and G_3 representing an excited configuration of neurons of the first level

In order to determine the general configuration of neurons of the first level, which is demanded to realize the given task, a graph G'^+ represented that configuration should be determined. For this purpose the expression G_1^+, G_2^+ and G_3^+ are composed together in such a way that identical elements $b_i{}^k($, corresponding to nodes of the graphs in Fig. 8 and Fig. 9, coincide. This operation is related with the operation of overlaying graphs G_1, G_2 and G_3 shown in Fig. 8 and 9. As a result of the composition operation performed on expressions G_1^+, G_2^+ and G_3^+, one obtains the following collective expression G'^+.

$$G'^+ = {}^0(b_4{}^1(z_2e_1b_3, z_1e_2b_3, z_1e_3[e_1]b_4, z_2e_3b_4, z_2e_2[e_3]b_2, z_1e_1[e_3]b_2{}^2(\\
z_1e_2b_4, z_2e_2[e_3]b_1, z_2e_3b_1, z_1e_3b_3, z_1e_1[e_2]b_3, z_2e_1b_3{}^3(z_2e_1[e_3]b_3,\\
z_2e_3b_3, z_2e_2b_2, z_1e_3[e_1]b_1, z_1e_2b_1, z_1e_1b_1{}^4(z_1e_1b_2, z_1e_2b_2, z_2e_2b_2,\\
z_2e_1b_3, z_2e_3b_3, z_1e_3b_4)^4)^3)^2)^1)^0$$

(4)

On the basis of the expression G'^+ one can draw the graph G' being a picture of a part of the first level which participate in executing a given task.

Taking into account the bracken edges appearing in the graphs G_1, G_2 and G_3, or those terms in the symbolic expression G_1^+, G_2^+ and G_3^+ in which a symbol $[e_j]$ is included, we can determine the graph $\tilde{G}'^+$ that represent a configuration of neurons on the second level. From the symbolic expressions G_1^+, G_2^+ and G_3^+ we obtain:

$$\tilde{G}_1^+ = {}^0(e_1{}^1(b_2z_1e_2, b_4z_1e_3, b_3z_2e_3)^1)^0$$
$$\tilde{G}_2^+ = {}^0(e_2{}^1(b_2z_2e_3, b_4z_2e_3)^1)^0$$
$$\tilde{G}_3^+ = {}^0(e_3{}^1(b_4z_1e_1, b_3z_1e_1)^1)^0$$

(5)

These expressions are combined together to obtain only one symbolic expression $\tilde{G}'^+$,

$$\tilde{G}'^+ = {}^0(e_1{}^1(b_3z_2e_3, b_4z_1e_3, b_2z_1e_2{}^2(b_2z_2e_3, b_4z_2e_3{}^3(b_4z_1e_1, b_3z_1e_1)^3)^2)^1)^0$$

(6)

On the basis of the expression $\tilde{G}'^+$ one can draw the graph $\tilde{G}'$ representing transitions between active states of neurons on the second level.

4   A Design of Logical Structure of the Hardware

The departure point for the synthesis of logical structure of the hardware which models the above considered levels of neurons are the symbolic expressions G'^+ (4) and $\tilde{G}'^+$ (6). We assume that the expression G'^+ represents part $\langle B \rangle$ of the hardware $\langle A' \rangle$, and the expression $\tilde{G}'^+$ represents part $\langle E \rangle$ of $\langle A' \rangle$. At first we shall describe a method of determining the logical structure of the part $\langle B \rangle$. We shall apply conventional logic, i.e. we shall use AND, OR, NOT gates and J-K flip-flops for describing the logical structure. To be able to univocally determine on the basis of the expression G'^+ the logical structure of part $\langle B \rangle$ on the level of single gates and flip-flops, this expression needs a further transformation into an expression G'^*. To realize the essential point of this transformation, let us look at the term $b_4{}'(z_2 e_1 b_3, \ldots$ of the expression G'^+. This term tells us that from the node b_4 of the graph G' there comes out an edge aiming at the node b_3, described by a pair of symbols $z_2 e_1$. The notation $b_4{}'($ can be regarded as a symbol of a flip-flop within the logical structure of the part $\langle B \rangle$. In the expression describing the logical structure, any flip-flop circuit will be denoted by a bold face b_i letter. Now, if we transform the term $b_4{}'(z_2 e_1 b_3)$ in the expression (4) to the form $\ldots b_4{}'([[b_4 z_2]e_1]b_3;$ then the notation $[b_4 z_2]$ will represent for us the AND gate with input signals b_4 and z_2. Similarly, the notation $[[b_4 z_2]e_1]$ should be treated as a double input gate AND whose one input is fed by a signal from gate $[b_4 z_2]$ and the signal e_1 coming from part $\langle E \rangle$. In accordance with the remarks formulated above, the notation $b_4{}'([[b_4 z_2]e_1]b_3, \ldots$ will be a representation of a logical circuit displayed in Fig. 10.

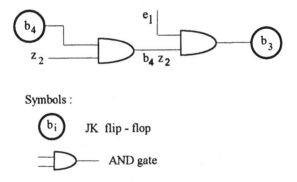

Symbols :

b_i JK flip - flop

    AND gate

Fig. 10. A logical circuit created on the basis of the expression $b_4{}'([[b_4 z_2]e_1]b_3, \ldots$

To give an example, let us consider a fragment of the expression G'^+ (4)

$$\ldots b_4{}'(z_2 e_1 b_3, z_1 e_2 b_3, \ldots \tag{7}$$

After a transformation of the expression G'^+ into G'^* this fragment will assume the following form:

$$G'^* = {}^0(b_4{}'(\{[[b_4 z_2]e_1],[[b_4 z_1]e_2]\} b_3, \ldots \tag{8}$$

In the expression G'^{*} the notation of the type $\{[...],[...]\}b_3$ denotes an OR gate whose inputs are displayed in square bracket, with output directed toward a flip-flop denoted by b_3. For example, the notation $\{[[b_4z_2]e_1],[[b_4z_1]e_2]\}b_3$ represents a logical circuit shown in Fig. 11.

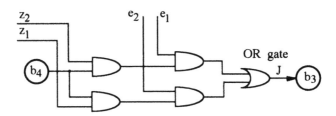

Fig. 11. A logical circuit derived from the expression $b_4{}'(\{[[b_4z_2]e_1],[[b_4z_1]e_2]\}b_3$

With regard to what has been said above in a symbolic expression describing the logical structure of part $\langle B \rangle$ one should take into account the outputs directed toward the part $\langle E \rangle$. These outputs are denoted in the expression G'^{+} (4) by the symbol $[e_r]$ and should be appeared in the symbolic expression G'^{*}. For example, the term $b_4{}'(...,z_2e_2[e_3]b_2,z_1e_1[e_3]\ b_2{}^2(...$ appearing in the expression G'^{+} (4) is transformed into the following form:

$$\{[[b_4z_2]e_2]e_3,[[b_4z_1]e_1]e_3\}b_2. \tag{9}$$

Having taken into consideration the part $b_4{}'(...b_2{}^2($ of G'^{+} (4) we obtain the following part of the expression G'^{*},

$$G'^{*} = {}^0(b_4{}'(\{[[b_4z_2]e_1],[[b_4z_1]e_2]\}b_3,\{[[b_4z_1]e_3]e_1,[[b_4z_2]e_2]\}b_4,$$
$$\{[[b_4z_2]e_2]e_3,[[b_4z_1]e_1]e_3\}b_3{}^2(... \tag{10}$$

On the basis of the expression G'^{*} (10) one can draw a block diagram of logical circuit that realizes all transitions of the part B of the hardware from the state b_4 to the remaining states, i.e. b_2 and b_3. This block diagram is presented in Fig. 12.

In order to derive a logical block diagram of the part $\langle E \rangle$ of the hardware $\langle A' \rangle$ the expression $\tilde{G}'^{+}$ (6) is transformed to an expression $\tilde{G}'^{*}$ following the same rules as applied at transforming the expression G'^{+} (4). Consequently, the term $e_1{}'(b_3z_2e_3$ in the expression $\tilde{G}'^{+}$ (6) will be given the form $e_1{}'([[b_3z_2]e_1]e_3,...$ within the expression $\tilde{G}'^{*}$. The symbolic expression $\tilde{G}'^{*}$ has the form given below:

$$\tilde{G}'^* = {}^0(e_1{}^1([[b_3z_2]e_1]e_3,[[b_4z_1]e_1]e_3,[[b_2z_1]e_1]e_2{}^2([[b_4z_2]e_2]e_3,$$
$$[[b_2z_2]e_2]e_3{}^3([[b_3z_1]e_3]e_1,[[b_4z_1]e_3]e_1)^3)^2)^1)^0 \tag{11}$$

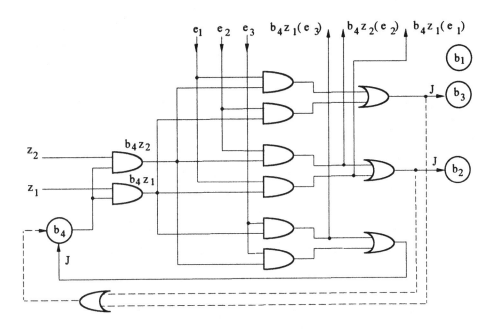

Fig. 12. A logical circuit realizing the transition from state b_4 to states b_2 and b_3

In this expression the terms of the type $[[b_iz_j]e_s]e_3$ represent strictly specified AND gates presented in the structure of the part $\langle B \rangle$. In consequence these gates should not be included any more in the structure of the part $\langle E \rangle$. Thus, in the structure of part $\langle E \rangle$ there appear only flip-flops e_r, and transitions between these flip-flops will be included into the structure of the part $\langle B \rangle$. On the basis of the symbolic expression $\tilde{G}'^*$ we are then in a position to uniquely define the logical structure of the part $\langle E \rangle$, as well as this part of the structure of the part $\langle B \rangle$ that affects the performance of $\langle E \rangle$. A block diagram of the logical structure of part $\langle E \rangle$, sketched on the basis of the expression $G'^*(11)$, has been displayed in Fig. 13. In this figure logical circuits cancelling the active states of flip-flops e_r are drawn with dashed lines.

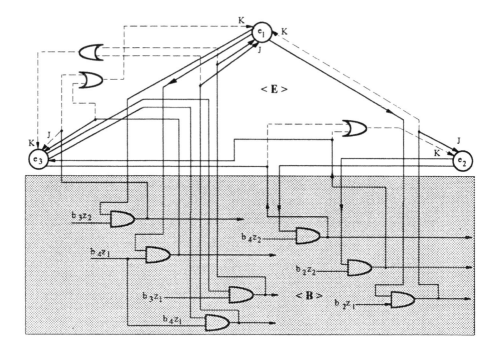

Fig. 13. A block diagram of the logical structure of part $\langle E \rangle$ of the hardware $\langle A \rangle$ including logical circuits of part $\langle B \rangle$ affecting transitions in $\langle E \rangle$

5   Conclusions

In this paper an approach to such hardware has been introduced as would be able to realize some tasks of artificial intelligence. A proposition has been submitted that designed hardware could achieve ability to executed those tasks if its organization and behavior would be comparable with the organization and behavior of the human brain. For this purpose, some tasks lain in the domain of artificial intelligence have been considered. On the basis of analysis of these tasks a hypothesis about organization and behavior of a part of human brain which participate in solving those tasks has been derived. Having that hypothesis a design of demanded hardware has been proposed. This paper introduced only an idea which requires further researches.

References

1. Arbib M.A., "The Metaphorical Brain", John Wiley and Sons, Inc., New York 1972.
2. Chew Lim Tan, Tong Seng Quah, Hoon Heng Teh, "An Artificial Neural Network that Models Human Decision Making", "Computer" March 1996, IEEE Press.

3. Kazimierczak J., "A Technique to Transform Programs into Circuits", Microprocessing and Microprogramming, vol. 22, pp. 125-140, North-Holland 1988.

4. Kazimierczak J., Łysakowska B., "Intelligent adaptive control of a mobile robot: The automaton with an internal and external parameter approach", Robotica, vol. 6, pp. 319-326, Cambridge 1988.

5. Kazimierczak J., "Acquisition and Representation of Knowledge on the Level of Programming Language for Automatic Programming", Proceedings of the ACM Computer Science Conference'93, Indianapolis, pp. 221-228, ACM Press, New York 1993.

6. Kazimierczak J., "Concept and Synthesis of an Operating System Nucleus Implemented in Computer Hardware", Proceedings of the ACM Computer Science Conference'97, St. Louis, Missouri, pp. 273-284, ACM Press, New York 1997.

7. Zvi Kohavi, "Switching and Finite Automata Theory", McGraw-Hill Book Comp., New York 1978.

Evolvable Hardware

Adaptive Equalization of Digital Communication Channels Using Evolvable Hardware

Masahiro Murakawa[1] Shuji Yoshizawa[1] Tetsuya Higuchi[2]

[1] University of Tokyo, 7-3-1, Hongo, Bunkyo, Tokyo, Japan
e-mail: murakawa@bios.t.u-tokyo.ac.jp
[2] Electrotechnical Laboratory, 1-1-4, Umezono, Tsukuba, Ibaraki, Japan

Abstract. This paper investigates the application of function-level Evolvable Hardware (EHW) to the adaptive equalization of digital communication channel. EHW is hardware that is built on programmable logic devices such as field programmable gate arrays. Its architecture can be reconfigured by using genetic learning to adapt to new, unknown environments in real time. We propose an EHW-based adaptive equalizer whose adaptive capability and fast processing speed make possible high-speed channel equalization. Simulation results show that the EHW-based equalizers have superior performance to traditional equalizers based on the linear transversal filter.

1 Introduction

During the last decade, the rapid progress in multimedia communications has increased the demand for high-speed mobile digital data transmission. One important problem in digital communications is channel equalization[1]. Channel equalization is a method of compensating for linear or non-linear channel distortion at the receiver's end to recover the original symbols.

In mobile digital communications, equalization is very difficult for the following two reasons. First, the channel acquires non-linear distortion by transmission of the signal through multiple paths. Multiple paths are caused by reflection of the transmitted radio waves by objects such as buildings. Second, the characteristics of the transmission channel vary continuously. For example, as the transmitter moves, environmental conditions such as landscape and the presence of buildings change, in turn changing the characteristics of the transmission channel. To absorb such changes and compensate for the channel distortion, the equalizer must alter the function while it works.

Conventionally, to overcome these difficulties, adaptive equalizers based on a linear transversal filter are used. They adopt a learning algorithm such as LMS[2] for on-line adaptation. But, when the non-linear channel distortion is very serious, the adaptive equalizer based on a linear transversal filter suffers from severe performance degradation.

In this paper, we propose the use of a function-level Evolvable Hardware (EHW)[3] for the adaptive equalizer. EHW is hardware that is built on programmable logic devices such as field programmable gate arrays (FPGAs). Its architecture can be reconfigured by using genetic learning to adapt to new, unknown environments in real time. The EHW has the following advantages:

- The execution speed of the evolved system is extremely fast because the result of adaptation to the environment is the hardware structure itself.
- The EHW can accomplish on-line adaptation through supervised learning.
- The function-level EHW has superior capability in synthesizing non-linear functions to the gate-level EHW.

Owing to these advantages, the function-level EHW is very suitable for adaptive equalization of mobile digital data transmission channels. Actually, simulation results show that EHW-based equalizers yield performance superior to that of traditional equalizers based on the linear transversal filter.

The rest of this paper is organized as follows: in section 2, the function-level EHW is explained in detail. In section 3, the adaptive equalizer based on the function-level EHW is introduced. Section 4 gives simulation results of its learning performance. In section 5, it is demonstrated that this equalizer can adaptively equalize time-varying channels. Section 6 concludes this paper.

2 Function-Level Evolvable Hardware

EHW is hardware that changes its hardware structure to adapt itself to the environment in which it is embedded[4]. To attain this goal, EHW utilizes field programmable gate arrays (FPGAs) and genetic algorithms (GAs)[5]. FPGAs are hardware devices whose architecture can be determined by downloading a binary string, called architecture bits. GAs are robust search algorithms that use multiple chromosomes (usually represented as binary strings) and apply a natural-selection-like operation to them to find better solutions. The basic idea of EHW is to regard the architecture bits of FPGAs as chromosomes of GAs and to find suitable hardware structure by using GAs, as shown in Fig. 1. Every chromosome is downloaded into FPGAs for evaluation while the EHW is learning.

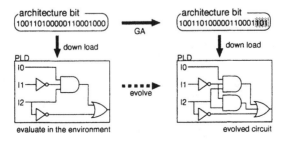

Fig. 1. Evolvable Hardware

Most research on EHW have a common problem that the evolved circuit size is small. Such hardware evolutions are based on primitive gates such as ANDs or ORs; we call the evolution at this level *gate-level* evolution. Consequently, gate-level evolutions cannot derive strong enough functions for practical applications.

To overcome this difficulty, recently a *function-level* EHW has been proposed[3]. The hardware evolutions are based on high-level hardware functions (e.g., adder, subtractor, sine generator, etc.) instead of primitive gates (e.g., AND or OR gates) as in the gate-level evolution. It was successfully demonstrated in Ref. 3 that non-linear discrimination problems such as the two-spiral problem can be solved by the function-level EHW.

Detailed descriptions of the function-level EHW are given below.

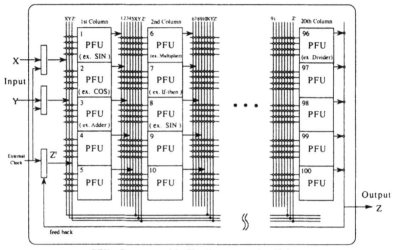

PFU : Programmable Floating processing Unit

Fig. 2. The FPGA model for function-level EHW

2.1 The FPGA model for the function-level EHW

We use the FPGA model shown in Fig. 2 to accomplish function-level evolution. The FPGA model consists of 20 columns, each containing five programmable floating processing units (PFUs). Each PFU can implement one of the following seven functions: an adder, a subtractor, an if-then, a sine generator, a cosine generator, a multiplier, or a divider. The selection of the function to be implemented by a PFU is determined by chromosome genes given to the PFU. Constant generators are also included in each PFU. Columns are interconnected by crossbar switches. The crossbars determine the inputs to PFUs.

This FPGA model assumes two inputs and one output. The data handled in the FPGA are floating-point numbers.

2.2 Genetic learning

Genetic learning in function-level evolution determines the PFU functions and the interconnection among PFUs. This means that a hardware function which is more suitable for the environment is discovered through genetic learning. The genetic operators used in function-level evolution are reproduction and mutation; crossover is not used.

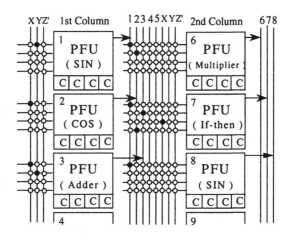

Fig. 3. Part of the FPGA model

Chromosome representation The variable-length chromosome GA (VGA) is used here. The VGA makes possible faster GA execution and larger circuit evolution [6].

An allele in a chromosome consists of a PFU *number*, a PFU *function*, and *input operand(s)*. Using Fig. 3 as an example, $(6, \times, 1, 2)$ is an allele that specifies the hardware function executed at the sixth PFU in the second column. The function is a multiplication using operands coming from the first PFU (i.e., sine) and the second PFU (i.e., cosine) in the first column. The output of the sixth PFU will be $sin(Y) \times cos(X)$.

A chromosome might have an allele configuration as follows:

$$(1, sin, Y), (2, cos, X), \cdots, (6, \times, 1, 2), \cdots, (96, \div, 90, 91)$$

representing a hardware function implemented by a full FPGA such as the one in Fig. 2.

Multiple chromosomes are prepared as a population for a GA. By repetitive applications of GA operations to the population, the most desirable hardware function is gradually synthesized.

Reproduction The roulette wheel reproduction rule is used. Through the elitist strategy, the chromosome with the best fitness value is always reproduced.

When EHW is used for pattern classification problems, the fitness value of each chromosome is determined by n/N, where N is the number of training patterns and n is the number of classifications made correctly by the EHW.

Mutation After the reproduction, every chromosome undergoes one mutation. There are three kinds of mutations.

1. Mutation of an operand
   Using Fig. 3 as an example, if (8,sin,1) is mutated to (8,sin,2), the new output is $sin(cos(X))$.
2. Mutation of a function
   If (8,sin,1) is mutated to (8,cos,1), the new output is $cos(sin(Y))$.
3. Insertion of a new allele
   Suppose the fourth PFU in the first column is not used. If (8,sin,1) is mutated to (8,sin,4), a new allele having 4 as its PFU number will be inserted in the chromosome.

3 Adaptive Equalization Using the Function-Level EHW

High-speed communications channels are often impaired by linear and non-linear channel distortion and additive noise. To obtain reliable data transmission in such communications systems, adaptive equalizers are required.

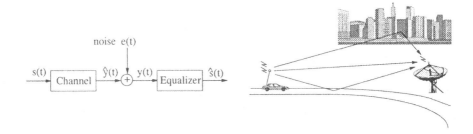

Fig. 4. Model of data transmission system

Fig. 5. Mobile communication

The digital communications system we consider here is illustrated in Fig. 4, in which a binary sequence s(t) is transmitted through a dispersive channel and then corrupted by additive noise e(t). Particularly in digital mobile communications, the channel can be influenced by environmental conditions such as landscape and the presence of buildings(Fig. 5).

The transmitted symbol s(t) is assumed to be an independent sequence taking values of either 1 or -1 with an equal probability. The task of the equalizer is to recover the transmitted symbols based on the channel observation y(t).

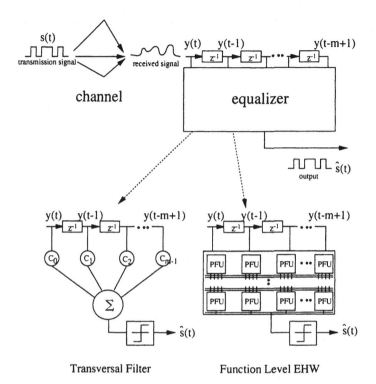

Fig. 6. Adaptive equalizers

Existing adaptive equalization techniques for the time-varying channel employ a linear transversal filter. In Fig. 6, the estimate of the transmitted symbol s(t) is given by:

$$\hat{s}(t) = sgn\left(\sum_{i=0}^{m-1} c_i y(t-i)\right) = sgn(c^T y(t)) \tag{1}$$

$$c = [c_0\ c_1 \cdots c_{m-1}]^T \quad y = [y(t)\ y(t-1) \cdots y(t-m+1)]^T \tag{2}$$

where c_i is the coefficient of the filter and the integer m is known as the order of the equalizer.

Depending on whether the equalizer knows the originally transmitted sequence s(t) or not, the equalization is characterized respectively as trained adaptation or blind equalization. The most widely used algorithm for linear trained adaptation equalizers is the Least Mean Squares (LMS) algorithm. In the LMS algorithm, the coefficient c is adjusted to match the channel characteristics during the adaptation period.

However, if the non-linear channel distortion is too severe to ignore, the adaptive equalizer based on a linear transversal filter suffers from severe performance degradation. In this paper we propose the use of the function-level EHW for the adaptive equalizer. The communications system that employs the EHW-based trained adaptation equalizer is shown in Fig. 6. The transmitter sends a *known training sequence* to the receiver, and the receiver adjust the EHW-based equalizer so that it reproduces the correct transmitted symbols.

The EHW-based equalizer has three advantages. First, the execution speed of the equalizer is extremely fast because the result of adaptation to the environment is the hardware structure itself. Second, the EHW-based equalizer can accomplish on-line adaptation. Third, the non-linear operators in the function-level EHW make possible non-linear equalization.

4   Learning Performance of the EHW Based Equalizer

We first simulated the learning performance of the proposed EHW based equalizer. The transfer function of the channel was given by $G(z) = 1.0 + 1.5z^{-1}$ and a zero-mean white Gaussian noise was added to the output of the channel. The order of the equalizer was 2 (m = 2).

A training set of eight data points was generated at every generation. The fitness value of each individual was determined by $n/8$, where n was the number of correct classifications by the EHW. The bit-error-rate (BER) was defined as the ratio of misclassified to correct symbols at the output of the best-of-generation individual. The BER was evaluated at every generation based on 10^4 random input symbols. The population size was 100.

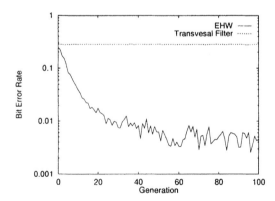

Fig. 7. Learning performance of the EHW-based equalizer (SNR: 21 dB)

Fig. 7 shows the learning performance of the EHW-based equalizer for a signal-to-noise ratio (SNR) of 21 dB. The solid curve shows a learning curve obtained by averaging 100 independent runs. The broken line shows a learning curve of the transversal-filter-based equalizer, whose total number of training sequences was the same as that of the EHW. As can been seen, the EHW-based

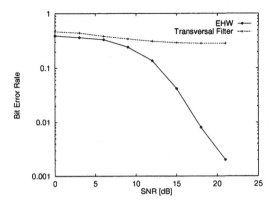

Fig. 8. Bit error rate of the EHW-based equalizer versus SNR

equalizer gives a much lower BER than that achieved by the transversal-filter-based equalizer. This is due to the ability of the function-level EHW to synthesize non-linear functions.

Fig. 8 shows the BER versus SNR achieved by the EHW-based equalizer at generation 100. We found that a significant improvement in the BER can be achieved by the EHW-based equalizer, especially at a high SNR.

5   Adaptive Equalization of Time-Varying Channels

We also simulated the performance of the EHW-based adaptive equalizer for time-varying channels.

In real communication systems, the characteristics of the channel are usually time-varying. For example, in digital mobile communications the channel is altered by environmental conditions such as landscape and the presence of buildings. Hence the adaptive equalizers are required to follow such changes and compensate for the channel distortion. In order to meet this requirement, the EHW-based adaptive equalizer is trained with a known sequence transmitted with a regular interval.

The channel used in the simulations is the non-linear channel shown in Fig. 9. The transmitted sequence is passed through a linear channel whose transfer function is $G(z) = 1 + d_1 z^{-1}$, and the output of the channel is added to the nonlinear harmonics. The value of the gain coefficients d_2, d_3, and d_4 determine how severe the nonlinear distortion will be. Such non-linear channel models are frequently encountered in data transmission over digital satellite links. The linear transversal-filter-based adaptive equalizer cannot compensate for such non-linear channel distortion.

We simulated the BER achieved by the EHW based adaptive equalizer whose order was 2 with the condition that the coefficient d_1 changed on generation. Two

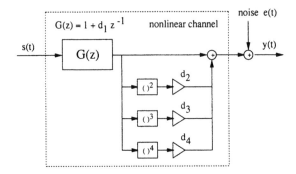

Fig. 9. Non-linear transmission channel used in the simulations

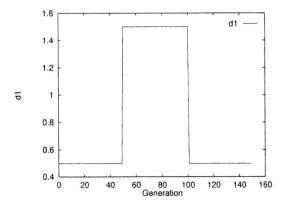

Fig. 10. Time-varying channel (d_1 changed drastically)

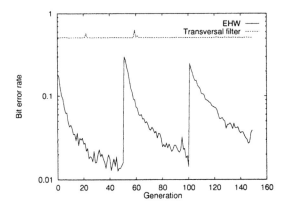

Fig. 11. Adaptive equalization of the EHW-based equalizer (SNR: 20 dB)

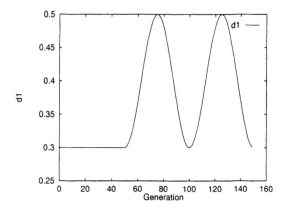

Fig. 12. Time-varying channel (d_1 changed gradually)

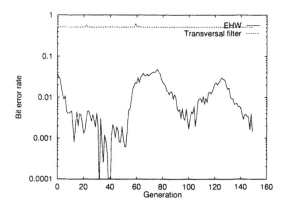

Fig. 13. Adaptive equalization of the EHW-based equalizer (SNR: 20 dB)

simulations were performed for the case in which d_1 changed drastically and the case in which d_1 changed gradually. In the simulations, the coefficients were set to $d_2 = 0.6, d_3 = 0.5$, and $d_4 = 0.4$. The population size was 100.

Fig. 10 shows the coefficient d_1 that changed drastically with generation, while Fig. 12 shows the d_1 that changed gradually. The BERs achieved by the EHW-based equalizer for these channels of Fig. 10 and Fig. 12 are shown in Fig. 11 and Fig. 13, respectively. As can be seen from the latter pair of figures, the EHW-based equalizers have the ability to follow both drastic and gradual environmental change. This is due to the adaptive capability of the GAs.

6 Conclusion

This paper has proposed an adaptive equalizer based on the function-level EHW. The EHW-based equalizer can be used for the trained adaptive equalization of linear and nonlinear communication channels. The results of simulations demonstrate the superiority of the EHW-based equalizers in BER performance over traditional equalizers based on the linear transversal filter. Especially at a high SNR, a two-magnitude improvement in the BER can be achieved by the EHW-based equalizer. This is owing to the ability of the function-level EHW to synthesize non-linear functions. We have also shown that the EHW based equalizers are capable of following both the drastic and gradual environmental change by virtue of the adaptive capability of the GAs.

Considering the application of the proposed equalizer to the real world, further simulations are required, based on a precise model of the communications channel. Also needed is examination to determine to what extent the EHW-based equalizer can cope with the speed of environmental change.

References

1. Proakis J. , "Digital Communications," Prentice Hall Inc., 1988.
2. Widrow B. , "Adaptive Signal Processing," Prentice Hall Inc., 1985.
3. Murakawa M. et al., "Hardware Evolution at Function Level," Proc. of Parallel Problem Solving from Nature (PPSN'96), 1996.
4. Higuchi T. et al., "Evolvable Hardware with Genetic Learning," Massively Parallel Artificial Intelligence (ed. H. Kitano), MIT Press, 1994.
5. Goldberg D., "Genetic Algorithms in Search, Optimization, and Machine Learning," Addison Wesley, 1989.
6. Kajitani I. et al., "Variable Length Chromosome GA for Evolvable Hardware," Proc. of the 1996 IEEE International Conference on Evolutionary Computation (ICEC'96), pp.443-447, 1996.

An Evolved Circuit, Intrinsic in Silicon, Entwined with Physics

Adrian Thompson*

COGS, University of Sussex, Brighton, BN1 9QH, UK

Abstract. 'Intrinsic' Hardware Evolution is the use of artificial evolution — such as a Genetic Algorithm — to design an electronic circuit automatically, where each fitness evaluation is the measurement of a circuit's performance when physically instantiated in a real reconfigurable VLSI chip. This paper makes a detailed case-study of the first such application of evolution directly to the configuration of a Field Programmable Gate Array (FPGA). Evolution is allowed to explore beyond the scope of conventional design methods, resulting in a highly efficient circuit with a richer structure and dynamics and a greater respect for the natural properties of the implementation medium than is usual. The application is a simple, but not toy, problem: a tone-discrimination task. Practical details are considered throughout.

1   Introduction

This paper describes a case-study in intrinsic hardware evolution: the use of artificial evolution — such as a Genetic Algorithm — to design a circuit automatically, where each fitness evaluation is the measurement of a circuit's performance when physically instantiated in a real reconfigurable VLSI chip. The term 'intrinsic' is used simply to indicate that the circuits are always tried out 'for real' rather than in simulation [1]. However, my dictionary also gives the following meanings to the word: *genuine, inherent, belonging to the point at issue.* I suggest that the point at issue with intrinsic hardware evolution is to allow the genuine inherent physical behaviour of the silicon be used freely, rather than just using hardware as a fast implementation of an idealised simulation or designer's model. I aim to show this through an example.

The following sections consider the first ever [13] intrinsically evolved FPGA configuration in great detail — there are interesting issues at every turn. The results speak for themselves, so I will save until later the underlying theory which motivated the rather unconventional approach taken. Then, in an extended discussion section, these ideas will be portrayed in the light of the experimental results that demonstrate their significance.

---

* Email: adrianth@cogs.susx.ac.uk
WWW: http://www.cogs.susx.ac.uk/users/adrianth/

2 The Evolvable Hardware

The Xilinx XC6216 [15] Field Programmable Gate Array (FPGA) [11] is a reconfigurable VLSI chip particularly suitable for evolutionary work. It will soon be commercially available — the work reported here was carried out on a β-test version. A simplified representation of the device is shown in Fig. 1. It has

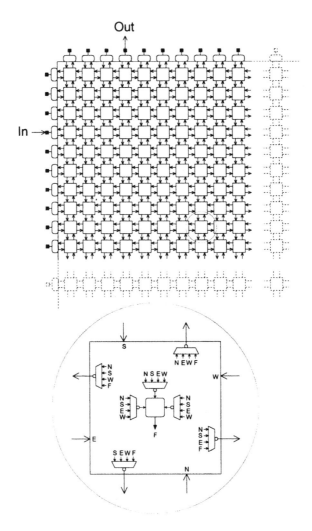

Fig. 1. A simplified view of the XC6216 FPGA. Only those features used in the experiment are shown. **Top:** A 10 × 10 corner of the 64 × 64 array of cells; **Below:** the internals of an individual cell, showing the function unit at its centre. The symbol ⌐▱ represents a multiplexer — which of its four inputs is connected to the output (via an inversion) is controlled by the configuration memory. Similar multiplexers are used to implement the user-configurable function F.

an array of 64 × 64 reconfigurable cells, each of which is connected to its four neighbours: North, East, West and South (NEWS) as shown. There is also a hierarchical arrangement of wires spanning 4, 16 and 64 cells, but these were not used in this experiment. Each cell contains a function unit that can be configured to perform any boolean function of two inputs, or multiplexer functions of three inputs. Each of a function unit's three inputs (not all of which are necessarily used) can be configured to be sourced by any of the four NEWS neighbours. The output of a cell in each of the NEWS directions can be configured to be driven either by the output F of its function unit, or by the signal arriving at one of the other NEWS faces. This allows a cell to connect some of its NEWS neighbours directly together at the same time as performing a function; a cell can 'route across itself' in some directions while giving the output of function F in others. The cells are configured independently (they do not all perform the same function), so even using only the nearest-neighbour links a very large range of possible circuits can be implemented.

Around the periphery of the array of cells are Input/Output Blocks (IOBs) and pads that interface the signals at the edge of the array to the pins of the chip. This is done in a more complex and flexible way than shown in the figure; all that is important here is that the chip was configured with a single input and a single output as shown. The choice of input and output positions was made before the experiment started, and then kept fixed. The unused IOBs simply appeared as inputs of a constant value. Only a 10 × 10 corner of the chip was used, and the unused cells were also configured just to produce a constant value. There are numerous other features of the device that were not used, and have not been mentioned.

At any time, the configuration of the chip is determined by the bits held in an on-chip memory, which can be written from software running on a host computer. No configuration of the cells can cause the device to be damaged — it is impossible to connect two outputs together, for instance, because all internal connections are uni-directional. So an evolutionary algorithm can be allowed to manipulate the configuration of the real chip without the need for legality constraints or checking. Here, we directly encode the configuration bits for the 10 × 10 corner — determining how the four outputs of each cell are derived and what function is performed by each function unit — onto a linear bit-string genotype of length 1800 bits. This was done in a raster fashion, reading cell-by-cell from left to right along each row, and taking the rows from bottom to top.

3 The Experiment

The task was to evolve a circuit — a configuration of the 10 × 10 corner of the FPGA — to discriminate between square waves of 1kHz and 10kHz presented at the input. Ideally, the output should go to +5V as soon as one of the frequencies is present, and 0V for the other one. The task was intended as a first step into the domains of pattern recognition and signal processing, rather than being an application in itself. One could imagine, however, such a circuit being used to demodulate frequency-modulated binary data received over a telephone line.

It might be thought that this task is trivially easy. So it would be, if the circuit had access to a clock or external resources such as RC time-constants by which the period of the input could be timed or filtered. It had not. Evolution was required to produce a configuration of the array of 100 logic cells to discriminate between input periods *five orders of magnitude* longer than the input ⇒ output propagation time of each cell (which is just a few nanoseconds). No clock, and no off-chip components could be used: a continuous-time recurrent arrangement of the 100 cells had to be found which could perform the task entirely on-chip. Many people thought this would not be possible.

The evolutionary algorithm was basically a conventional generational Genetic Algorithm (GA) [3]. The population of size 50 was initialised by generating fifty random strings of 1800 bits each. After evaluation of each individual on the real FPGA, the next generation was formed by first copying over the single fittest individual unchanged (elitism); the remaining 49 members were derived from parents chosen through linear rank-based selection, in which the fittest individual of the current generation had an expectation of twice as many offspring as the median-ranked individual. The probability of single-point crossover was 0.7, and the per-bit mutation probability was set such that the expected number of mutations per genotype was 2.7. This mutation rate was arrived at in accordance with the Species Adaptation Genetic Algorithm (SAGA) theory of Harvey [4], along with a little experimentation.

The GA was run on a normal desktop PC interfaced to some simple in-house electronics[2] as shown in Fig. 2. To evaluate the fitness of an individual, the hardware-reset signal of the FPGA was first momentarily asserted to make certain that any internal conditions arising from previous evaluations were removed. Then the 1800 bits of the genotype were used to configure the 10×10 corner of the FPGA as described in the previous section, and the FPGA was enabled. At this stage, there now exists on the chip a genetically specified circuit behaving in real-time according to semiconductor physics.

---

[2] TECHNICAL ELECTRONICS NOTES: The FPGA and its interface to the PC, the tone generator, and the analogue integrator all reside comfortably on a single full-length card plugging into the AT (ISA) Bus of the PC. The analogue integrator was of the basic op-amp/resistor/capacitor type, with a MOSFET to reset it to zero [7]. A MC68HC11A0 micro-controller operated this reset signal (and that of the FPGA), and performed 8-bit A/D conversion on the integrator output. A final accuracy of 16 bits in the integrator reading was obtained by summing (in software) the result of integration over 256 sub-intervals, with an A/D conversion followed by a resetting of the analogue integrator performed after each sub-interval. The same micro-controller was responsible for the generation of the tone.

Locations in the configuration memory of the FPGA and in the dual-port RAM used by the the micro-controller could be read and written by the PC via some registers mapped into the AT-Bus I/O space. The XC6216 requires some small but non-trivial circuitry to allow this — schematics are available from the author.

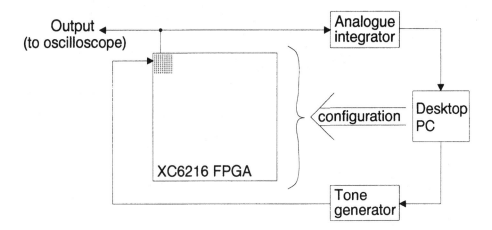

Fig. 2. The experimental arrangement.

The fitness of this physically instantiated circuit was then automatically evaluated as follows. The tone generator drove the circuit's input with five 500ms bursts of the 1kHz square-wave, and five of the 10kHz wave. These ten test tones were shuffled into a random order, which was changed every time. There was no gap between the test tones. The analogue integrator was reset to zero at the beginning of each test tone, and then it integrated the voltage of the circuit's output pin over the 500ms duration of the tone. Let the integrator reading at the end of test tone number t be denoted i_t ($t=1,2,\ldots 10$). Let S_1 be the set of five 1kHz test tones, and S_{10} the set of five 10kHz test tones. Then the individual's fitness was calculated as:

$$\text{fitness} = \frac{1}{10}\left|\left(k_1 \sum_{t \in S_1} i_t\right) - \left(k_2 \sum_{t \in S_{10}} i_t\right)\right| \quad \text{where} \begin{cases} k_1 = 1/30730.746 \\ k_2 = 1/30527.973 \end{cases} \quad (1)$$

This fitness function demands the maximising of the difference between the average output voltage when a 1kHz input is present and the average output voltage when the 10kHz input is present. The calibration constants k_1 and k_2 were empirically determined, such that circuits simply connecting their output directly to the input would receive zero fitness. Otherwise, with $k_1 = k_2 = 1.0$, small frequency-sensitive effects in the integration of the square-waves were found to make these useless circuits an inescapable local optimum.

It is important that the evaluation method — here embodied in the analogue integrator and the fitness function Eqn. 1 — facilitates an evolutionary pathway of very small incremental improvements. Earlier experiments, where the evaluation method only paid attention to whether the output voltage was above or below the logic threshold, met with failure. It should be recognised that to evolve non-trivial behaviours, the development of an appropriate evaluation technique can also be a non-trivial task.

4 Results

Throughout the experiment, an oscilloscope was directly attached to the output pin of the FPGA (see Fig. 2), so that the behaviour of the evolving circuits could be visually inspected. Fig. 3 shows photographs of the oscilloscope screen, illustrating the improving behaviour of the best individual in the population at various times over the course of evolution.

The individual in the initial random population of 50 that happened to get the highest score produced a constant +5V output at all times, irrespective of the input. It received a fitness of slightly above zero just because of noise. Thus, there was no individual in the initial population that demonstrated any ability whatsoever to perform the task.

After 220 generations, the best circuit was basically copying the input to the output. However, on what would have been the high part of the square wave, a high frequency component was also present, visible as a blurred thickening of the line in the photograph. This high-frequency component exceeds the maximum rate at which the FPGA can make logic transitions, so the output makes small oscillations about a voltage *slightly below* the normal logic-high output voltage for the high part of the square wave. After another 100 generations, the behaviour was much the same, with the addition of occasional glitches to 0V when the output would otherwise have been high.

Once 650 generations had elapsed, definite progress had been made. For the 1kHz input, the output stayed high (with a small component of the input wave still present) only occasionally pulsing to a low voltage. For the 10kHz input, the input was still basically being copied to the output. By generation 1100, this behaviour had been refined, so that the output stayed almost perfectly at +5V only when the 1kHz input was present.

By generation 1400, the neat behaviour for the 1kHz input had been abandoned, but now the output was mostly high for the 1kHz input, and mostly low for the 10kHz input... with very strange looking waveforms. This behaviour was then gradually improved. Notice the waveforms at generation 2550 — they would seem utterly absurd to a digital designer. Even though this is a digital FPGA, and we are evolving a recurrent network of logic gates, the gates are not being used to 'do' logic. Logic gates are in fact high-gain arrangements of a few transistors, so that the transistors are usually saturated — corresponding to logic 0 and 1. Evolution does not 'know' that this was the intention of the designers of the FPGA, so just uses whatever behaviour these high-gain groups of transistors happen to exhibit when connected in arbitrary ways (many of which a digital designer must avoid in order to make digital logic a valid model of the system's behaviour). This is not a digital system, but a continuous-time, continuous valued dynamical system made from a recurrent arrangement of high-gain groups of transistors — hence the unusual waveforms.

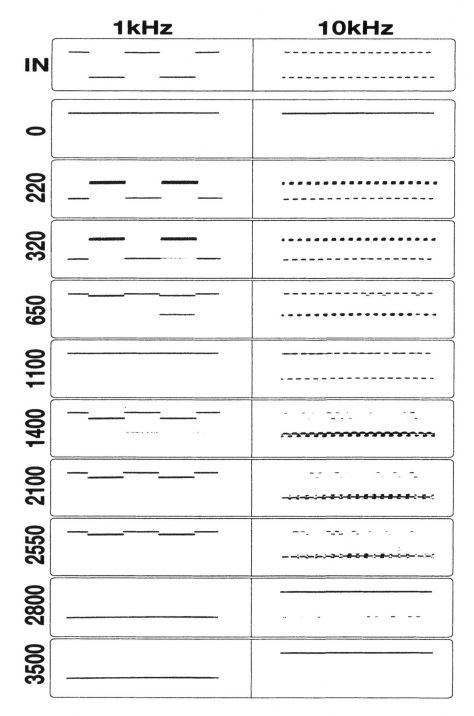

Fig. 3. Photographs of the oscilloscope screen. **Top:** the 1kHz and 10kHz input waveforms. **Below:** the corresponding output of the best individual in the population after the number of generations marked down the side.

By generation 2800, the only defect in the behaviour was rapid glitching present on the output for the 10kHz input. Here, the output polarity has changed over: it is now low for the 1kHz input and high for 10kHz. This change would have no impact on fitness because of the absolute value signs in the fitness function (Eqn. 1); in general it is a good idea to allow evolution to solve the problem in as many ways as possible — the more solutions there are, the easier they are to find.

In the final photograph at generation 3500, we see the perfect desired behaviour. In fact, there were infrequent unwanted spikes in the output (not visible in the photo); these were finally eliminated at around generation 4100. The GA was run for a further 1000 generations without any observable change in the behaviour of the best individual. The final circuit (which I will arbitrarily take to be the best individual of generation 5000) appears to be perfect when observed by eye on the oscilloscope. If the input is changed from 1kHz to 10kHz (or vice-versa), then the output changes cleanly between a steady +5V and a steady 0V without any perceptible delay.

Graphs of maximum and mean fitness, and of genetic convergence, are given in Fig. 4. These graphs suggest that some interesting population dynamics took place, especially at around generation 2660. The experiment is analysed in depth from an evolution-theoretic perspective in a companion paper [5], so I will not dwell on it here. Crucial to any attempt to understand the evolutionary process which took place is the observation that the population had formed a genetically converged 'species' *before* fitness began to increase: this is contrary to conventional GA thinking, but at the heart of Harvey's Species Adaptation Genetic Algorithm (SAGA) [4] conceptual framework.

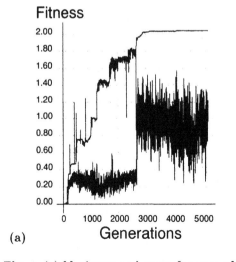

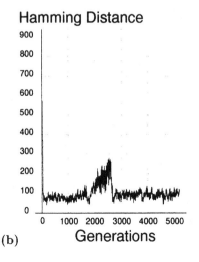

(a)   (b)

Fig. 4. (a) Maximum and mean fitnesses of the population at each generation. (b) Genetic convergence, measured as the mean Hamming distance between the genotypes of pairs of individuals, averaged over all possible pairs.

The entire experiment took 2-3 weeks. This time was dominated by the five seconds taken to evaluate each individual, with a small contribution from the process of calculating and saving data to aid later analysis. The times taken for the application of selection, the genetic operators, and to configure the FPGA were all negligible in comparison. It is not known whether the experiment would have succeeded if the individuals had been evaluated for shorter periods of time — fitness evaluations should be just accurate enough that the small incremental improvements in performance that facilitate evolution are not swamped by noise. An exciting aspect of hardware evolution is that very high-speed tasks can be tackled, for instance in the pattern recognition or signal processing domains, where fitness evaluation — and hence evolution — can be very rapid. The recognition of audio tones, as in this experiment, is a long duration task in comparison to many of these, because it is reasonable to expect that the individuals will need to be evaluated for many periods of the (slow) input waveforms, especially in the early stages of evolution. The author was engaged in a different project while the experiment was running, so it consumed no *human* time.

5 Analysis

The final circuit is shown in Fig. 5; observe the many feedback paths. The lack of modularity in the topology is unsurprising, because there was no bias in the genetic encoding scheme in favour of this.

Parts of the circuit that could not possibly affect the output can be pruned away. This was done by tracing all possible paths through the circuit (by way of wires and function units) that eventually connect to the output. In doing so, it was assumed that all of a function unit's inputs could affect the function unit output, even when the actual function performed meant that this should not theoretically be the case. This assumption was made because it is not known exactly how function units connected in continuous-time feedback loops actually *do* behave. In Fig. 6, cells and wires are only drawn if there is a connected path by which they could possibly affect the output, which leaves only about half of them.

To ascertain fully which parts were actually contributing to the behaviour, a search was conducted to find the largest set of cells that could have their function unit outputs simultaneously clamped to constant values (**0** or **1**) without affecting the behaviour. To clamp a cell, the configuration was altered so that the function output of that cell was sourced by the flip-flop inside its function unit (a feature of the chip which has not been mentioned until now, and which was not used during evolution): the contents of these flip-flops can be written by the PC and can be protected against any further changes. A program was written to randomly select a cell, clamp it to a random value, perform a fitness evaluation, and to return the cell to its un-clamped configuration if performance was degraded, otherwise to leave the clamp in place. This procedure was iterated, gradually building up a maximal set of cells that can be clamped without altering fitness.

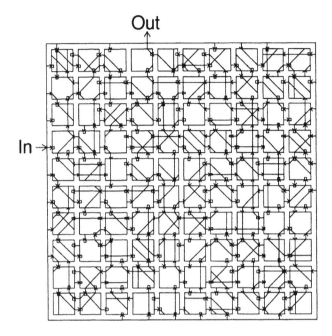

Fig. 5. The final evolved circuit. The 10 × 10 array of cells is shown, along with all connections that eventually connect an output to an input. Connections driven by a cell's function output are represented by arrows originating from the cell boundary. Connections into a cell which are selected as inputs to its function unit have a small square drawn on them. The actual setting of each function unit is not indicated in this diagram.

In the above automatic search procedure, the fitness evaluations were more rigorous (longer) than those carried out during evolution, so that very small deteriorations in fitness would be detected (remember there is always some noise during the evaluations). However, there was still a problem: clamping some of the cells in the extreme top-left corner produced such a tiny decrement in fitness that the evaluations did not detect it, but yet by the time *all* of these cells of small influence had been clamped, the effect on fitness was quite noticeable. In these cases manual intervention was used (informed by several runs of the automatic method), with evaluations happening by watching the oscilloscope screen for several minutes to check for any infrequent spikes that might have been caused by the newly introduced clamp.

Fig. 7 shows the functional part of the circuit that remains when the largest possible set of cells has been clamped without affecting the behaviour. The cells shaded gray cannot be clamped without degrading performance, *even though there is no connected path by which they could influence the output* — they were not present on the pruned diagram of Fig. 6. They must be influencing the rest of the circuit by some means other than the normal cell-to-cell wires: this probably takes the form of a very localised interaction with immediately neighbouring

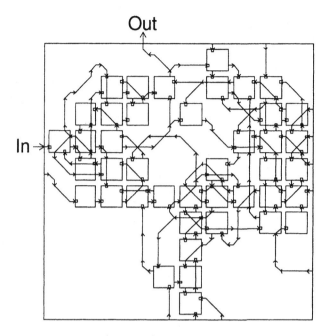

Fig. 6. The pruned circuit diagram: cells and wires are only drawn if there is a connected path through which they could possibly affect the output.

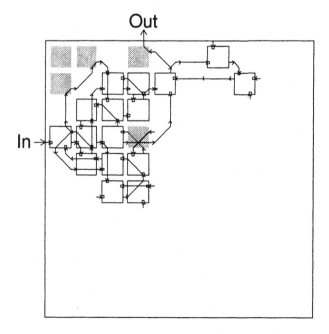

Fig. 7. The functional part of the circuit. Cells not drawn here can be clamped to constant values without affecting the circuit's behaviour — see main text.

components. Possible mechanisms include interaction through the power-supply wiring, or electromagnetic coupling. Clamping one of the gray cells in the top-left corner has only a small impact on behaviour, introducing either unwanted pulses into the output, or a small time delay before the output changes state when the input frequency is changed. However, clamping the function unit of the bottom-right gray cell, which also has two active connections routed through it, degrades operation severely even though that function output is not selected as an input to any of the NEWS neighbours: it doesn't go anywhere.

This circuit is discriminating between inputs of period 1ms and 0.1ms using only 32 cells, each with a propagation delay of less than 5ns, and with no off-chip components whatsoever: a surprising feat. Evolution has been free to explore the full repertoire of behaviours available from the silicon resources provided, even being able to exploit the subtle interactions between adjacent components that are not directly connected. The input/output behaviour of the circuit is a digital one, because that is what maximising the fitness function required, but the complex analogue waveforms seen at the output during the intermediate stages of evolution betray the rich continuous-time continuous-value dynamics that are likely to be internally present.

In [12] it was shown that in GAs like the one used here, there can be a tendency for circuits to evolve to be relatively unaffected by genetic mutations, on average. (This effect was first noticed in a different context [2, 8], and only occurs significantly in engineering GAs under particular — but common — conditions.) Depending on the genetic encoding scheme, this can have a variety of consequences for the phenotype, including graceful degradation in the presence of certain hardware faults. For our circuit evolved here, however, increasing the proportion of the possible mutations that do not reduce fitness may result in *decreasing* the number of cells implicated in generating the behaviour. So it may be no accident that the functional core of cells seen in Fig. 7 is small.

So far, we have only considered the response of the circuit to the two frequencies it was evolved to discriminate. How does it behave when other frequencies of square wave are applied to the input? Fig. 8 shows the average output voltage (measured using the analogue integrator over a period of 5 seconds) for input frequencies from 31.25kHz to 0.625kHz. When the case temperature of the FPGA is 31.2°C (as it was, ±5°C, during evolution), then for input frequencies ≥ 4.5kHz the output stays at a steady +5V, and for frequencies ≤ 1.6kHz at a steady 0V. Thus, the test frequencies (marked F1 and F2 in the figure) are correctly discriminated with a considerable margin for error. As the frequency is reduced from 4.5kHz, the output begins to rapidly pulse low for a small fraction of the time; as the frequency is reduced further the output spends more time at 0V and less time at +5V, until finally resting at a steady 0V as the frequency reaches 1.6kHz. These properties might be considered 'generalisation.'

Fig. 8 also shows the circuit's behaviour when hot or cold. The high temperature was achieved by placing a 60W light-bulb near the chip, the low temperature by opening all of the laboratory windows on a cool breezy evening. Varying the temperature moves the frequency response curve to the left or right, so once

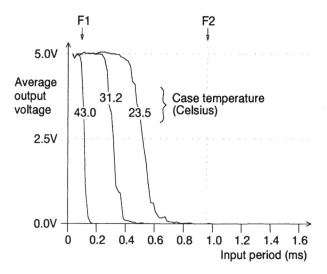

Fig. 8. The frequency response of the final circuit, measured at three different temperatures. F1 and F2 are the two frequencies that the circuit was evolved to discriminate; in fact, for ease of implementation, they happen to be of period 0.096ms (10.416kHz) and 0.960ms (1.042kHz) respectively, rather than exactly 10kHz and 1kHz as mentioned in the main text.

the margin for error is exhausted the circuit no longer behaves perfectly to discriminate between F1 and F2. In the examples given here, at 43.0°C the output is not steady at +5V for F1, but is pulsing to 0V for a small fraction of the time. Conversely, at 23.5°C the output is not a steady 0V for F2, but is pulsing to +5V for a small fraction of the time. This is not surprising: the only time reference that the system has is the natural dynamical behaviour of the components, and properties such as resistance, capacitance and propagation delays are temperature dependent. The circuit operates perfectly over the 10°C range of temperatures that the population was exposed to during evolution, and no more could reasonably be expected of it. We'll return to the issue of evolving temperature stability in the discussion that follows.

6 Discussion

The idea of enabling sophisticated behaviour to arise from an unusually small number of electronic components by allowing them to interact more freely than is customary dates back at least as far as Grey Walter's electromechanical 'tortoises' in 1949 (when the active components were thermionic valves and relays)[6]. More recently, Mead's philosophy for analogue neural VLSI has been to exploit the behaviours that semiconductor structures naturally exhibit, rather than choosing a set of functions and *then* trying to implement them in hardware [9]. In 'Pulse-stream' neural networks, the use of continuous-time dynamics has been demonstrated to release new power from a digital substrate [10].

The core principle that these ideas approach is to look for an efficient composition of electronic components selected from a set of *physical* (not abstract) resources, such that their coupled natural behaviours collectively give rise to the required overall system behaviour. In this paper, we have seen evolution do exactly that. A 'primordial soup' of reconfigurable electronic components has been manipulated according to the overall behaviour it exhibits, and on no other criterion, with no constraints imposed upon the structure or its dynamical behaviour other than those inherent in the resources provided.

For a human to design such a system on paper would require the set of coupled differential equations describing the detailed electronic and electromagnetic interactions of every piece of metal, oxide, doped silicon, etc., in the system to be considered at all stages of the design process. Because this is not practical, the structure and dynamical behaviour of the system must be constrained to make design tractable. The basic strategy is to: (1) Break the system into smaller parts that can be understood individually. (2) Restrict the interactions between these parts so *that* can be understood. (3) Apply 1 and 2 hierarchically, allowing design at increasing levels of abstraction.

Thus, conventional design always requires constraints to be applied to the circuit's spatial structure and/or dynamical behaviour. Evolution, working by judging the effects of variations applied to the real physical hardware, does not. That is why the circuit was evolved without the enforcement of any spatial structure, such as limitations upon recurrent connections, or the imposition of modularity, and without dynamical constraints such as a synchronising clock or handshaking between modules.[3] This sets free all of the detailed properties of the components to be used in developing the required overall behaviour. It is reasonable to claim that the evolved circuit consequently uses significantly less silicon area than would be required by a human designer faced with the same problem, but such assertions are always open to attack from genius designers.

The outstanding problem with allowing evolution a free hand to exploit the resources is that the evolving circuits can become tailored too specifically to the exact conditions prevailing during evolution. For instance, our example circuit was shown to be using subtle interactions between adjacent components on the silicon; surely if this evolved configuration were used with another, nominally identical, FPGA chip then it would no longer work? Every chip has slightly different propagation delays, capacitances, etc., and the circuit could have come crucially to rely on those of the particular chip on which it was evolved. To investigate this question, the final population at generation 5000 was used to configure a *completely different* 10×10 region of the same FPGA chip (Fig. 9).

When used to configure this new region, the individual in the population that was fittest at the old position deteriorated by $\approx 7\%$. However, there was another individual in the population which, at the new position, was within 0.1% of perfect fitness. Evolution was allowed to continue at the new position, and after only 100 generations had recovered perfect performance. When this new

---

[3] See [14, 13] for an expansion of this argument and earlier experiments specifically designed to explore the feasibility of such unconstrained evolution.

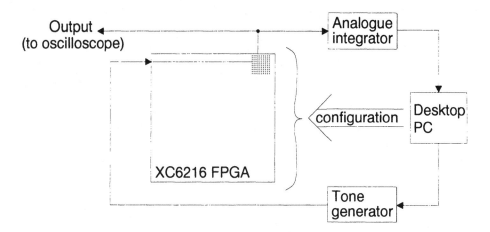

Fig. 9. Moving the circuit to a different region of the FPGA.

population was moved back to the *original* region of silicon, again the transfer reduced the fitness of the individual that used to be fittest, but there was another individual in the population that behaved perfectly there.

Recall that the circuit works perfectly over the 10°C range of temperatures to which the population was exposed during evolution. This, together with the ease with which evolution was observed to adapt the circuit to work on a new region of silicon, suggests a unified solution to the problem of evolving circuits with engineering tolerances. The plan is to have ~10 nominally identical FPGA chips, selected from separate batches (so as to be as different from each-other as possible), held at different temperatures using Peltier-effect heat pumps, and with a range of permissible power-supply voltages. To evaluate an individual's fitness, it will be tested on each of the FPGAs and given a score according to its ability to perform under all of these conditions. This will not slow down evolution because the FPGAs can operate in parallel. The hope is that evolution will produce a configuration that works at any permissible temperature and power-supply voltage and for any FPGA of that type. Success is not certain, because it will not be possible to expose the evolving circuits to every possible combination of conditions, but there is good reason to think that it can be made easier for evolution to generalise than to specialise. If the unconstrained efficient evolutionary exploitation of resources can be made an engineering practicality, the pay-offs will be great.

7 Conclusion

When an evolutionary fitness evaluation is the judging of the physical behaviour of a reconfigurable electronic device, evolution can be allowed to explore the whole space of possible configurations. Much of this space is beyond the scope of conventional design methods, so concepts of what electronic circuits can look like need to be broadened. The benefit of such a step is that evolution can then utilise the available resources more efficiently, with richer circuit structure, dynamical behaviour, and respect for the natural physical behaviour of the medium. This has been demonstrated by a simple, but not toy, experiment — the first direct 'intrinsic' evolution of an FPGA configuration.

Acknowledgements: This work was funded by the School of Cognitive & Computing Sciences. Special thanks to the Xilinx Development Corporation, John Gray, Phil Husbands, Dave Cliff, Inman Harvey, Giles Mayley and Tony Hirst.

References

1. H. de Garis. Growing an artificial brain with a million neural net modules inside a trillion cell cellular automaton machine. In *Proc. 4th Int. Symp. on Micro Machine and Human Science*, pp211–214, 1993.
2. M. Eigen. New concepts for dealing with the evolution of nucleic acids. In *Cold Spring Harbor Symposia on Quantitative Biology*, vol. LII, 1987.
3. D.E. Goldberg. *Genetic Algorithms in Search, Optimisation & Machine Learning*. Addison Wesley, 1989.
4. I. Harvey. Species Adaptation Genetic Algorithms: A basis for a continuing SAGA. In F. J. Varela and P Bourgine, eds, *Towards a Practice of Autonomous Systems: Proc. 1st Eur. Conf. on Artificial Life*, pp346–354. MIT Press, 1992.
5. I. Harvey and A. Thompson. Through the labyrinth evolution finds a way: A silicon ridge. In this volume: *Proc. 1st Int. Conf. on Evolvable Systems: From Biology to Hardware 1996 (ICES96)*, Springer-Verlag LNCS.
6. O. Holland. Grey Walter: the pioneer of real artificial life. In *ALife V: Proc. 5th Int. Workshop Synthesis and Simulation of Living Systems*. MIT Press, 1996.
7. P. Horowitz and W. Hill. *The Art of Electronics*. Cambridge University Press, 2nd edition, 1989.
8. M.A. Huynen and P. Hogeweg. Pattern generation in molecular evolution: Exploitation of the variation in RNA landscapes. *J. Mol. Evol.*, 39:71–79, 1994.
9. C.A. Mead. *Analog VLSI and Neural Systems*. Addison Wesley, 1989.
10. A.F. Murray. Analogue neural VLSI: Issues, trends and pulses. *Artificial Neural Networks*, 2:35–43, 1992.
11. J.V. Oldfield and R.C. Dorf. *Field Programmable Gate Arrays: Reconfigurable logic for rapid prototyping and implementation of digital systems*. Wiley, 1995.
12. A. Thompson. Evolutionary techniques for fault tolerance. In *Proc. UKACC Int. Conf. on Control 1996 (CONTROL'96)*, pp693–698. IEE Conference Publication No. 427, 1996.
13. A. Thompson. Silicon evolution. In J. R. Koza et al., eds, *Proc. of Genetic Programming 1996 (GP96)*, pp444–452. MIT Press, 1996.
14. A. Thompson, I. Harvey & P. Husbands. Unconstrained evolution and hard consequences. In E. Sanchez & M. Tomassini, eds, *Towards Evolvable Hardware: The evolutionary engineering approach*, pp136–165. Springer-Verlag LNCS 1062, 1996.
15. Xilinx, Inc. XC6200 Advanced product specification V1.0, 7/96. In *The Programmable Logic Data Book*. 1996. See http://www.xilinx.com.

Through the Labyrinth Evolution Finds a Way: A Silicon Ridge

Inman Harvey and Adrian Thompson*

School of Cognitive and Computing Sciences,
University of Sussex,
Brighton, BN1 9QH, UK

Abstract. Artificial evolution is discussed in the context of a successful experiment evolving a hardware configuration for a silicon chip (a Field Programmable Gate Array); the real chip was used to evaluate individual configurations on a tone-recognition task. The evolutionary pathway is analysed; it is shown that the population is genetically highly converged and travels far through genotype space. Species Adaptation Genetic Algorithms (SAGA) are appropriate for this type of evolution, and it is shown how an appropriate mutation rate was chosen. The role of junk on the genotype is discussed, and it is suggested that neutral networks (paths through genotype space via mutations which leave fitness unchanged) may be crucial to the effectiveness of evolution.

1 Introduction

To evolve silicon hardware successfully one must understand the limitations and possibilities of hardware; but one must also understand the limitations and possibilities of artificial evolution. This paper focuses on these latter concerns, and illustrates them with data taken from the first ever successful experiment evolving circuit designs directly onto a Field Programmable Gate Array for the purpose of signal recognition, using intrinsic evolution.

Newcomers to the field of hardware evolution tend to draw their ideas about artificial evolution from the Evolutionary Algorithm (EA) literature. Over the last 20 years a body of practice has accumulated, applying algorithms based upon evolutionary methods to a wide range of optimisation problems. It will be suggested here that hardware evolution will typically require a subtly but very significantly different methodology from that portrayed in much of the EA literature. We claim that the usual fears of premature convergence of the population are groundless; that entrapment at local optima does not usually take the form that is conventionally pictured; and neutral networks through genotype space should be encouraged through the artificial equivalent of potentially useful junk DNA.

---

* Emails: inmanh, adrianth @cogs.susx.ac.uk
WWW: http://www.cogs.susx.ac.uk/users/inmanh, adrianth

Section 2 of this paper will summarise the hardware experiment, the pattern of fitnesses exhibited during the run, and the final evolved hardware configuration. The degree of genetic convergence in the population is plotted, showing that a high degree of convergence is achieved almost immediately and then maintained. This runs counter to conventional EA expectations, where pains are often taken to prevent this on the (usually false) assumption that such convergence implies there will be no further evolution. In Section 3 the background assumptions of mainstream EAs are examined; Section 4 suggests why this often leads to false assumptions about genetic convergence.

In Section 5 the concepts of neutrality and neutral networks within a fitness landscape are introduced. Section 6 discusses the role and usefulness of junk or redundancy in genotypes, and suggests that some forms of junk should be encouraged in hardware evolution.

Having given a theoretical context, we then return to the data from the real experiment in Section 7, where the genetic variation is plotted. The movement of the population through genotype space is plotted in Section 8, and we discuss the speed of this movement during evolution. We give a baseline comparison with a similar population not under directed selection. There are indications that when under selection, the population 'searches' along high-fitness ridges or neutral networks within the fitness landscape. In Section 9 we are able to give a sketch of the fitness landscape, and relate it to the evolutionary pathway.

Before drawing general conclusions, Section 10 discusses the philosophy behind Species Adaptation Genetic Algorithms (SAGA), where GAs are adapted to the sort of conditions that will usually hold in hardware evolution: the continued evolution, without entrapment, of a genetically converged population or 'species'.

2   The Hardware Evolution experiment

In a companion paper [26] Thompson describes in detail the intrinsic hardware evolution directly onto a Field Programmable Gate Array (FPGA) of a circuit which is required to distinguish between two input tones of 1kHz and 10kHz. The natural timescale of signals within the FPGA, used without any clocking, is of the order of nanoseconds, hence this task is difficult with the limited resources made available. Here the experiment is summarised in sufficient detail to give a context for the analysis of the dynamics of the evolutionary process.

A genotype of 1800 bits directly encoded the functions and pattern of connections of 100 logic blocks, or 'cells', within the FPGA: 18 bits for each cell. A population of size 50 was initialised randomly. In a generational genetic algorithm (GA) each genotype was used to reconfigure the FPGA which was then evaluated at the task. The next generation was generated by first copying over the elite member unchanged; the remaining 49 members were derived from parents chosen through linear rank-based selection, in which the fittest individual of the current generation had an expectation of twice as many offspring as the median-ranked individual. Single-point crossover probability was 0.7, and the

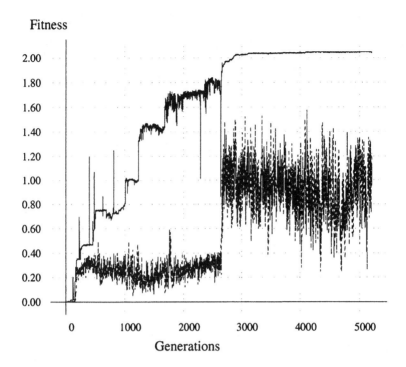

Fig. 1. The maximum and mean fitnesses of the population are plotted over 5220 generations. The dramatic rise in mean fitness occurs at generation 2660.

per-bit mutation rate was set such that the expected number of mutations per genotype was 2.7.

Evolution was continued for 5220 generations, with full genetic data saved every 10 generations. In Figure 1 the maximum and mean fitnesses are plotted, showing a dramatic increase in the mean at around 2660 generations, and the maximum fitness reaching a plateau at around 3000 generations (there is a small but definite further improvement shortly after generation 4000). This fitness corresponds to near-perfect performance at discriminating between the two input tones, signalling the result by a high or low output signal.

The fittest hardware design of generation 5000 is shown in Figure 2, where the arrows show the genetically specified pattern of connections between each of the 100 available cells (though not the functions performed by each cell). Systematic testing demonstrates that it is only a group of cells in the top left corner, between the input and output nodes, that is functional in the performance of the task. All of the other cells can be simultaneously clamped to fixed values without affecting the behaviour. The relevant functional cells are shown on the right in Figure 2. Some 2/3 of the cells, left blank in the right-hand figure, are irrelevant in the context of the functional part; hence mutations altering such irrelevant connections will be neutral with respect to fitness, except possibly for some immediately around the periphery of the functional part.

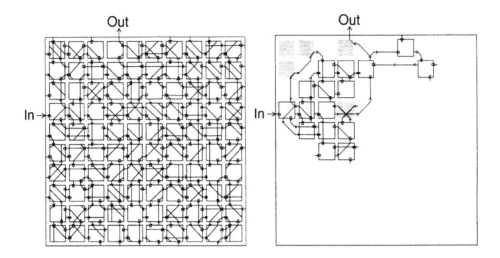

Fig. 2. On the left is shown the genetically specified configuration of the FPGA after evolution for 5000 generations. On the right is shown the subset of these connections which is functional, in the sense that the hardware still functions appropriately when the rest of the chip is clamped to fixed values.

This implies that for this particular functional design, up to 2/3 of the 1800 bits could be mutated without affecting the fitness. One can also realistically expect that there may be many other unrelated functional designs with a comparable number of redundant bits. Thus it would be a conservative estimate to suggest that out of the 2^{1800} possible points in genotype space — binary genotypes of length 1800 — at least 2^{1200} points represent hardware designs successful at the task.

In Figure 3 we plot on the left side the genetic convergence within the population as evolution progresses. This is measured as the average Hamming distance between pairs of genotypes drawn from the population. In the initial random population of genotypes of length 1800, this average is around 50% or 900 bits, but it can be seen that genetic convergence to below the 100 level is rapid, occurring within the first 45 generations. There is a temporary climb to above the 200 level after 2000 generations, which then falls back at around 2660 generations: the same time as the sudden rise in mean fitness shown in Figure 1.

The rapid convergence is not surprising, because something similar happens even in the absence of selective forces, merely through random genetic drift. In [1] it is shown that with a population of size N, with zero mutation, uniform recombination of n binary loci, and random selection of parents the mean convergence time to zero variation is approximately $1.4N(0.5 log_e n + 1.0)^{1.1}$ generations. In the presence of mutation the convergence is not to a zero level of variation, but to a higher balance between selection/drift and mutation; this is reached considerably sooner. To give a baseline comparison to the hardware

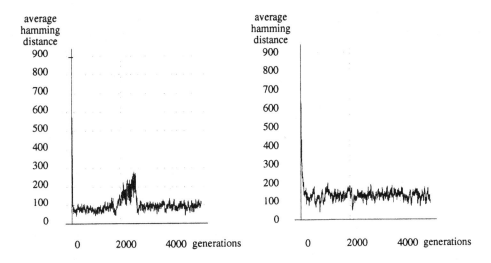

Fig. 3. Plot of genetic convergence within the population (measured as average Hamming distance between pairs of genotypes) against generations. On the left, from the evolving hardware; on the right, with fitnesses randomly allocated.

evolution example, an evolutionary experiment was run with the identical GA conditions, save only that the fitness of each genotype was allocated randomly at each test instead of being based on performance at the tone-discrimination task. The convergence statistics are shown on the right of Figure 3. In this case it takes some 220 generations to drop down to the 150 level; the average value is then maintained somewhat above the 100 mark.

Almost all of the improvement in fitness occurs after genetic convergence; the same phenomenon was discussed in the context of a different set of evolutionary experiments in [11]. For many users of GAs this is unexpected; the following sections will attempt to explain their surprise.

3   A Broad Picture of Evolutionary Algorithms

Historically there have been at least four main strands or flavours of evolutionary algorithms (EAs). The most prominent flavour is Genetic Algorithms (GAs) founded in the 1960s by John Holland [14, 9] where genotypes typically are strings of characters, often binary. Since the 1980s a prominent offshoot of GAs has been Genetic Programming (GP) [2, 21] where genotypes are typically simplified computer programs in the form of tree structures. Independent flavours include Evolution Strategies (ES) [23, 24] where the genotypes include real numbers; and Evolutionary Programming (EP) [7, 6]. Many concerns are shared across these paradigms; in this paper we shall focus on GAs, but the conclusions have wider applicability.

A common assumption is made within the GA community, explicitly or implicitly, that there are two distinct phases within an evolutionary run. The second and final stage is referred to as 'convergence', and the first stage is the transition from the (often randomised) starting population to the converged state.

There is a potential ambiguity in the word 'converged' which is regrettably often unrecognised. One meaning is that of genetic convergence: the state of the population where every genotype is similar to every other one. A second meaning is that the genetic search has converged onto its final resting place, perhaps a local optimum; often a fitness graph is shown with the fitness increasing to some asymptotic value, reaching it at 'the point of convergence'. It is frequently assumed by members of the GA and GP community that the two different types of convergence occur at one and the same time; hence the ambiguity is often considered to be of little importance.

There are some particular constrained circumstances in which this does indeed happen, but in general it does not. The evolutionary pathway of hardware design studied in depth in this paper is a clear case where the time of genetic convergence does not correspond to the time of convergence to a fitness optimum. In the general evolutionary case a population has a high degree of genetic convergence — it is a 'species' — with the degree of convergence maintained by a balance between selective pressures and genetic operators. Selective pressures act to increase convergence; genetic operators such as mutation act to increase diversity.

If there is any initial imbalance in favour of selection between these competing pressures, then a transitory period of genetic convergence occurs, often very rapidly. This does not in general mean that further evolution ceases.

4 Why is there this confusion about convergence?

Many GA implementations do fit into the particular constrained circumstances in which genetic convergence does indeed (exceptionally) imply no further evolutionary progress. When using a GA to optimise a function of n variables, a common strategy would be to use m bits to code for each variable on a genotype of length nm; the encoding of each variable to m-bit precision could be in binary or using a Gray code [9].

In such a case every locus on the genotype is functional — literally in that a change of a bit normally changes the value of the function that is being optimised. Fitness functions typically used in testing GAs (e.g. most of the De Jong test suite [9]) fall into this category. A fitness landscape can be visualised for such functions. In the 2-dimensional version ($n = 2$) the landscape consists of one or more hills in a bounded region of terrain where the x and y axes correspond to the two variables. The varying height of the landscape corresponds to the fitness, i.e. the value of the function for corresponding values of the two variables. Optimisation involves finding the highest hills.

Such a landscape can be generalised in two ways: firstly by extending the dimensionality to values of $n > 2$; secondly by recognising that where variables are each encoded discretely by m bits on the genotype, there are precisely 2^{mn}

genotypes, corresponding to 2^{mn} corners of a hypercube, each of which can be allocated a fitness. This hypercube visualisation recognises that genotype space is a graph or lattice of connected points; one needs a mathematician's abstract attitude to high dimensional spaces to visualise this in more than 3 dimensions. One of the messages of this paper is that one should be cautious in extending intuitions from low dimensions too readily to these high dimensional lattices where they may no longer be valid. Since a change of any one bit — a move from one corner of the hypercube to a neighbouring corner — results in a change of fitness, we call this a *non-neutral* landscape. The well-known N-K fitness landscapes [19] fit into this category.

A genetically converged population is represented by a number of nearby (in Hamming distance) points on the lattice given by the corners of the hypercube. Such a population under selection, reproduction, recombination and mutation will soon move upwards on a non-neutral fitness landscape until it arrives at a local optimum. Only if selective forces are low, or mutation rates small, will this be a protracted process.

So in these circumstances of non-neutral fitness landscapes the conventional expectation is generally correct; convergence in both senses does indeed happen at much the same time. The population converges on a local optimum, perhaps not the global optimum, and gets stuck there. *However*, much of artificial evolution, including specifically hardware evolution, will not fit this pattern of a non-neutral fitness landscape.

5 Neutrality

It is sometimes useful to borrow (and adapt) the notion of a *phenotype* from biology; in artificial evolution the phenotype is the instantiated design that a given genotype encodes — the end-product. It is open to the experimenter to choose the level of description in terms of which phenotypes are defined. For instance in the context of the FPGA hardware described here, phenotypes can be described in terms of (a) chip configuration or (b) input/output behaviour or (c) fitness score. Under options (a) or (b), two genotypes may be of equal fitness either because they encode the same phenotype, or because they encode two different phenotypes which happen to be equally fit. What counts as a change of phenotype under one description may be identical under a different description.

In this paper we are here taking option (c), the fitness score, and hence we are ignoring for present purposes changes of chip configuration which do not alter the fitness. In this context it should be recalled that under conventional conditions of population genetics, in a population of size N fitness differences between individuals that make a change in their expected number of offspring of the order of $1/N$ or less are effectively neutral (i.e. the differences are 'invisible' to selective pressure) [5, 20].

A neutral network of a fitness landscape is defined as a set of connected points of equivalent fitness, each representing a separate genotype; here *connected* means that there exists a path of single (neutral) mutations which can traverse the network between any two points on it without affecting fitness.

If a genetic encoding has a number of completely redundant loci, which are never used to determine the phenotype under any circumstance, then such loci automatically generate neutral networks. If a non-neutral genetic encoding (one which generates a non-neutral fitness landscape), with binary genotypes of length n, is modified by the addition of g extra redundant loci, then each phenotype will now be represented by 2^g points in genotype space instead of just one. These points will form a connected neutral network. However nothing will have been gained by this exercise — we shall term this type of redundancy *useless junk*.

In contrast, it is possible to add redundancy, and generate neutral networks, so as to transform the evolvability of a fitness landscape. One example [18] can transform a fitness landscape of any degree of ruggedness to a different, but related, landscape with a single global optimum; there will no longer be any local optima at which to get trapped. To do this an original encoding using n loci is transformed into one using $(1 + 2n)$. The first 'deciding' bit determines which of the following two groups of n bits is to be considered or ignored; the chosen one of these two groups is then interpreted according to the original encoding. Since for any given genotype one group of n bits is currently not being expressed, neutral mutations are possible within this group until the position of the global optimum is encoded (but not yet expressed) therein. A single mutation of the first deciding bit then generates the global optimum, without any intermediate points of lower value having been encountered.

This example is of theoretical rather than practical interest; the neutral, unselected search within the unexpressed group is no better than random search. So a landscape on which one can get trapped in a local optimum has been transformed into one in which one will never get trapped — but it will take too long (through random search) to find the goal. There are parallels between this impractical scheme and the practical operation of gene duplication plus mutation, which may be a powerful factor in natural evolution [22]. The single deciding bit acts as a form of gene switch, a concept which can be developed for generating neutral networks in artificial evolution.

This theoretical example introduces the notion of *potentially useful junk*.[2] This refers to loci on a genotype which *within the current context* of the rest of the genotype are functionless (mutations make no difference), but given different values elsewhere on the genotype they may well become significant.

In the evolved hardware which is the subject of this paper, it can be seen that the encoding allows for plenty of potentially useful junk, and consequently many neutral networks. The junk parts of the genotype where the potential for future usefulness is highest code for the periphery of the functional part of the FPGA, as shown in Figure 2. The bigger this periphery the more potential there may be. In the current experiments the only connections allowed were between neighbouring cells in the 2-D layout, but the chip does have the potential

---

[2] "Garbage you throw out; junk is what you store in the attic in case it might be useful one day" — attributed to S. Brenner, with reference to junk DNA.

for 'Magic wires' and 'fastlanes' which connect non-neighbouring cells, and in effect give a higher dimension topology for the chip than the simple 2-D layout. Use of these would tend to create a larger periphery to the functional part of the chip, with potentially useful effects on evolutionary pathways. With the direct encoding used in this experiment this would also possibly introduce new epistatic interactions between loci far apart on the genotype; whether or not this is desirable is currently not known.

6 Is it likely that junk is useful?

Neutral networks may be of use if they allow escape from what would otherwise be a local optimum, and this to happen within a reasonable timescale which must be much less than that of random search. The place of such networks is discussed in the context of RNA evolution in [17]. An analysis of transitions between one neutral network and the next is given in [8]. It appears that neutral networks do indeed have a beneficial effect within RNA fitness landscapes, in the sense that higher fitnesses are thereby achieved. In artificial evolution neutral networks have generally been ignored, an exception being [15].

Is this beneficial effect likely to hold true for landscapes with neutral networks such as those of hardware evolution? It is certainly not true of *all* landscapes with neutral networks; for instance the examples above with useless junk, and those with a single global optimum only reachable through random search. But there is reason to hope that much junk is potentially useful junk — though ultimately this depends on the underlying problem and on the method of encoding on the genotype. Where potentially useful junk can be generated by a suitable choice of encoding, this should be encouraged.

One relatively simple way to encourage potentially useful junk in hardware evolution is to make available many more components than are actually required at the end of the day. This was the case here, and the following analysis lends support to the hypothesis that this junk may well have been useful. Once a high fitness feasible design has been achieved, if this uses most of the available components then few mutations will leave it unaffected. If other regions of the associated neutral network exist where phenotypes are unaffected by a larger number of mutations (corresponding to a plateau rather than a ridge in the fitness landscape) then selection will tend to move the population as a whole towards such areas [3, 16, 25]. This can imply a tendency for designs that achieve the same functionality with less components to be favoured over those with more [26]; nevertheless the availability of more components may well make intermediate stages of evolution much easier by providing useful junk.

7 Genotype usage through evolutionary history

Those loci on genotypes that are currently under selection will tend to be conserved within the population at the selected values, over many generations. Other

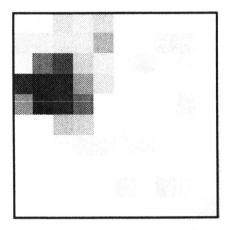

Fig. 4. For each block of 18 bits on the genotype (corresponding to one of the 10×10 cells on the FPGA), the amount of variation is calculated in the population between generations 1000 and 5000. This is plotted on a similar layout to that of Figure 2, with places of least variation darkest.

loci that are currently junk will not be so constrained; one should however be aware of the vagaries of genetic drift in small populations. Genetic drift refers to the consequences of sampling error in reproduction in small populations even when a locus is not under directed selection. The proportions of an allele at a locus will vary stochastically, or drift, until a state of all 0s or all 1s is reached; this state will generally be maintained in the medium term, until it is dislodged by mutation. With the current parameter values, 'medium term' means hundreds of generations; to see loci that are not under selection being dislodged by mutation, we must view the history over thousands of generations.

We can plot variance in alleles at each of the 1800 loci in the population between generations 1000 and 5000. This can be averaged over each block of 18 bits, which corresponds to the configuration of a single cell on the FPGA with the direct encoding used. In Figure 4 this is displayed graphically in the same layout as in Figure 2. It can be seen that there is a strong correspondence between the darker areas representing least variance, and the functional parts of the chip, as shown in Figure 2. Analysis of the genetic variation can thus be a useful tool in helping to predict which parts of the chip may be pruned away as redundant. The fact that this picture contains a lot of grey, rather than simply black and white, suggests that there has indeed been much exploration of neutral networks.

8 Saga through genotype space

The converged population is not stuck, but moves through genotype space; when it is not climbing a fitness slope it will be drifting along a neutral network. Whereas drift at a single locus may soon reach an absorbing barrier (all 0s or all

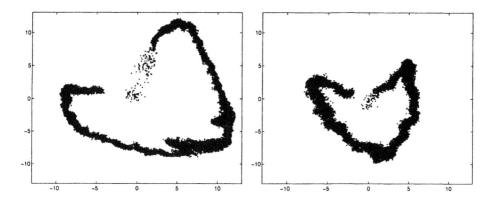

Fig. 5. The 'Saga' of the population starting from a scattered initial generation in the centre, and then in each case proceeding clockwise with a converged population; on the left data from the hardware evolution, on the right a comparison run with fitnesses randomly allocated. Every 10th generation all 50 genotypes in the population are projected onto 1st and 2nd Principal Components (PCs) derived from the population movement through genotype space over 5220 generations. The PCs are different on left and right, but the same scale has been used.

1s), neutral networks are frequently so enormous that in practice one can drift interminably. One way of visualising this movement is by projecting the 1800-dimensional genotype space onto just 2 dimensions. We choose the First and Second Principal Components of the movement of the genotype 'centroid' of the population through the 5220 generations to define the particular projection; this automatically gives a maximum spread to the displayed pathway. In Figure 5 on the left is shown every 10th generation plotted with this projection.

No special significance should be given to the gross shape of the pathway; any process of random drift or 'drunkard's walk' will give a somewhat similar path. In Section 2 we mentioned a baseline comparison where the identical GA was used, but allocating fitnesses at random without reference to the task. The pathway for this run is shown on the right.

The same scale is used on both sides of Figure 5. The population on the left is generally more converged, and moves faster, than the one on the right. In Figure 6 the speed of movement (through genotype space) of the centroids of each population is plotted, and can be seen to be some 3 times faster under selection than in the random case.

In Figure 5 on the left, there are two periods when the population appears to spread out; at "4 o'clock" and again at "5 o'clock". The latter period, leading up to the transition around generation 2660 (visible in the mean fitnesses in Figure 1), shows the population for any one generation spread out into a cigar shape under this projection. In Figure 7 on the left is shown the elite member of generation 2550 and the rest of the population.

This projection has been calculated so as to emphasise the spread of geno-

417

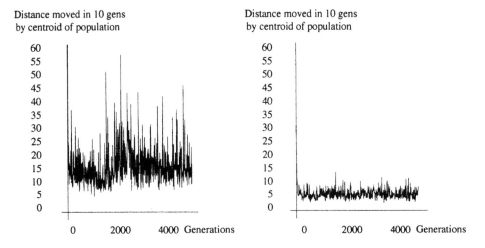

Fig. 6. Plots of how fast the 'centroids' of the populations shown in Figure 5 move; on the left with evolving hardware, on the right with randomly allocated fitnesses.

types from the population history or saga. The corollary of this is that genotypes that are in no way associated with this saga, such as random genotypes, are likely to have only very small components in the direction of these axes, and will appear near the origin or centre of this projection (as indeed the initial random populations do in Figure 5).

In general the elite member of any snapshot is always furthest from the origin of this projection. Non-elite members of a population tend to be mutations derived from the elite of earlier generations; mutations tend to make the genotype 'more random' — have less association with the path of the population — and hence are nearer the origin than the elite.[3] This can be seen on the right hand side of Figure 7, where the 'comet's tail' of mutants streams away from the elite member towards the origin; these mutants were generated by taking 50 copies of the elite, and applying mutations to each at rates varying linearly between 0 and 6.12 expected mutations per 1800 bits, to produce a spread roughly comparable with the left hand side. On the left, with the data from hardware evolution, the comet's tail points off to one side of the origin, indicating that these genotypes are not solely the product of random mutation, but rather of mutation plus selection. One can conclude that the population is spread out along high value ridges in the fitness landscape, which is where even fitter regions may be found. This corresponds to the picture put forward by Eigen et. al. in [4], pages 6886–6887:

---

[3] This follows from the fact that in high-dimensional spaces two random vectors are almost always nearly orthogonal; the projection of one onto the other (dot product) can be expected to be small.

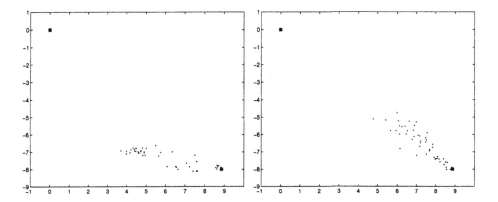

Fig. 7. The same projection is used here as in Figure 5, except it is scaled up and the origin (marked) is now at the top left. In the left figure all 50 genotypes of generation 2500 of hardware evolution are plotted; the elite member is marked at bottom right. On the right the same elite member is shown with 49 other genotypes mutated at random without selection; it can be seen that the population forms a 'comet's tail' pointing towards the origin in this latter case. Here the same PCs are used for the projections on the left and the right.

"In conventional natural selection theory, advantageous mutations drove the evolutionary process. The neutral theory introduced selectively neutral mutants, in addition to the advantageous ones, which contribute to evolution through random drift. The concept of quasi-species shows that much weight is attributed to those slightly deleterious mutants that are situated along high ridges in the value landscape. They guide populations toward the peaks of high selective values."

9   A picture of the landscape

Some indication of the nature of the fitness landscape can be found by plotting the fitnesses of each snapshot taken of the population during evolution. In Figure 8 the fitnesses of each individual in the population are ranged in order within each snapshot, and then ranged alongside each other to cover the 5220 generations (with one snapshot every 10th generation). Necessarily this landscape does not show *all* of the fitness terrain around the current population, but only that part actually sampled (with noise) by genotypes generated through evolution.

Throughout the run there is a small part of the population, usually around 10 out of 50, which has zero or near-zero fitness. After generation 2660 there is a high plateau on which most of the population lies; this is an indication of how much neutral mutation is possible at this stage. For a period leading up to this plateau there is a narrow ridge which holds the current elite, with almost all

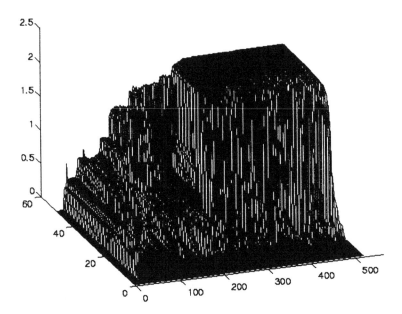

Fig. 8. The fitness landscape, as sampled by the population. Over 500 snapshots are displayed from left to right (one every 10 generations). Within each such snapshot the 50 members of the population are ranked (from back to front of the figure) according to fitness, which is plotted on the vertical scale. After the initial stages a narrow ridge leads up to a flat plateau which starts around snapshot 266, which is generation 2660.

the population considerably less fit. This period corresponds to the increase in genetic diversity in the population (Figures 3 and 5). If the GA had not used the strategy of elitism, this type of ridge might not have occurred; this speculation has not been checked by experiment.

10   Use of SAGA

Normal GAs, indeed most other EAs, are not designed to be used with the genetically converged populations that we have seen here. The Species Adaptation Genetic Algorithm, or SAGA [12, 10, 13], is designed expressly for this purpose. The differences from normal GAs are subtle but significant, and the core requirements of SAGA have been used in the hardware evolution described here.

These can be briefly summarised. Evolution is directed by selection exploiting differences in fitness causes by variations in the genetic make-up of the population. Conventional GAs and GP usually assume that all or nearly all of the variation is that of the initial random population, and hence evolution will cease when this initial variation is exhausted; mutation is thus considered a background genetic operator, and recombination is usually assumed to be the primary operator. In SAGA the reverse is true; of these two genetic operators, mutation is primary, and recombination (though useful) is secondary.

The effectiveness of evolutionary search in a converged species is largely determined by the balance between selection and mutation. Zero mutation leads to stasis, and too much mutation for the selective pressure leads to loss of whatever useful genetic information may have accumulated previously in the genotypes (the 'error catastrophe'); an optimum must lie between these two extremes, and in fact is the maximum rate which still lies below the error catastrophe. In [4] it is shown that in an asexual population with a large population size, on a fitness landscape consisting of a single peak on an otherwise flat plain, the error catastrophe occurs at a rate of $m = \ln(\sigma)$ expected mutations per genotype; where σ is the *superiority* parameter of the master sequence — the factor by which a fittest member on the peak outbreeds the average members on the plain.

For the rank-based selection scheme used here (which includes elitism), $\ln(\sigma) \sim 1$; other selection schemes with significantly different selection pressures need adjustment accordingly. Adding recombination to an asexual population, and reducing population size to the small figures we use such as 50, makes relatively little difference to the impact of the error catastrophe on the effectiveness of evolutionary search [13]. The change from the rugged single-peak landscape to realistic fitness landscapes may make more difference, but always such as to increase the critical error rate.

What *does* make a significant difference, however, is the presence of redundancy or junk in the genotype. If the target mutation rate is, as in our case, 1 mutation per genotype on the assumption of no redundancy, in the presence of junk this should be increased so as to give an expected 1 mutation per *nonredundant* part of the genotype. Usually one does not know in advance what proportion will be junk — it may alter during evolution — but estimates can be checked as evolution progresses. In this case the mutation rate used of 2.7 bits mutated per genotype would be safe (i.e. less than the error catastrophe) if at least 63% of the genotype was redundant; results suggest that this figure was a reasonable one to use. For further details of SAGA [12, 10, 13] should be consulted. A fundamental aspect of the SAGA approach is that of incremental evolution: after success at some level of complexity future evolution at more complex related tasks can continue from the current converged population, rather than starting again with a random population.

11 Conclusion

The picture of artificial evolution presented here differs significantly from that in the main body of EA literature. We have demonstrated with an example of evolution of the hardware configuration on a real silicon chip that fitness improvement can indeed take place within a genetically converged population. With this different approach there is no need to worry about premature convergence, and thus small population sizes may be perfectly acceptable.

An attempt has been made to promote a different perspective on fitness land-scapes. Whereas one *can* create such landscapes with isolated hills which allow a population to get trapped at local optima, it is suggested that landscapes associated with many real problems are *not* of this nature. If there is redundancy of the right kind — 'potentially useful junk' — then neutral networks percolating through genotype space may eliminate most local optima (perhaps all that are likely to be encountered) except for global optima. Though some neutral networks may not be useful, it may be possible to encourage the presence of the practical kind by allowing for many more components to be available for hardware evolution than will be necessary in the final design.

There is a lot more to evolution than meets the eye, and naive models and metaphors may lead to poor decisions in the design of evolutionary algorithms, or prejudice against reasonable decisions. In the context of much contemporary EA practice, the use of a small population of size 50, with genotypes of length 1800 bits, continued for 5000 generations with a genetically converged population on a hard real-world problem, would seem to many to be folly. With the different perspective of SAGA, and consideration of the role of junk and neutral networks, it seems more plausible. The actual result achieved, on a real silicon chip, supports the choice of method.

References

1. H. Asoh and H. Muehlenbein. On the mean convergence time of evolutionary algorithms without selection and mutation. In H.-P. Schwefel Y. Davidor and R. Männer, editors, *Parallel Problem Solving from Nature (PPSN III), Lecture Notes in Computer Science 866*, pages 88–97. Springer-Verlag, 1994.
2. N. L. Cramer. A representation for the adaptive generation of simple sequential programs. In J.J. Grefenstette, editor, *Proceedings of an International Conference on Genetic Algorithms and Their Applications*, Hillsdale NJ, 1985. Lawrence Erlbaum Associates.
3. M. Eigen. New concepts for dealing with the evolution of nucleic acids. In *Cold Spring Harbor Symposia on Quantitative Biology, vol. LII*, 1987.
4. M. Eigen, J. McCaskill, and P. Schuster. Molecular quasi-species. *Journal of Physical Chemistry*, 92:6881–6891, 1988.
5. D.S. Falconer. *Introduction to Quantitative Genetics, 3rd edition*. Longman, 1989.
6. D. B. Fogel. *System Identification through Simulated Evolution*. Ginn Press, Needham, MA, 1991.
7. L.J. Fogel, A.J. Owens, and M.J. Walsh. *Artificial Intelligence through Simulated Evolution*. John Wiley, New York, 1966.
8. C.V. Forst, C. Reidys, and J. Weber. Evolutionary dynamics and optimization: Neutral networks as model-landscapes for RNA secondary-structure folding-landscapes. In F. Moran, A. Moreno, J.J. Merelo, and P. Cachon, editors, *Advances in Artificial Life: Proceedings of the Third European Conference on Artificial Life (ECAL95). Lecture Notes in Artificial Intelligence 929*, pages 128–147. Springer Verlag, 1995.
9. D. E. Goldberg. *Genetic Algorithms in Search, Optimization and Machine Learning*. Addison-Wesley, Reading MA, 1989.

10. I. Harvey. Evolutionary robotics and SAGA: the case for hill crawling and tournament selection. In C. Langton, editor, *Artificial Life III, Santa Fe Institute Studies in the Sciences of Complexity, Proc. Vol. XVI*, pages 299–326. Addison Wesley, 1993.

11. I. Harvey, P. Husbands, and D. T. Cliff. Genetic convergence in a species of evolved robot control architectures. In S. Forrest, editor, *Genetic Algorithms: Proceedings of Fifth Intl. Conference*, page 636, San Mateo CA, 1993. Morgan Kaufmann.

12. I. Harvey. Species adaptation genetic algorithms: A basis for a continuing SAGA. In F. J. Varela and P. Bourgine, editors, *Toward a Practice of Autonomous Systems: Proceedings of the First European Conference on Artificial Life*, pages 346–354. MIT Press/Bradford Books, Cambridge, MA, 1992.

13. I.R. Harvey. *The Artificial Evolution of Behaviour*. PhD thesis, COGS, University of Sussex. Also available via www on http://www.cogs.susx.ac.uk/users/inmanh/, 1995.

14. J. Holland. *Adaptation in Natural and Artificial Systems*. University of Michigan Press, Ann Arbor, USA, 1975.

15. T. Hoshino and M. Tsuchida. Manifestation of neutral genes in evolving robot navigation. Presented at Artificial Life V, Nara, Japan, May 1996.

16. M. Huynen and P. Hogeweg. Pattern generation in molecular evolution: Exploitation of the variation in RNA landscapes. *J. Mol. Evol.*, 39(7):1–79, 1994.

17. M.A. Huynen, P.F. Stadler, and W. Fontana. Smoothness within ruggedness: The role of neutrality in adaptation. *Proc. Natl. Acad. Sci. USA*, 93:397–401, January 1996.

18. N. Jakobi. Encoding scheme issues for open-ended artificial evolution. In Voigt, H.-M., et al. (eds.), *Parallel Problem Solving From Nature IV: Proc. of the Int. Conf. on Evolutionary Computation*, Vol. 1141 of *LNCS*. Heidelberg: Springer-Verlag. Forthcoming.

19. S. A. Kauffman. *Origins of Order: Self-Organization and Selection in Evolution*. Oxford University Press, 1993.

20. M. Kimura. *The Neutral Theory of Molecular Evolution*. Cambridge University Press, 1983.

21. J. R. Koza. *Genetic Programming*. MIT Press/Bradford Books, Cambridge MA, 1992.

22. S. Ohno. *Evolution by Gene Duplication*. Allen and Unwin, London, 1970.

23. I. Rechenberg. *Evolutionsstrategie: Optimierung techniser Systeme nach Prinzipien der biologischen Evolution*. Frommann-Holzboog Verlag, Stuttgart, 1973.

24. H.-P. Schwefel. *Numerical Optimization of Computer Models*. John Wiley, Chichester, U.K., 1981.

25. A. Thompson. Evolutionary techniques for fault tolerance. In *Proceedings of the UKACC International Conference on Control 1996 (CONTROL'96)*, pages 693–698. IEE Conference Publication Number 427, 1996.

26. A. Thompson. An evolved circuit, intrinsic in silicon, entwined with physics. In this volume: *Proceedings of The First International Conference on Evolvable Systems: From Biology to Hardware, 1996 (ICES96)*, Springer-Verlag.

Hardware Evolution System Introducing Dominant and Recessive Heredity

Tomofumi Hikage[†], Hitoshi Hemmi[††], and Katsunori Shimohara[†]

[†]NTT Human Interface Laboratories
1-2356 Take Yokosuka-Shi
Kanagawa 238-03, JAPAN
[††]Evolutionary Systems Department,
ATR Human Information Processing Research Laboratories,
2-2 Hikaridai, Seika-cho, Soraku-gun, Kyoto, 619-02, JAPAN
hikage@nttcvg.hil.ntt.jp, hemmi@hip.atr.co.jp, katsu@nttcvg.hil.ntt.jp

Abstract. This paper proposes a new hardware evolution system – a new AdAM (Adaptive Architecture Methodology), that introduces dominant and recessive heredity through diploid chromosomes in order to increase genetic diversity. Dominant and recessive heredity is implemented by two techniques: one node of a tree-structured chromosome can have two sub-trees corresponding to alleles: Dominant or recessive attributes of a new pair of sub-trees is decided randomly. Simulations using the artificial ant problem show that the new AdAM is superior to the old one in adaptability and robustness in the face of a changeable environment.

1 Introduction

This paper proposes a new AdAM (Adaptive Architecture Methodology) system that introduces genetic diversity to enhance system adaptability and robustness. In order to achieve genetic diversity, we graft dominant and recessive heredity onto the original AdAM [3].

Hardware evolution will enable a system to refine its hardware structure and behavior dynamically. Hardware evolution is now becoming a promising research field given recent progress in evolution methodology and reconfigurable hardware like the FPGA (Field Programmable Gate Array).

In order to create complex adaptive systems, hardware evolution and software evolution are indispensable. Hardware evolution is most effective when the designer can't grasp all conditions at the design stage. In particular, hardware evolution or evolvable hardware will greatly advance of the realization of autonomous robots suitable for uncertain and fluctuating environments.

One approach is to directly treat the configuration data of the FPGA as chromosomes and evolve them using the Genetic Algorithm [2]. Unfortunately, since the configuration data is low-level, it is difficult to apply this approach to large-scale and complex applications.

We are aiming to establish an evolutionary methodology for hardware and to create a computational framework to guide evolutionary hardware. For that purpose, we employ HDL (Hardware Description Language), which provides the new AdAM with the same advantages in terms of both software and hardware high-level description and flexibility, as it does the original AdAM.

HDL programs, which are converted into hardware behavior, are automatically generated and forced to evolve. In the sense that programs evolve as chromosomes, AdAM can be regarded as a form of software evolution. However, we can deal with hardware behavior and structure as pheno-types by using HDL-based programs. Emphasis should be paid to the fact that we treat hardware the same as software.

Dominant and recessive heredity is part of the new AdAM to increase genetic diversity, so that the system can achieve much more adaptability and robustness in changeable and fluctuating environments. To implement dominant and recessive heredity, nodes of tree-structured chromosome are extended so as to have a pair of sub-trees which represent rewriting rules as alleles. Since this creates diploids, only one node of every chromosome is randomly selected to have a pair of sub-trees, and genetic operations are controlled to generate only diploids. Whether a rewriting rule is dominant or recessive is decided randomly as each rule pair appears.

In this paper, we compare the new AdAM to the old one to verify adaptability and robustness in changeable and fluctuating environments through simulations on the artificial ant problem.

2 How to Model Dominant and Recessive Heredity

Most multi-cellular organisms are diploid with a set of diploid chromosomes, and exhibit dominant and recessive heredity; a haploid, by comparison, has only one set of chromosomes. Since we usually employ a model in which one individual has one "chromosome", we postulate that a diploid has a pair of chromosomes and a haploid has a chromosome to simplify the following discussion.

In the case of a haploid, a change in a chromosome immediately appears as a change in its pheno-type experiencing selection pressure. In general, most changes in a chromosome due to mutation or deletion are harmful, in the sense that the probability of destroying a gene that generates an effective function is higher than that of gaining a new gene with a superior function. In order for a haploid to acquire a new function effectively, gene duplication should occur before having mutation. Such duplication allows the haploid to have copies of a gene for an existing function in pheno-type, its chromosome can acquire a new gene by mutation while keeping the original gene. That is, gene duplication must precede mutation so haploid evolution takes a long time in general.

Genetic Programming (GP) and genetic algorithms (GA) usually employ a haploid model. This means that GP and GA have the same drawback mentioned above. In GP and GA, however, a crossover operation is introduced, with which two different chromosomes, i.e., individuals, exchange a part of their chromosomes to offset the drawback.

A diploid, on the other hand, has a pair of chromosomes that are different in most cases. One gene is, in pheno-type, dominant. Thus, when it is changed by mutation, its pheno-type suffers selection pressure according to its fitness for the environment. The other gene, the recessive one, does not appear in the pheno-type and can be changed irrespective of the pheno-type. That is, even if a change in a recessive gene yields an effective function, it does not immediately improve fitness.

A diploid does not need gene duplication before mutation, unlike a haploid, because the recessive gene is available even if the dominant one is destroyed by mutation. Moreover, if gene duplication occurs in either of the pair genes, and if crossover between the pair of chromosomes takes place, the dominant and recessive genes may be placed in the same chromosome. Thus, a diploid has much more diversity and a higher possibility of acquiring a new gene than a haploid, while still keeping the recessive gene [1].

A gene in a chromosome can be dominant or recessive, so a chromosome usually consists of a mix of dominant and recessive genes. However, system becomes too complex and complicated if we employ an equivalent chromosome structure. Thus, when we mimic diploid chromosomes and dominant-recessive heredity, chromosomes are randomly defined as dominant or recessive, for simplicity. In addition, since a chromosome in our system consists of a node tree as described later, we model diploid and dominant-recessive heredity with one chromosome, in which only one node has two sub-trees corresponding to alleles, instead of a pair of chromosomes.

3   New AdAM as an Autonomous Hardware Evolutionary System

3.1   Outline of the System

Figure 1 is the system block diagram of the new AdAM. It is the same as that of the old AdAM except chromosome expression and some genetic operations that reflect dominant-recessive heredity.

The HDL that we use is SFL (Structured Function description Language), developed for PARTHENON (Parallel Architecture Refiner THEorized by Ntt Original coNcept) system - a kind of CAD/CAM system for ASIC LSIs. AdAM is an evolutionary system based on the autonomous generation of HDL programs. SFL is a high-level language like C and so offers high-level descriptions that are easy to understand.

Production rules to generate chromosomes are a set of SFL grammar expressed by BNF (Backus-Naur Form) which work as rewriting rules. By applying such rewriting rules repeatedly, a chromosome is produced as a node tree in which each node corresponds to a rewriting rule. As a chromosome is expressed as a node tree, genetic operations can be assigned to the chromosome. Figure 2 shows a part of an SFL grammar definition (BNF expression). In Figure 2, the rX.m in round brackets expresses rule-number, where X is category-number

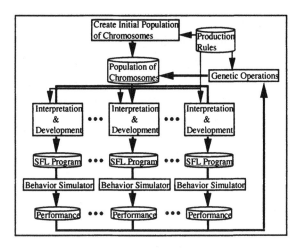

Fig. 1. System Block Diagram

```
(r0.0)  sfl_desc  ->  seq0_mod_def
(r1.0)  seq0_mod_def -> empty
(r1.1)  seq0_mod_def -> seq0_mod_def mod_def
(r2.0)  mod_def -> K_MODULE mod_name K_LBRACE seq0_submod_type_OR_fc_type_def
         seq0_mod_component_def seq0_mod_ctrl_pin_arg_def seq0_stage_and_task_def
         seq0_mod_act_with_ctrl_pin_def seq0_stage_act_process_def K_RBRACE
(r3.0)  mod_name -> nm_mod
                •
                •
                •
```

Fig. 2. A part of SFL grammar definition (BNF expression)

and m is sub-number. For example, r1.0 expresses rule-number of seq0_mod_def
→ empty, where 1 is category-number of seq0_mod_def and 0 is the sub-number
of empty. This rule rewrites seq0_mod_def into empty. Equally, r1.1 expresses
another rewriting rule of category seq0_mod_def.

A chromosome – a tree structure of rewriting rules based on SFL grammar –
is automatically interpreted and developed into an SFL program that describes
hardware behavior. Each non-terminal symbol, shown in small letters in Figure 2,
has a rewriting rule. The capital letters in the figure are terminal symbols that
have no rewriting rules. Non-terminal symbols must be rewritten as terminal
symbols. Figure 3 indicates a simple SFL program.

```
module jk{
        instrin Jin,Kin;
        instrout Q,invQ;
        stage_name LATCH { task RUN(); }
        instruct Jin generate LATCH.RUN();
        instruct Kin generate LATCH.RUN();
        stage LATCH {
                state_name J_state,K_state;
                first_state J_state;
                state J_state any{
                        Kin:    par{
                                goto K_state;
                        }
                        else:   Q();
                }
                state K_state any{
                        Jin:    par{
                                goto J_state;
                        }
                        else:   invQ();
                }
        }
}
```

Fig. 3. Example of SFL program

This example program describes JK latch. A description of the module is that the description unit of SFL starts at K_MODULE(module) and finishes at K_RBRACE(}) according to BNF expression guidelines.

Overall system behavior is as follows: First, the initial chromosome population is randomly created according to the production rules. Each chromosome generates an individual SFL program through interpretation and development. Individual SFL programs, i.e., individual hardware behaviors, are simulated and evaluated by a given fitness function. Individuals with good performance are selected so that they become a parent of the next generation. Genetic operations are applied to the chromosomes of the parents to generate the chromosomes of the next generation. Such evolutionary processes eventually evolve individuals that encode for desirable hardware behaviors.

3.2 Chromosome with Dominant-Recessive Heredity

Figure 4 indicates a chromosome with dominant-recessive heredity.

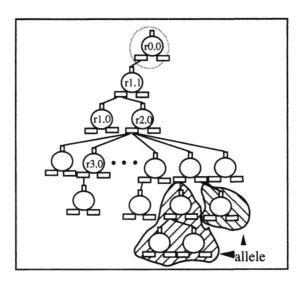

Fig. 4. A chromosome with dominant-recessive heredity

As mentioned before, we model dominant-recessive heredity with one chromosome by assigning one node to have two sub-trees corresponding to alleles. As shown in figure 4, a chromosome has a tree-structure of nodes each of which has a rule-number of a rewriting rule. In figure 4, two sub-trees are delineated by shading. One sub-tree is dominant; the remaining one is recessive.

Which node has two sub-trees is randomly determined when a chromosome is generated as follows: First, a node with rule-number r0.0 is generated. Category-number 0 expresses sfl_desc, and sfl_desc is rewritten to seq0_mod_def. The non-terminal symbol seq0_mod_def is assigned category-number 1 and has two sub-numbers 0 and 1. A node with a rule-number that has sub-numbers more than one, like seq0_mod_def, can be the node expressing dominant-recessive heredity, and the node used to express dominant-recessive heredity is randomly selected among all nodes with sub-number more than one. In the selected node, rules whose sub-numbers are randomly selected from those in the same category are allocated to two sub-trees of the node, and then which sub-tree is dominant is also determined randomly. If a node is not selected for dominant-recessive heredity, one of rules with sub-numbers is selected randomly. For example, a node next from r0.0 is generated by r1.0 or r1.1.

Moreover, especially by reproduction, one of genetic operations, described later, it happens that two sub-trees for alleles change to have both dominant or recessive. In the case, dominant or recessive of two sub-trees is also determined randomly.

3.3 Genetic Operations

Along with introducing dominant-recessive heredity into a chromosome, we define and employ the following genetic operations. These genetic operations have to be controlled so that chromosomes for diploids are always generated.

- Mutation: This operation changes the sub-number of a rule in a node, e.g., from rX.1 to rX.2 or rX.3, or changes the constant information of a node or the node information of constant number.
- Duplication / Deletion: This operation duplicates or deletes a node by changing the number of times the recursive rewriting rule is applied.
- HL (HapLoid)-type crossover: This operation exchanges the sub-numbers of two nodes, which have the same category-number, but each of which belongs to a different chromosome. As the only information exchanged is the sub-number of the same category number, each chromosome is kept grammatically correct. Figure 5 indicates the HL-type crossover operation.

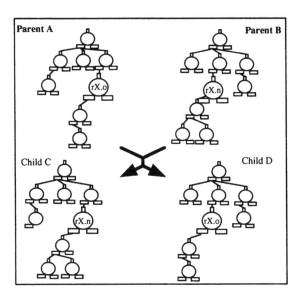

Fig. 5. Example of HL-type Crossover operation

- DL (DipLoid)-type crossover: This operation is newly introduced along with dominant-recessive heredity. It exchanges the sub-numbers of two nodes, which have the same category-number, but each of which belongs to either of two sub-trees corresponding to alleles in the same chromosome. Therefore, a part of recessive gene may be included in a dominant sub-tree, and vice versa.

– Reproduction: This operation is also newly introduced, and is the operation most characteristic of dominant-recessive heredity. It exchanges the sub-trees of two nodes that have the same category-number, but each of which belongs to a different chromosome. For example, the chromosomes owned by parent A and B yield child C and D, as shown in figure 6.

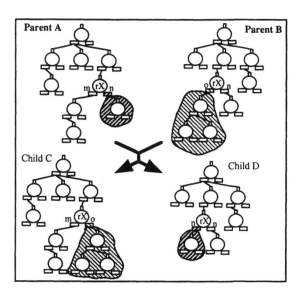

Fig. 6. Example of Reproduction with dominant and recessive heredity

Reproduction is to fuse two "germ cells", that is, the germ cells of parent A and B are fused to yield child C and child D. The dominant or recessive tags of the sub-trees of parent A and B are already determined before reproduction. In figure 6, X indicates category-number, and m, n and o indicate sub-numbers. If we assume that sub-number m and o are dominant for sub-number n, the sub-number n can not appear in pheno-type at the parent's generation. However, at the child's generation, child D has two recessive sub-numbers in its two sub-trees, that is, both sub-trees are recessive. In that case, the recessive sub-number appears in pheno-type.

The operations described above prevent any genetic operation from generating lethal chromosomes. Thus, efficient and effective evolution without waste can be achieved, and the SFL programs resulting from genetic operations can also satisfy the SFL program definition requirements.

4 Experiment

We compared the new AdAM to the old AdAM by examining their performance in terms of adaptability and robustness under environmental fluctuation; both were applied to the artificial ant problem. This problem is to evolve the hardware behavior of artificial ants so that they collect food faster and more effectively. The best artificial ant should collect the most food in the fewest steps. Figure 7 indicates the environment of the problem. The black squares are food. The experiment environment is a torus. The artificial ant starts at the top left corner.

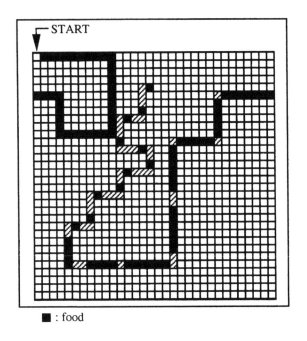

Fig. 7. Experiment Environment I

4.1 Experiment Conditions

Experiment conditions are as follows.

- Population size
  - 256 individuals
- Individuals/environment
  - 1 individual
- Selection Method
  - Roulette Model
  - Elite strategy
    * Elite size : 20

- Genetic operation

  - mutation
  - duplication
  - deletion
  - crossover(HL-type and DL-type)
  - reproduction

- I/O Facilities of Artificial Ant (see the figure 8)

  - Input pin : 5 pins
  - Output pin : 2 pins

- Fitness function

  - Fitness = Score + limit - Step + 1
    Score : units of food collected
    limit : Maximum step number allowed in this system (350 steps)
    Step : move from one square to another
    If all food is collected, AdAM is stopped.

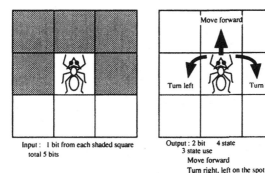

Fig. 8. Artificial Ant

4.2 Environmental Fluctuation

The effectiveness of the new AdAM was challenged by dynamically changing the food location. The food location was changed to form a model environmental fluctuation. Figure 9 indicates the changed environment.

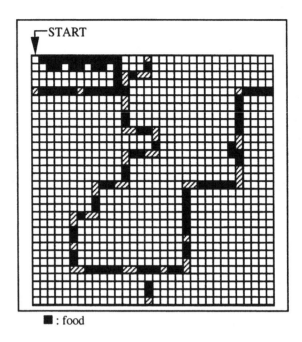

Fig. 9. Experiment Environment II

After individuals reaching maximal fitness appeared in environment I, the environment was changed to environment II. The maximal fitness was 293 points in environment I and 275 points in environment II.

In environment I, the best individual (artificial ant) uses only simple state transition. For the environment given, the ant does not need to judge the priority of inputs; it can collect the food quickly by moving forward.

In environment II, the ant must assess the priority for each input to collect all the food. If the ant simply moves forward, it will not be able to collect all the food within the step limit. That is, in environment II, the ant must acquire the behavior of assessing input priority in addition to those behaviors learned in environment I; environment II is more difficult and complicated than environment I.

5 Result

Figure 10 shows the result of the new AdAM, while figure 11 shows that of the old one; the environment was switched from I to II after individuals of maximum fitness emerged. Although the old AdAM achieved maximal fitness faster than the new one in environment I, the old AdAM spent more generations to get over 200 points than the new AdAM after the change in the environment. The new AdAM's mean of fitness values and standard deviation are higher than those of the old AdAM. That is, individuals in the new AdAM were better in performance as a whole than those of the old AdAM. The new AdAM is superior to the old AdAM in terms of genetic diversity because its standard deviation value was bigger.

Figure 12 shows the performance of the new AdAM, while figure 13 shows that of the old AdAM in environment II. The old AdAM evolved individuals that could get high fitness faster than the new AdAM in environment II. However, the old AdAM needed more generations to develop excellent individuals than the new AdAM when the environment fluctuated.

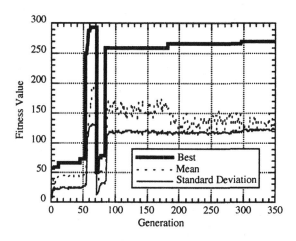

Fig. 10. Result of the new AdAM in fluctuating environment (I to II)

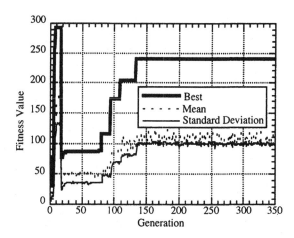

Fig. 11. Result of the old AdAM in fluctuating environment (I to II)

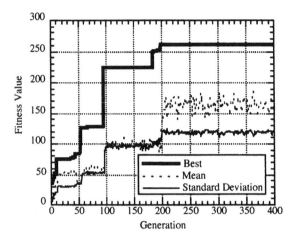

Fig. 12. Result of the new AdAM in environment II

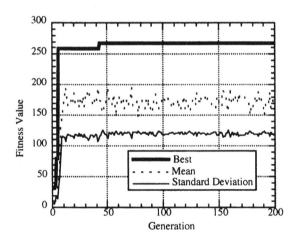

Fig. 13. Result of the old AdAM in environment II

6 Conclusion

We have proposed the new AdAM as an autonomous hardware evolution system. It offers dominant-recessive heredity to achieve genetic diversity which enhances system adaptability and robustness against changeable environments. A new chromosome model and several genetic operations were also proposed in order to implement diploid and dominant-recessive heredity into the system. In terms of the richness of chromosome expression and the number of genetic operations possible, we can say that the new AdAM achieves genetic diversity.

We compared the new AdAM to the old one by verifying system adaptability and robustness to environment changes by simulating the behavior of an artificial ant colony. Although the old system evolved individuals that got maximal points faster than the new one, the new system offered much faster adaptation than the old one when the environment was changed. That is, individuals of the new system are robust against environment changes.

Simulation results suggest that progressive evolution, in which evolution takes place stepwise to match environment changes, is much more effective and useful as a computational framework for hardware evolution. We will continue to refine the proposed system through investigations into a new chromosome expressed by a general graph and/or progressive evolution as further studies.

References

1. Bruce Albers et al. *MOLECULAR BIOLOGY OF THE CELL SECOND EDITION.* KYOIKUSHA, 1993.
2. Tetsuya Higuchi, Tatsuya Niwa, Toshio Tanaka, Hitoshi Iba, Hugo de Garis, and Tatsumi Furuya. Evolvable hardware – genetic based generation of electric circuitry at gate and hardware description language (HDL) levels. Technical Report 93–4, Electorotechnical Laboratory, Tsukuba, Ibaraki, Japan, 1993.
3. Hemmi Hitoshi, Jun'ichi Mizoguchi, and Katsunori Shimohara. Development and evolution of hardware behaviors. In *Artificial Life IV.* MIT Press, 1994.

CAM-Brain: A New Model for ATR's Cellular Automata Based Artificial Brain Project

Felix Gers   **Hugo de Garis**

ATR, Human Information Processing Laboratories
2-2 Hikari-dai, Seika-cho, Soraku-gun, Kyoto 619-02, Japan.
Phone: +81-774-95-1079 , Fax : +81-774-95-1008 ,
email: flx@hip.atr.co.jp & degaris@hip.atr.co.jp
URL : http://www.hip.atr.co.jp/~flx
& http://www.hip.atr.co.jp/~degaris

Abstract. This paper introduces a new model for ATR's CAM-Brain Project, which is far more efficient and simpler than the older model. The CAM-Brain Project aims at building a billion neuron artificial brain using "evolutionary engineering" technologies. Our neural structures are based on Cellular Automata (CA) and grow/evolve in special hardware such as MIT's "CAM-8" machine. With the CAM-8 and the new CAM-Brain model, it is possible to grow a neural structure with several million neurons in a 128 M cell CA-space, at a speed of 200 M cell-updates per second. The improvements in the new model are based on a new CA-implementation technique, on reducing the number of cell-behaviors to two, and on using genetic encoding of neural structures in which the chromosome is initially distributed homogeneously over the entire CA-space. This new CAM-Brain model allows the implementation of neural structures directly in parallel hardware, evolving at hardware speeds.

Keywords

Artificial Brains, Evolutionary Engineering, Neural Networks, Genetic Algorithms, Genetic Encoding, Cellular Automata, Cellular Automata Machine (CAM-8), Evolvable Hardware.

1   Introduction

The aim of ATR's CAM-Brain Project is to build/grow/evolve brain-like-systems, into which human knowledge can be inserted as initial conditions. The brain-like structures are based on cellular automata (CA) and grow inside cellular automata machines (CAMs). Evolutionary engineering techniques, combined with huge computing power, may then be able to evolve such brain-like systems with a level of functionality simular to that of biological brains. The expectation is that it will be much easier to investigate the behavior of engineering-based brains than biological brains, and that the engineering solutions will reveal insights into the working mechanisms of the evolved systems, which might then be rediscovered in biological brains.

In the CAM-Brain Project, the CAs act as the medium in which to conduct the growth, and in a second step, the neural signaling, of brain-like structures.

To apply evolutionary engineering techniques to the creation of huge neural systems, e.g. an artificial insect brain, we need a simulation tool that has to satisfy two conditions. First of all, it must be possible to simulate most of the important characteristics of biological neural networks. Secondly, the evolution times must be manageably short. The bottle-neck in evolutionary algorithms is the fitness measurement time, i.e. the time needed to simulate the systems. Hence it must be possible to simulate these systems on massively parallel hardware.

With CAs as a simulating tool, both demands can be met. Artificial neurons and neural networks can be modeled at any level of detail, because CA cells can symbolize a molecule in a biological cell membrane, a whole neuron, or any structure in between. Since CAs are only locally connected, they are ideal for implementation on purely parallel hardware. CAs are both structurally flexibile and inherently parallel. We use MIT's Cellular Automata Machine CAM-8 [1] for our CAM-Brain modeling.

With the new CAM-Brain model that is introduced in this paper, we have overcome many of the deficiencies of the old model. Nevertheless, the old model did at least demonstrate the realizability of evolvable neural structures based on CAs [4].

The improvements of the new model are based on several ideas, plus the use of a higher level language to generate our CAs and transform them into an executable rule-base. The new model is implemented on MIT's CAM-8 machine, using a kind of CA that we call a "partitioned" CA. Partitioned CAs use a signal-based cell interaction, instead of exchanging complete cell-states. (See section 3.1 for details.) We introduce a new technique for the genetic encoding of neural structures, based on the idea of initially distributing the network's chromosome homogeneously over the whole CA-space. We have reduced the number of CA cell interactions to two when modeling the signaling of artificial neurons.

These features of the new CAM-Brain model can be consistently combined, so that the complete model, including growth, signaling and evolution, can be implemented in 2 bytes of memory per cell, and with a small number of logic gates. The simplifiction means that it may be possible to implement the new model directly in hardware.

After describing the old model, (the first work in the field of CA-based brain-like structures [4]), the following three sections explain the improvements of the new model in more detail.

2 The Old CAM-Brain Model

The old model was based on Codd [3], i.e. the CA trails representing the axons and dendrites were 3 cells wide, with 2(4) outer "sheath" cells (in the 2D(3D) version), and an inner cell which was used to transmit the growth signals in the growth phase, and the neural signals in the signaling phase. As will become clear from the perspective of the new model, the essential feature of the old model was that the state of each cell contained NO information concerning the direction of flow of signals. It was the insertion of such directional information into the new model which enabled a revolution in the simplicity and efficiency of its implementation. In the old model, the sheath cells guided the signal flow. At junctions in the signal path, gating cells (turn left, turn right) were employed to direct neural signals to their receiving neurons. Collisions between CA trails became complex, too complex for writing high level general rules to govern their dynamics. Hence new CA state transition rules had to be found empirically, i.e CAM-Brain would run until a missing rule was found. It would then prompt the human operator for the desired next state, store the new rule automatically into the rule-base, and continue running until the next missing rule. The old model of 3D CAM-Brain required nearly 4 million rules. This large number of rules resulted from the 24 symmetry rotations of each 3D rule generated empirically. This empirical approach worked, but was slow. It took 2 years to generate all the 3D rules in the old model.

We now give a brief overview of some of the details of the old model. We do not spend a lot of time on this, because such details have already been presented in many publications, e.g. [4].

The CA based network of the old model consists of two types of neuron, excitatory and inhibitory. At first, neuron bodies are grown from an initial pattern, as shown in figure 1. The starting positions of the seed cells can be either user specified or placed under genetic control.

To each grown neuron a "chromosome" block of random growth instruc-

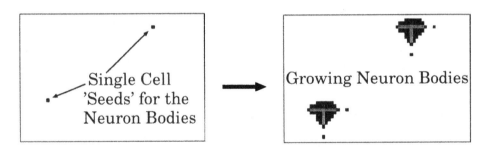

Fig. 1. The growth of the neuron bodies (right) from an initial pattern (left).

tions is attached. These chromosome blocks are stored in a subspace "under" the actual CA-space as shown in figure 2.

With these growth instructions, each neuron grows an axonic and a dendritic tree, which later will be used to conduct neural signals. A collision of a dendritic and an axonic branch results in the formation of a synapse. The synapses are not weighted. The weighting is expressed indirectly via the length of the dendritic trail that a signal has to pass through. A signal decreases linearly in strength while traveling along a dendrite, so the weighting can be adapted by evolving the length of the dendritic trail.

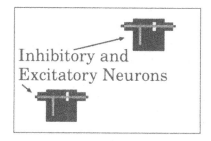 

Fig. 2. The neurons (left) with their individual chromosomes (right). A chromosome contains 6 different kinds of growth signals : grow straight, turn (left, right), split (left, right) and T-split. Every CA cell of a chromosome represents one growth signal.

Each neuron grows its dendrite and axon with its individual chromosome information, until either all the chromosome information is used, or there is no more growth possible in the given CA-space. Alternatively, the chromosome can be wrapped around, so that the available space determines when growth stops. Figure 3 shows examples of early and saturated growths.

To measure the fitness of a grown circuit, signal values are extracted from user chosen points, and are used to control some process. The quality of the control becomes the fitness of the circuit. A conventional genetic algorithm is then applied to a population of circuits. de Garis has shown, the evolvability of the model (for details, see [4]).

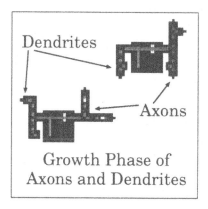 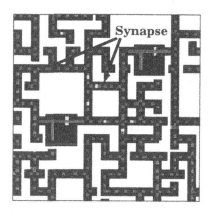

Fig. 3. (Left) The growth phase of the network continues until all CA trails are connected through synapses or are blocked. (Right) A network, fully grown with the old CAM-Brain model.

3 The New CAM-Brain Model

The old CAM-Brain model is a pioneering work. It shows the realizability of evolvable neural networks based on CA and the possibility of combining evolved neural net modules to build artificial brains. As with most pioneering works, the old model opens new directions for research, but still contains deficiencies to be overcome. Some of the deficiencies of the old CAM-Brain model are

- inefficient use of CA-space, (based on Codd's model [3], and i.e. the use of sheath (and gating) cells),
- restrictions on growable structures of the axonic and dendritic trees, due to its "fractal" growth model,
- the large number of CA-states and CA-rules,
- low (cellular) level of CA rule generation rather than using high level "abstract" rules, (empirical, rather than automatic CA rule generation, which caused the slow rate of rule generation (i.e. 2 years instead of 20 seconds) and inflexibility with respect to changing the model),
- the necessity of symmetry rotations for each rule (4 rotations in 2D and 24 in 3D).

In the new model we overcome these problems and add ideas from the old model to the new. The next subsection (3.1) gives a general description of the CA-type we use. We introduce a CA type which differs somewhat from the more usual form of CA. We call this variant "partitioned CA". In the following three subsections (3.2 - 3.3) we describe how the deficiencies of the old model are overcome, and how these improvements are incorporated into the new model. We call this new CAM-Brain model the "CoDi-model" for neural signaling. We also show how it is possible to distribute the chromosome initially over the CA-space, an idea which has several advantages over the old model.

3.1 General Description of our CAs

When designing our CA-based neural networks for the new model, our objective is to implement them directly in evolvable hardware, e.g. FPGAs, content addressable memory chips, etc. Therefore, we try to keep the CAs as simple as possible, by having a small number of bits to specify the state, keeping the CA rules few in number, and having few cellular neighbors.

We use a von Neumann neighborhood [2], i.e. a cell in a 2D space looks at its four nearest neighbors and its own state [1]. At present our CA is implemented on a 2D space. We plan to extend it to 3D CA-spaces soon, since the general approach taken in our model is independent of the dimensionality of the CA-space. We use two different CAs for the growth phase and the signaling phase in our neural networks. The cell interactions are similar in both CAs. The main difference between the two phases, and hence the two CAs, is that the type of a cell can only change during the growth phase, e.g. a blank cell can change to

---

[1] In general this situation is refered to as a "neighborhood" with 5 neighbors.

an axon cell. (See later sections for details.) Cell types do not change during the signaling phase, although cells do change their activity values and exchange neural signals. Since the cell-to-cell interactions are similar for the 2 phases, the same CA can be used in both. This might be useful, when a future learning algorithm needs to grow new connections, depending on the activity in already existing axonic or dendritic connections.

Partitioned CAs The states of our CAs have two parts, which are treated in different ways. The first part of the cell-state contains the cell's type. The second part serves as an interface to the cell's neighborhood. It contains the cell's activity and the input signals from the neighbors. Characteristic of our CA is that only part of the state of a cell is passed to its neighbors, namely the signal and then only to those neighbors specified in the fixed part of the cell state. We call our CAs "partitioned", because the state is partitioned into two parts, the first being fixed and the second is variable for each cell (see figure 4).

The advantage of this partitioning-technique is that the amount of informa-

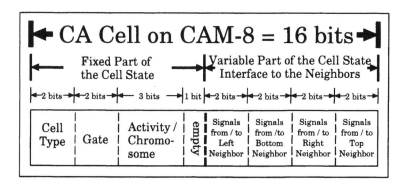

Fig. 4. State representation in the new CAM-Brain model, which uses 15 of the 16 bits in the CAM-8 CA-cell. During the growth phase 3 of the bits are used to store the chromosome's growth instructions. The same 3 bits are later used to store the activity of a neuron cell during the signaling phase.

tion that defines the new state of a CA cell is kept to a minimum, due to its avoidance of unnecessary information-exchange. However CAs with many complex behaviors are usually designed with complete exchange of cell states, thus ensuring redundant information exchange. The old CAM-Brain model used such a complex CA with full exchange of state [4].

The old CAM-Brain CA needs 8 bits and 32 K rules to grow its neural networks. Even with the small von Neumann neighborhood, i.e. 5 neighbors, a 40 bit string of neighborhood information is necessary to determine the new cell state. If this CA is not compressed and transformed into a number of smaller CAs (to fit into a 16 bit look-up-table (LUT)), it is not possible to implement it on MIT's CAM-8 machine [10]. This compression and transformation makes the execution of the CA very slow and inefficient, especially if the down-loading time for new CA rule-sets is large compared to the execution time. For a CA implementation in FPGAs, down-loading a new rule-set means reconfiguring the FPGA completely. This technique is unacceptably slow in the case of FPGAs and other programmable hardware devices, especially when 3D circuits become realizable[11].

Another possibility to reduce the number of states in CAs is to use non-uniform CAs [8]. In non-uniform CAs, different rule-sets are used simultaneously, depending on the position of a cell in the CA-space. This non-uniformity can reduce the number of CA-states, and hence the amount of information exchanged between CA cells. The reduction of CA-state size causes a reduction in the total number of CA-rules, which, for a non-uniform CA, is the sum of the numbers of rules in all rule-sets. Non-uniform CAs can always be transformed into uniform CAs by adding CA-states and corresponding CA-rules. The distinct smaller rule-sets of non-uniform CAs can be stored sequentially in the larger rule-table of the corresponding uniform CAs. To distinguish the rule-sets of the non-uniform CA, in this sequence, new states are added to the state-set. This extension of the state size allows cells to use the part of the rule-table which corresponds to their "local" rule-set in non-uniform CAs.

The transformation from non-uniform CAs to uniform CAs removes the advantages of non-uniform CAs for the sake of uniformity. On the other hand, uniformity might be necessary to run a model on prospective parallel hardware [9].

Using partitioned CAs offers the advantages of non-uniform CAs by reducing information transfer, even though it is uniform. The cells pass the same reduced amount of data to their neighbors as the cells of non-uniform CAs. They do not transmit information about their own cell type. This corresponds to the fact that cells in non-uniform CAs do not know which rule-set their neighbors use. In this way the partitioning technique reduces the amount of information transfer that is necessary for a cell to determine its new state, e.g. by looking it up in a look-up-table (LUT). As a consequence, the necessary LUT gets significantly shorter than the corresponding LUT of traditional CAs. The same reasoning is valid for the storage of CA cell rules in programmable hardware. Partitioned CAs simplify the hardware programming.

This is a crucial advantage for the implementation of CAs into parallel hardware. Large LUTs or complex programs cannot be stored locally, and hence are non-parallel. Using the partitioning-technique, it becomes possible to use CAs with complex interactions for our new CAM-Brain model, and to simulate neural networks directly in parallel hardware.

The next three subsections describe the new CAM-Brain model and its implementation as a partitioned-CA in detail.

3.2 Codi-Model for the Signaling of Neural Structures in CA

As mentioned in section 2, in the 2D version of the old model, transportation of information through the CA was inspired by Codd [3]. Neural signals pass through 3-cell wide trails, which are gridded on a 6-cell grid (see figure 3). The neuron bodies consist of $11 \cdot 11 = 121$ cells (see figure 2). The result is a CA-space with at least 25% of the CA-cells unused. Two thirds of the occupied cells are sheath or blocker cells, and remain constant during the whole signaling phase of the neural network. This is inefficient use of the CA-space. An efficient CA neuron model should satisfy the following criteria.

- Its CA trails should be only one cell wide. For example, it could use signaling techniques similar to the model proposed by von Neumann [2].
- The grid should be dispensable, so that signal trails can be dense without disturbing each other.
- Contiguous signaling should be possible.
- Neuron bodies should be as small as possible, e.g. one cell.

We succeeded in satisfying all these criteria and incorporated them into a single model for the signaling of neural structures. We named the model "CoDi" based on the essential features of its axonic and dendritic cell behavior, i.e. to "collect" or to "distribute" the cellular values from/to the CA cells' neighbors. The CoDi-model works with four basic types of CA cells, neuron body, axon, dendrite and blank. Blank cells represent empty space. They do not participate in any cell interaction during the signaling of the neural network.

Neuron bodies consist of one CA cell. The neuron body cells collect neural signals from the surrounding dendritic cells and apply an internally defined function to the collected data. In the CoDi model the neurons sum the incoming signal values and fire after a threshold is reached. This behavior of the neuron bodies can be modified easily to suit a given problem. The output of the neuron bodies is passed on to its surrounding axon cells.

Axonic cells distribute data originating from the neuron body. Dendritic cells collect data and eventually pass it to the neuron body. These two types of cell-to-cell interaction cover all kinds of cell encounters. Axonic and dendritic cells do not need to know anything about the cell type of their neighbors. They behave in the same way towards any type of neighbor. Therefore it is not necessary to distinguish between the different cases of cell-to-cell interactions, e.g. axon-neuron, axon-dendrite, dendrite-dendrite etc, so that designing separate rules for each case is unnecessary. The two basic interactions cover every case, and they can be expressed simply, using a small number of rules. (This simplicity is essential when we come to implement CAM-Brain directly in hardware.) To make axonic and dendritic signal trails (which can be dense in the CA-space) without blockers or sheaths, the four basic cell types need further specification. To the two bits which specify the basic cell type, we added two more bits to set the "gate" of a cell. (See figure 4.) A neuron cell uses this gate to store its orientation, i.e. the direction in which the axon is pointing. In an axon cell, the gate points to the neighbor from which the neural signals are received. An axon cell accepts input only from this neighbor, but makes its own output available to all its neighbors. In this way axon cells distribute information. The source of information is always a neuron cell. Dendritic cells collect information by accepting information from any neighbor. They give their output, (which is the sum of their inputs) only to the neighbor specified by their own gate. In this way, dendrite cells collect and sum neural signals, until the final sum of collected neural signals reaches the neuron cell.

Each axon and dendrite cell "belongs" to exactly one neuron cell. This configuration of the CA-space is guaranteed by the preceding growth phase (section 3.3). Figure 5 shows the signaling phase with the CoDi-model. Each neuron is given two dendritic trees and two axonic trees. One of the axonic trees distributes inhibitory signals of the neuron, the other distributes excitatory signals. A neuron has two dendritic trees to maximize the amount of information that can be passed to the neuron body in each time step.

We use neural signals with signal-values in the range from -1 to 2 (= 2

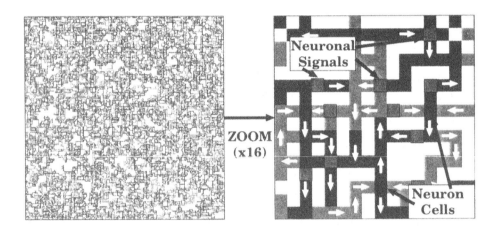

Fig. 5. (Left) A CA-space with $256 \cdot 256$ cells in the signaling phase with the CoDi-model. (Right) A zoom (x16) with five neuron cells (black) with two dendrites and two axons each. The arrows inside the axonic (dark grey) signal trails and dendritic (light grey) signal trails indicate the direction of information flow during the signaling phase. In the growth phase the signals in the dendritic trails travel the other way, because they are emitted by the neuron. To improve the visualization of neural trails, the underlying chromosome is gridded with a checker-board pattern of "grow-straight" instructions. This causes the neural trails to grow on a 2-cell grid.

bits). Signals can be contiguous. This is not possible in the old model, where neural signals travel in groups of three cells (triplets). Two of these three cells determine the direction of the signal movement. In the new model, the direction of motion is stored in the underlying signal trails. The signals are one cell wide and can fill a trail completely, as is the case in the growth phase.

Synapses The old CAM-Brain model uses synapses to transfer neural signals from axons to dendrites. The new model does not need synapses, because dendrite cells that are in contact with an axonic trail (i.e. have an axon cell as neighbor) collect the neural signals directly from the axonic trail. This results from the behavior of axon cells, which distribute to every neighbor, and from the behavior of the dendrite cells, which collect from any neighbor.

However axon-axon and dendrite-dendrite interactions of cells that do not belong to the same neuron are not possible. This is because axon cells take their information only from the neighboring axon cell that is closer to the "source" neuron on their axonic trail. Dendritic cells behave in the opposite way, by passing signals to the dendritic neighbor that is closer to the neuron or to the neuron cell itself. This configuration of the CA-space results from the neural growth phase.

These two simple behaviors of axon and dendrite cells (i.e. collect and distribute), which are exact opposites, allow axonic and dendritic trails to be dense in the CA-space, without interacting in an unintended way. The CoDi-model remains functional even if every cell in the CA-space is used.

3.3 Genetic Encoding of Neural Networks in CA with a Distributed Chromosome

Biological Systems have to self-assemble. The genetic encoding of biological brains is indirect, sophisticated and incomplete. There is not enough genetic information in vertebrates to encode the detailed connectivity of their brain circuits.

However, these characteristics of biological brain growth need not apply to electronic brains. Electronic circuits do not have to grow. They can be copied into hardware as initial settings. The genetic encoding can be direct. For example, a neural circuit can be stored as a spatial pattern in bitmap format. The genetic information can be complete, so that a given chromosome always produces exactly the same neural circuit and vice versa.

For our CAM-Brain models, we have kept the biological feature of growing a system. We choose a genetic encoding that is indirect, but very different from biological genetic encoding. We wanted our encoding to be one to one, so that a given chromosome leads to exactly one neural network and vice versa. In this way we ensure that a chromosome has a unique fitness value and hence do not have to spend time calculating an average fitness for each chromosome.

The advantage of a grown system is that the structural integrity of the basic components (in our case the neurons) is preserved in the presents of mutations (or alternations in the chromosome). All non-blank cells in the CA-space can be assigned to exactly one neuron, (the neuron from which they were grown). This is the case for any chromosome. Though the connectivity of the neural network may be unsuitable for a given problem after the growth according to a arbitrary chromosome, a consistent structure of the neural network is always guaranteed. The importance of this feature becomes apparent when we compare a) the effect of genetic operators on a neural system that grows and b) neural systems that are mapped directly to their chromosomes. In the first case a mutation leads to a small, local change in the direction of a signal trail. In the case of direct mapping, a mutation and the resulting local change of a cell, which may belong to a signal trail of a complex axonic tree, can lead to disconnection of a large amount of circuitry. This circuitry would then be disconnected from its source neuron, and loose the integration into the basic structure of the neural network model. The situation gets worse, if we look at the way new neural connections come into being. The probability that a new and useful signal trail evolves if all cells in the CA-space are mutated independently (as is the case for direct mapping), are very small. However a mutation in a grown neural network (for example from "turn right" to "split right") leads to the growth of a new neural signal trail that is consistently integrated into the structure of the source neuron and hence the whole network. This feature of a grown system reduces the search space for possible CA-patterns to the patterns that represent consistent neural structures. The selection of an indirect genetic encoding is a logical consequence of the decision to grow our networks. The principle of growth and indirect genetic encoding has already been used in the old CAM-Brain model. But the way the growth of the neural networks is genetically encoded in the old model (see section OldCAM-BrainModel) has several disadvantages.

- When the spatial distance between neuron bodies can vary, one has to ensure that chromosomes do not overlap, by establishing buffer zones in the chromosome subspace (see figure 2). This forces a tradeoff between wasting memory and the freedom to position the neurons. Additionally, in a neural network, it is always the case that some neurons have more frequent and longer connections than others. These "bigger" neurons have to "wrap around" their chromosomes, i.e they run short of chromosome-growth-signals. While the "smaller" neurons, in an area where the neuron density is high, may not need all their chromosome information.
- A growth signal that comes to a junction has to split into two identical signals in order to supply all branches of its axonic and dendritic trees. This produces a kind of "fractal growth" and hence restricts the range of possible growth patterns.
- The position of the chromosome information in the CA-subspace has no connection with the position of the circuitry that grows from it. So, it is not possible to use a local learning algorithm that can change a local connection and rewrite the chromosome according to that change.

The new CAM-Brain model solves these problems by distributing the chromosome initially throughout the CA-space.

Each neuron no longer has its own chromosome. Instead, every cell in the CA-space contains one instruction of the chromosome, i.e. one growth instruction, so that the chromosome belongs to the network as a whole.

The neuron bodies can be positioned arbitrarily in the CA-space. (See figure 6.) Neuron cells send two kinds of growth signals to their neighbors, either "grow

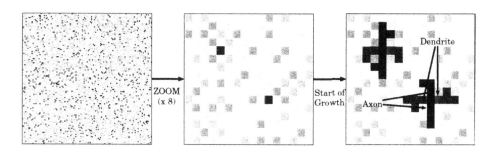

Fig. 6. (Left) An initial pattern of 128 · 128 CA cells with a neuron-density of 5%. The (black) neuron cells are randomly positioned. (Middle) A zoom (x8) with two neuron cells on a random (2-cell gridded) chromosome. (Right) The beginning of the growth phase, after three CA-steps.

a dendrite", or "grow an axon". These signals are passed to the direct neighbors of the neuron cell according to its chromosome information. The blank neighbors, which receive a neural growth signal, turn into either an axon cell or a dendrite cell. (See figure 6 (right).) The just grown cells, then receive the next growth signal from the neuron body cell, which is sending them continuously. To decide in which directions axonic or dendritic trails should grow, the grown cells look at their chromosome information, that encodes the growth instructions, of which there are 6 types (grow straight, turn left or right, split left forward or right forward and block growth, see figure 7). Using its chromosome instructions and knowing from which direction the growth signals come, a cell can determine the neighbors to which it transmits the incoming growth signals. (See RHS of figure 7.) If, for example, a growth signal comes from the left neighbor to a cell whose chromosome instruction is "split right", it will pass the growth signal to its right and bottom neighbors. The model can easily be expanded to other types of growth signal, e.g. a T-split.

After a cell is grown, it accepts growth signals only from the direction from which it received its first signal. This "reception direction" information is stored in the "gate" position of each cell's state.

Figure 8 shows two cut-outs of neural networks at the end of a growth phase. The growth phase stops when growth is saturated, i.e. when no more trails (as specified by the instructions of the network's chromosome) can be formed.

In most of our figures, we have used gridded neural structures, in order to render the axonic and dendritic trails more visible. To force neuron trails to grow on a 2-cell grid, the underlying chromosome is initialized in a "checker-board" pattern. The "black" squares of the checker-board are filled with "grow-straight" instructions and the white squares are filled with arbitrary growth instructions.

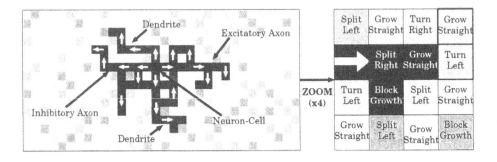

Fig. 7. (Left) A neuron in the CoDi-model with two dendrites and two axons. The arrows inside the axonic and dendritic signal trails indicate the direction of information flow during the growth phase. The neuron grows on a random (2-cell gridded) chromosome in the CA-space. (Right) A signal trail grows according to the underlying chromosome instructions. The trail continues to grow over the two striped cells in the next two time steps. The branch of the trail that leads downward is blocked by the blocker instruction in the chromosome.

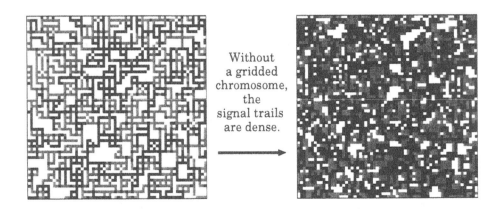

Fig. 8. Two cut-outs of neural networks at the end of the growth phase. (Left) A network grown from a gridded chromosome. (Right) A network grown without gridded chromosome instructions.

Since gridding is only a visualization trick, an unconstrained evolutionary algorithm can grow trails which are dense (the RHS of figure 8 shows an example of an ungridded neural network).

Beside the "grow straight", "turn left" etc. instructions, there are "block-growth" instructions in the chromosome. They cause a cell in an axonic or dendritic trail to block the passage of the incoming growth signals to its neighbors, so the branch of the trail stops growing. With a high density of blocker instructions in a chromosome, an area in the CA-space can be transformed into a more or less impenetrable "wall". Figure 9 shows two "walls" with different

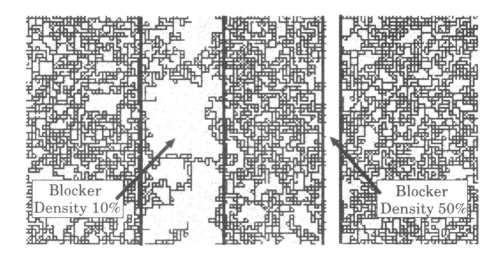

Fig. 9. The use of "walls" to modularize neural circuits in the new CAM-Brain model. The density of blocker instructions on the chromosome determines the probability of trail-growth through a wall and hence the penetrability of the wall.

densities of blocker instructions in the two regions. A combination of walls can partition a neural network into more or less independent modules. This modular partitioning will probably play a major role when the time comes to assemble evolved modules to make an artificial brain. On the CAM-8 with 128 M cells, it is possible to evolve several tens of thousands of modules with 100 2D neurons each.

The grown circuitry and the chromosome are connected locally, so that a local learning algorithm can make local changes in the phenotype and the genotype of the network. This enables the use of Lamarckian evolution (with local learning algorithms) in addition to Darwinian evolution.

The distributed chromosome technique of the new CAM-Brain model makes maximum use of the available CA-space and enables the growth of any type of network connectivity. It also allows local learning to be combined with the evolution of grown neural networks.

3.4 A CA-Model Directly Evolvable in Hardware Must be Simple

Why is the simplicity of our CA-model important? We built CAs to model brain-like structures. Biological brains, the highly optimized results of millions of years of evolution, are obvious sources of inspiration for our work.

Even the tiny brains of insects consist of about a million neurons with complex connectivity. In order to model a 3D neuron with CAs in a way that preserves the essential functionality of real neurons, we estimate that we need at least a cube of 10^3 CA cells. So the minimum size of CA-space to model an insect brain is 10^9 cells. To update a billion CA cells at acceptable speeds will require the use of parallel hardware. The ideal case for CA updating is to give each CA cell its own processor, so that the whole CA-space is updated, in parallel, in one clock cycle.

To build parallel hardware with 10^9 independent CA processors (for a small mammalian brain it would be 10^{12} CA processors), the components must be very simple in their internal logic, with very little memory. Large look-up-tables (LUTs), as are common in traditional CA implementations, will be impossible. Hence, for a CA to be executable on parallel hardware, the rules must be few enough to be expressible within the limited number of logic gates per CA processor inside the parallel hardware. For this reason we tried to keep our new CAM-brain model as simple as possible.

The CA state format of the new CAM-Brain model has 2 bits for the basic cell type, 2 bits for the gate, and 3 bits for the chromosome information. The chromosome is only used in the growth phase. In the signaling phase, the bits of the chromosome information position of the state are used to store the cells' activities. Together there are 7 bits of internal cell state. (See figure 4.) The growth signals and activity signals are encoded in 2 bits. Thus every CA cell receives 2 bits · 4 neighbors = 8 bits of signal information. So altogether, a cell in our model needs 15 bits of memory. As the CAM-8 hardware works with 16 bits per CA cell, it is possible to update our new CAM-Brain CA-space in one machine cycle of the CAM-8. Since the new model does not need sub-cells (a CAM-8 feature allowing more than 16 bits of state per CA cell), it can use the full CAM-8 CA-space of 128 M cells with an update speed of 200 M cells per second.

Considering the fact that each CA cell uses 2 bytes of memory per cell, and that the underlying principles of our CA-rules are expressible with a small number of logic gates, we are hoping to be able to implement the new CAM-Brain model on a large FPGA, so that the neural structures evolve directly in hardware, at hardware speeds.

4 Conclusions

This paper introduced a new CAM-Brain model for the evolution of brain-like structures, based on cellular automata.

The new model, as the old, separates the growth and signaling phases. The underlying CAs for both phases are based on partitioned-CA, which will simplify an implementation in parallel hardware.

For the growth phase, we introduced a distributed-chromosome technique, which allows us to grow arbitrary neural structures, and enables the combined use of local learning and evolutionary engineering.

The neural signaling is realized with the CoDi-model, which allows maximum use of the given CA-space and reduces drastically the number of necessary CA-states and CA-rules. All presented techniques can be extended to 3D CA-spaces without difficulty.

The new CAM-Brain model can be executed on MIT's "CAM-8" hardware with several million 2D neurons in a 128 M cell CA-space, at a speed of 200 M cell-updates per second. In a 3D CA-space it will be possible to grow more then 100.000 3D neurons on the CAM-8.

The simplicity of the new model, and the use of 2 bytes of memory per CA cell allow an implementation in programmable hardware. Hence, with the new CAM-Brain model, it will be possible to evolve brain-like structures directly in hardware, at hardware speeds (a necessity, if we are to evolve neural structures with millions of neurons [12]).

References

1. Toffoli, T. & Margolous, N. *'Cellular Automata Machines'*, MIT Press, Cambridge, MA , 1987.
2. von Neumann, J. *'Theory of Self-Reproducing Automata'*, ed. Burks A.W.University of Illinois Press, Urbana , 1966
3. Codd, E.F. *'Cellular Automata'*, Academic Press, NY 1968.
4. de Garis, H. *'CAM-BRAIN: The Evolutionary Engineering of a Billion Neuron Artificial Brain by 2001 which Grows/Evolves at Electronic Speed Inside a Cellular Automata Machine (CAM)'*, in 'Towards Evolvable Hardware', Springer, Berlin, Heidelberg, NY , 1996.
5. Lloyd, S. *'A Potentially Realizable Quantum Computer'*, Science **261**, 1569-1571 1993.
6. Koza, J.R. *'Genetic Programming: On the Programming of Computers by the Means of Natural Selection'*, Cambrige, MA, MIT Press , 1992
7. Koza, J.R. & Bennet, F.H. & Andre, D. & Keane, M.M. *'Toward Evolution of Electronic Animals Using Genetic Programming'*, ALife V Conference Proceedings, MIT Press , 1996.
8. Sipper, M. *'Co-evolving Non-Uniform Cellular Automata to Perform Computations'*, Physica D **92**, 193-208 1996.
9. Carter, F. L. *'Molecular Electronic Devices'*, North-Holland, Amsterdam, NY, Oxford, Tokyo , 1986.
10. Gers, F. A. & de Garis, H. *'Porting a Cellular Automata Based Artificial Brain to MIT's Cellular Automata Machine 'CAM-8"*, (submitted).
11. Margolus, N. *'Crystalline Computation'*, (Preprint).
12. de Garis, H. *'One Chip Evolvable Hardware: 1C-EHW'*, (submitted).

Genetic Programming

Evolution of a 60 Decibel Op Amp Using Genetic Programming

Forrest H Bennett III[1]

John R. Koza[2]

David Andre[3]

Martin A. Keane[4]

1) Visiting Scholar, Computer Science Department,
Stanford University, Stanford, California 94305 USA
fhb3@slip.net

2) Computer Science Department, Stanford University

3) Computer Science Department, University of California, Berkeley, California

4) Econometrics Inc., 5733 West Grover
Chicago, IL 60630 USA

Abstract: Genetic programming was used to evolve both the topology and sizing (numerical values) for each component of a low-distortion, low-bias 60 decibel (1000-to-1) amplifier with good frequency generalization.

1. Introduction

In nature, complex structures are designed by means of evolution and natural selection. This suggests the possibility of applying the techniques of evolutionary computation in order to automate the design of complex structures.

The problem of circuit synthesis involves designing an electrical circuit that satisfies user-specified design goals. A complete design of an electrical circuit includes both its topology and the sizing of all its components. The *topology* of a circuit consists of the number of components in the circuit, the type of each component, and a list of all the connections between the components. The *sizing* of a circuit consists of the component value(s) of each component.

Evolvable hardware is one approach to automated circuit synthesis. Early pioneering work in this field includes that of Higuchi, Niwa, Tanaka, Iba, de Garis, and Furuya (1993a, 1993b); Hemmi, Mizoguchi, and Shimohara (1994); Mizoguchi, Hemmi, and Shimohara (1994); and the work presented at the 1995 workshop on evolvable hardware in Lausanne (Sanchez and Tomassini 1996).

The design of analog circuits and mixed analog-digital circuits has not proved to be amenable to automation (Rutenbar 1993). CMOS operational amplifier (op amp) circuits have been designed using a modified version of the genetic algorithm (Kruiskamp 1996; Kruiskamp and Leenaerts 1995); however, the topology of each op amp was one of 24 topologies based on the conventional human-designed stages of an op amp. Thompson (1996) used a genetic algorithm to evolve a frequency discriminator on a Xilinx 6216 reconfigurable processor in analog mode.

2. Genetic Programming

Genetic programming is an extension of the genetic algorithm described in John Holland's pioneering *Adaptation in Natural and Artificial Systems* (1975).

The book *Genetic Programming: On the Programming of Computers by Means of Natural Selection* (Koza 1992) provides evidence that genetic programming can solve, or approximately solve, a variety of problems. See also Koza and Rice 1992. Genetic programming (GP) starts with a primordial ooze of randomly generated computer programs composed of the available programmatic ingredients and then applies the principles of animal husbandry to breed a new (and often improved) population of programs. The breeding is done in a domain-independent way using the Darwinian principle of survival of the fittest, an analog of the naturally-occurring genetic operation of crossover (sexual recombination), and occasional mutation. The crossover operation is designed to create syntactically valid offspring programs (given closure amongst the set of ingredients). Genetic programming combines the expressive high-level symbolic representations of computer programs with the near-optimal efficiency of learning of Holland's genetic algorithm. A computer program that solves (or approximately solves) a given problem often emerges from this process.

The book *Genetic Programming II: Automatic Discovery of Reusable Programs* (Koza 1994a, 1994b) describes a way to evolve multi-part programs consisting of a main program and one or more reusable, parameterized, hierarchically-called subprograms.

Recent work in the field of genetic programming is described in Kinnear 1994, Angeline and Kinnear 1996, Koza, Goldberg, Fogel, and Riolo 1996.

Gruau's *cellular encoding* (1996) is an innovative technique in which genetic programming is used to concurrently evolve the architecture, weights, thresholds, and biases of neurons in a neural network.

3. Evolution of Circuits

Genetic programming can be applied to circuits if a mapping is established between the kind of rooted, point-labeled trees with ordered branches used in genetic programming and the line-labeled cyclic graphs encountered in the world of circuits.

Developmental biology provides the motivation for this mapping. The starting point of the growth process used herein is a very simple embryonic electrical circuit. The embryonic circuit contains certain fixed parts appropriate to the problem at hand and certain wires that are capable of subsequent modification. An electrical circuit is progressively developed by applying the functions in a circuit-constructing program tree to the modifiable wires of the embryonic circuit (and, later, to both the modifiable wires and other components of the successor circuits).

The functions are divided into four categories:

(1) connection-modifying functions that modify the topology of circuit (starting with the embryonic circuit), and

(2) component-creating functions that insert components into the topology of the circuit,

(3) arithmetic-performing functions that appear in arithmetic-performing subtrees as argument(s) to the component-creating functions and specify the numerical value of the component, and

(4) automatically defined functions in function-defining branches.

Each branch of the program tree is created in accordance with a constrained syntactic structure. Branches are composed from construction-continuing subtree(s) that continue the developmental process and arithmetic-performing subtree(s) that determine the numerical value of the component. Connection-modifying functions have one or more construction-continuing subtrees, but no arithmetic-performing subtrees. Component-creating functions have one construction-continuing subtree and typically have one arithmetic-performing subtree. This constrained syntactic structure is preserved by using structure-preserving crossover with point typing (Koza 1994a).

3.1. The Embryonic Circuit

The developmental process for converting a program tree into an electrical circuit begins with an embryonic circuit.

Figure 1 shows a one-input, one-output embryonic circuit that serves as a test harness for the evolving circuits. **USOURCE** is the incoming signal. **UOUT** is the output signal. There is a fixed 1,000 Ohm load resistor **RLOAD** and a fixed 1,000 Ohm source resistor **RSOURCE**. Because we are evolving an amplifier, there is also a fixed 1,000,000 Ohm feedback resistor **RFEEDBACK**, a fixed 1,000 Ohm balancing source resistor **RBALANCE_SOURCE**, and a fixed 1,000,000 Ohm balancing feedback resistor **RBALANCE_FEEDBACK**. This arrangement limits the possible amplification of the evolving circuit to the 1000-to-1 ratio (which corresponds to 60 dB) of the feedback resistor to the source resistor.

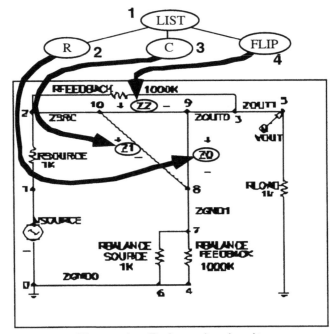

Figure 1   Embryonic circuit.

There are three modifiable wires **Z0**, **Z1**, and **Z2** arranged in a triangle so as to provide connectivity between the input, the output, and the balancing resistors. All of the above elements (except **Z0**, **Z1**, and **Z2**) are fixed and are not modified during the developmental process. At the beginning of the developmental process, there is a writing head pointing to (highlighting) each of the three modifiable wires. All development occurs at wires or components to which a writing head points.

The top part of this figure shows a portion of an illustrative circuit-constructing program tree. It contains a resistor-creating R function (labeled 2 and described later), a capacitor-creating C function (labeled 3), and a polarity-reversing FLIP function (labeled 4) and a connective LIST function (labeled 1). The R and C functions cause modifiable wires **Z0** and **Z1** to become a resistor and capacitor, respectively; the FLIP function reverses the polarity of modifiable wire **Z2**.

3.2. Component-Creating Functions

Each circuit-constructing program tree in the population contains component-creating functions and connection-modifying functions.

Each component-creating function inserts a component into the developing circuit and assigns component value(s) to the component. Each component-creating function has a writing head that points to an associated highlighted component in the developing circuit and modifies the highlighted component in some way. The construction-continuing subtree of each component-creating function points to a successor function or terminal in the circuit-constructing program tree.

Space does not permit a detailed description of each function that we use herein. See Koza, Andre, Bennett, and Keane (1996), and Koza, Bennett, Andre, and Keane (1996a, 1996b, 1996c, 1996d, 1996e) for details.

The two-argument resistor-creating R function causes the highlighted component to be changed into a resistor. The value of the resistor in kilo-Ohms is specified by its arithmetic-performing subtree.

The arithmetic-performing subtree of a component-creating function consists of a composition of arithmetic functions (addition and subtraction) and random constants (in the range −1.000 to +1.000). The arithmetic-performing subtree specifies the numerical value of a component by returning a floating-point value that is, in turn, interpreted, in a logarithmic way, as the value for the component in a range of 10 orders of magnitude (using a unit of measure that is appropriate for the particular type of component involved). The floating-point value is interpreted as the value of the component as described in Koza, Andre, Bennett, and Keane (1996, 1997).

Figure 2 shows a modifiable wire **Z0** connecting nodes 1 and 2 of a partial circuit containing four capacitors.

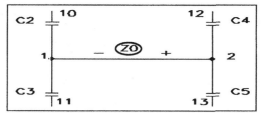

Figure 2 Modifiable wire Z0.

Figure 3 shows the result of applying the R function to the modifiable wire **Z0** of figure 2.

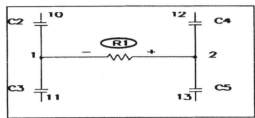

Figure 3   Result of applying the R function.

Similarly, the two-argument capacitor-creating C function causes the highlighted component to be changed into a capacitor. The value of the capacitor in nano-Farads is specified by its arithmetic-performing subtree.

The one-argument Q_D_PNP diode-creating function causes a diode to be inserted in lieu of the highlighted component, where the diode is implemented using a PNP transistor whose collector and base are connected to each other. The Q_D_NPN function inserts a diode using an NPN transistor in a similar manner.

There are also six one-argument transistor-creating functions (called Q_POS_COLL_NPN, Q_GND_EMIT_NPN, Q_NEG_EMIT_NPN, Q_GND_EMIT_PNP, Q_POS_EMIT_PNP, Q_NEG_COLL_PNP) that insert a transistor in lieu of the highlighted component. For example, the Q_POS_COLL_NPN function inserts a NPN transistor whose collector is connected to the positive voltage source.

The three-argument transistor-creating Q_3_NPN function causes an NPN transistor (model Q2N3904) to be inserted in place of the highlighted component and one of the nodes to which the highlighted component is connected. The Q_3_NPN function creates five new nodes and three new modifiable wires. There is no writing head on the new transistor. Similarly, the three-argument transistor-creating Q_3_PNP function causes a PNP transistor (model Q2N3906) to be inserted.

Figure 4 shows the result of applying the Q_3_NPN0 function, thereby creating transistor **Q6** in lieu of modifiable wire **Z0** of figure 2.

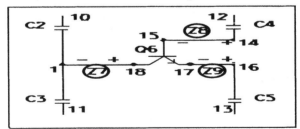

Figure 4   Result of applying the Q_3_NPN0 function.

3.3.   Connection-Modifying Functions

Each connection-modifying function in a program tree points to an associated highlighted component and modifies the topology of the developing circuit.

The one-argument polarity-reversing FLIP function attaches the positive end of the highlighted component to the node to which its negative end is currently attached and vice versa. After execution of the FLIP function, there is one writing head pointing to the component.

The three-argument SERIES division function creates a series composition consisting of the highlighted component, a copy of it, one new modifiable wire, and two new nodes (each with a writing head).

Figure 5 illustrates the result of applying the SERIES division function to resistor **R1** from figure 3.

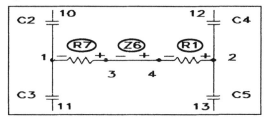

Figure 5   Result after applying the SERIES function.

The four-argument PSS and PSL parallel division functions create a parallel composition consisting of the original highlighted component, a copy of it, two new wires, and two new nodes. Figure 6 shows the result of applying PSS to the resistor **R1** from figure 3.

There are six three-argument functions (called T_GND_0, T_GND_1, T_POS_0, T_POS_1, T_NEG_0, T_NEG_1) that insert two new nodes and two new modifiable wires and make a connection to ground, positive voltage source, or negative voltage source, respectively. Figure 7 shows the results of applying the T_GND_0 function to the circuit of figure 3.

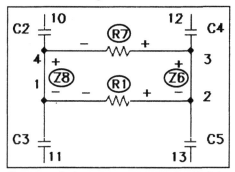

Figure 6   Result of the PSS parallel division function.

The three-argument PAIR_CONNECT_0 and PAIR_CONNECT_1 functions enable distant parts of a circuit to be connected together. The first PAIR_CONNECT to occur in the development of a circuit creates two new wires, two new nodes, and one temporary port. The next PAIR_CONNECT to occur (whether PAIR_CONNECT_0 or PAIR_CONNECT_1) creates two new wires and one new node, connects the temporary port to the end of one of these new wires, and then removes the temporary port.

The one-argument NOP function has no effect on the highlighted component; however, it delays activity on the developmental path on which it appears in relation to other developmental paths in the overall program tree.

The zero-argument END function causes the highlighted component to lose its writing head. The END function causes its writing head to be lost – thereby ending that particular developmental path.

The zero-argument SAFE_CUT function causes the highlighted component to be removed from the circuit provided that the degree of the nodes at both ends of the highlighted component is three (i.e., no dangling components or wires are created).

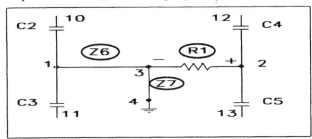

Figure 7   Result of applying the T_GND_0 function.

4.   Preparatory Steps

Our goal in this paper is to evolve a design for a 60 decibel amplifier with low distortion and low bias. Before applying genetic programming to circuit synthesis, the user must perform seven major preparatory steps, namely

(1) identifying the embryonic circuit that is suitable for the problem,

(2) determining the architecture of the overall circuit-constructing program trees,

(3) identifying the terminals of the to-be-evolved programs,

(4) identifying the primitive functions contained in the to-be-evolved programs,

(5) creating the fitness measure,

(6) choosing certain control parameters (notably population size and the maximum number of generations to be run), and

(7) determining the termination criterion and method of result designation.

The feedback embryo of figure 1 is suitable for this problem.

Since the embryonic circuit has three writing heads – one associated with each of the result-producing branches – there are three result-producing branches (called RPB0, RPB1, and RPB2) in each program tree. We decided to include two one-argument automatically defined functions (called ADF0 and ADF1) in each program tree and that there would be no hierarchical references among the automatically defined functions. Thus, there are two function-defining branches in each program tree. Consequently, the architecture of each overall program tree in the population consists of a total of five branches (two function-defining branches and three result-producing branches) joined by a LIST function.

The function sets are identical for all three result-producing branches of the program trees. The terminal sets are identical for all three result-producing branches.

For the three result-producing branches, the function set, $\mathcal{F}_{\text{ccs-rpb}}$, for each construction-continuing subtree is

$\mathcal{F}_{\text{ccs-rpb}}$ = {ADF0, ADF1, R, C, SERIES, PSS, PSL, FLIP, NOP, T_GND_0,
          T_GND_1, T_POS_0, T_POS_1, T_NEG_0, T_NEG_1,
          PAIR_CONNECT_0, PAIR_CONNECT_1, Q_D_NPN, Q_D_PNP,

Q_3_NPN0, ..., Q_3_NPN11, Q_3_PNP0, ..., Q_3_PNP11, Q_POS_COLL_NPN, Q_GND_EMIT_NPN, Q_NEG_EMIT_NPN, Q_GND_EMIT_PNP, Q_POS_EMIT_PNP, Q_NEG_COLL_PNP}.

For the three result-producing branches, the terminal set, $T_{ccs\text{-}rpb}$, for each construction-continuing subtree consists of

$T_{ccs\text{-}rpb} = \{$END, SAFE_CUT$\}$.

For the three result-producing branches, the function set, $F_{aps\text{-}rpb}$, for each arithmetic-performing subtree is,

$F_{aps\text{-}rpb} = \{+, -\}$.

For the three result-producing branches, the terminal set, $T_{aps\text{-}rpb}$, for each arithmetic-performing subtree consists of

$T_{aps\text{-}rpb} = \{\Re\}$,

where $\Re$ represents floating-point random constants from -1.0 to $+1.0$.

The terminal sets are identical for both function-defining branches (automatically defined functions) of the program trees. The function sets are identical for both function-defining branches.

For the two function-defining branches, the function set, $F_{ccs\text{-}adf}$, for each construction-continuing subtree is

$F_{ccs\text{-}adf} = F_{ccs\text{-}rpb} - \{$ADF0, ADF1$\}$.

For the two function-defining branches, the terminal set, $T_{ccs\text{-}adf}$, for each construction-continuing subtree is

$T_{ccs\text{-}adf} = T_{ccs\text{-}rpb}$.

For the two function-defining branches, the function set, $F_{aps\text{-}adf}$, for each arithmetic-performing subtree is,

$F_{aps\text{-}adf} = F_{aps\text{-}rpb} = \{+, -\}$.

For the two function-defining branches, the terminal set, $T_{aps\text{-}adf}$, for each arithmetic-performing subtree consists of

$T_{aps\text{-}adf} = \{\Re\} \cup \{$ARG0$\}$,

where ARG0 is the dummy variable (formal parameter) of the automatically defined function.

The evaluation of fitness for each individual circuit-constructing program tree in the population begins with its execution. This execution applies the functions in the program tree to the embryonic circuit, thereby developing the embryonic circuit into a fully developed circuit. A netlist describing the fully developed circuit is then created. The netlist identifies each component of the circuit, the nodes to which that component is connected, and the value of that component. Each circuit is then simulated to determine its behavior. The 217,000-line SPICE (Simulation Program with Integrated Circuit Emphasis) simulation program (Quarles et al. 1994) was modified to run as a submodule within the genetic programming system.

An amplifier can be viewed in terms of its response to a DC input. An ideal inverting amplifier circuit would receive a DC input, invert it, and multiply it by the amplification factor. A circuit is flawed to the extent that it does not achieve the desired amplification; to the extent that the output signal is not centered on 0 volts

(i. e., it has a bias); and to the extent that the DC response of the circuit is not linear.

Thus, for this problem, we used a fitness measure based on SPICE's DC sweep. The DC sweep analysis measures the DC response of the circuit at several different DC input voltages. The circuits were analyzed with a 5 point DC sweep ranging from −10 millvolts to +10 MV, with input points at −10 MV, −5 MV, 0 MV, +5 MV, and +10 MV. SPICE then simulated the circuit's behavior for each of these five DC voltages.

Fitness is then calculated from four penalties derived from these five DC output values. Fitness is the sum of the amplification penalty, the bias penalty, and the two non-linearity penalties.

First, the amplification factor of the circuit is measured by the slope of the straight line between the output for −10 MV and the output for +10 MV (i.e., between the outputs for the endpoints of the DC sweep). If the amplification factor is less than the target (which is 60 dB for this problem), there is a penalty equal to the shortfall in amplification.

Second, the bias is computed using the DC output associated with a DC input of 0 volts. There is a penalty equal to the bias times a weight. For this problem, a weight of 0.1 is used.

Finally, the linearity is measured by the deviation between the slope of each of two shorter lines and the overall amplification factor of the circuit. The first shorter line segment connects the output value associated with an input of −10 MV and the output value for −5 MV. The second shorter line segment connects the output value for +5 MV and the output for +10 MV. There is a penalty for each of these shorter line segments equal to the absolute value of the difference in slope between the respective shorter line segment and the overall amplification factor of the circuit.

Many of the circuits that are randomly created in the initial random population and many that are created by the crossover and mutation operations are so bizarre that they cannot be simulated by SPICE. Such circuits are assigned a high penalty value of fitness (10^8).

The population size, M, was 640,000. The crossover percentage was 89%; the reproduction percentage was 10%; and the mutation percentage was 1%. A maximum size of 300 points was established for each of the branches in each overall program. The other minor parameters were the default values in Koza 1994a (appendix D).

This problem was run on a medium-grained parallel Parsytec computer system consisting of 64 80 MHz Power PC 601 processors arranged in a toroidal mesh with a host PC Pentium type computer. The distributed genetic algorithm was used with a population size of $Q = 10,000$ at each of the $D = 64$ demes. On each generation, four boatloads of emigrants, each consisting of $B = 2\%$ (the migration rate) of the node's subpopulation (selected on the basis of fitness) were dispatched to each of the four toroidally adjacent processing nodes. See Andre and Koza 1996 for details.

5 . Results

The best circuit (figure 11) from generation 0 achieves a fitness of 986.1 and has nine transistors, three capacitors, and two resistors (in addition to the five resistors of the feedback embryo).

About 45% of the circuits of generation 0 cannot be simulated by SPICE. However, the percentage of unsimulatable programs drops to 8% by generation 1. Moreover, this percentage does not exceed 2% after generation 16 and does not exceed 1% after generation 58.

The best circuit (figure 12) from generation 49 achieves a fitness of 404.0 and has 16 transistors, no capacitors, and four resistors (in addition to the five resistors of the feedback embryo).

The best circuit (figure 13) from generation 109 has 22 transistors, no capacitors, and 11 resistors (in addition to the five resistors of the feedback embryo). It achieves a fitness of 0.178. Its circuit-constructing program tree has 40, 98, and 27 points, respectively, in its first, second, and third result-producing branches and 33 and 144 points, respectively, in its first and second automatically defined functions.

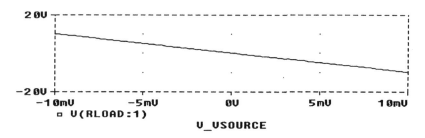

Figure 8   DC sweep of best circuit from generation 109.

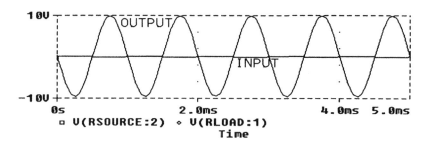

Figure 9   Time domain behavior of best of generation 109.

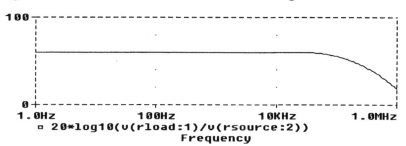

Figure 10   AC sweep for best circuit from generation109.

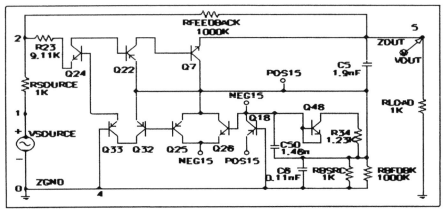

Figure 11 Best circuit from generation 0.

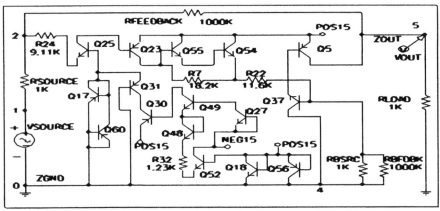

Figure 12 Best circuit from generation 49.

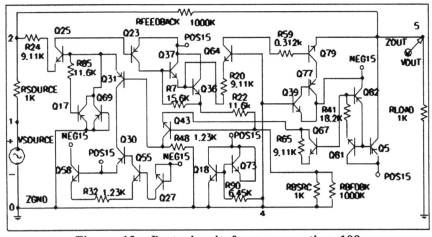

Figure 13 Best circuit from generation 109.

The amplification of an op amp can be measured from the DC sweep (figure 8). The amplification in decibels is 20 times the common logarithm of the ratio of the change in the output divided by the change of the input (i.e., 20 millivolts here) . The amplification is 60 dB here (i.e., 1,000-to-1 ratio). There is a bias of 0.2 volts. Notice the linearity of the DC sweep in this figure.

Figure 9 shows the time domain behavior of the best circuit from generation 109. The vertical axis shows voltage and ranges from −10 volts to +10 volts. The input is the 10 millivolt sinusoidal signal; however, this sinusoidal input signal appears as a nearly straight line because of the scale. At 1,000 Hz, the amplification is 59.7 dB; the bias is 0.18 volts; and the distortion is very low (0.17%).

The amplification of an op amp can also be measured from an AC sweep. Figure 10 shows that the amplification for the best circuit of generation 109. The amplification at 1,000 Hz is 59.7 dB.The flatband gain is 60 dB and the 3 dB bandwidth is 79, 333 Hz.

We then tested whether the genetically evolved 22-transistor best circuit from generation 109 provided more than 60 dB amplification when embedded in a test harness that is appropriate for testing amplification of up to 80 dB. The amplification is 80.15 dB when SPICE's DC sweep is applied. The amplification is 77.79 dB at 1,000 Hz in the time domain. When the AC sweep is applied, the circuit delivers over 80 dB of amplification from 1 Hz to 36 Hz; the circuit delivers over 77.86 dB from 36 Hz to 55,000 Hz.

6. Other Examples of Evolutionary Circuit Design Using Genetic Programming

The above techniques have recently been successfully applied to a variety of other problems of evolutionary circuit design.

6.1. Lowpass "Brick Wall" Filter

Genetic programming successfully evolved a design for a lowpass filter with passband below 1,000 Hz and a stopband above 2,000 Hz with requirements equivalent to that of a fifth order elliptic filter (Koza, Bennett, Andre, and Keane 1996a, 1996c). In some runs of this problem, the genetically evolved lowpass filter has a topology that is similar to that employed by human engineers. For example, in one run, a 100% compliant evolved circuit (figure 14) had the recognizable ladder topology of a Butterworth or Chebychev filter (i.e., a composition of series inductors horizontally with capacitors as vertical shunts). In another run, a 100%-compliant recognizable "bridged T" filter was evolved.

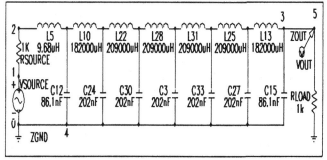

Figure 14   Genetically evolved ladder filter circuit.

6.2. Asymmetric Bandpass Filter

A difficult-to-design asymmetric bandpass filter with requirement equivalent to a tenth-order elliptic filter (Koza, Bennett, Andre, and Keane 1996d) was evolved. Figure 15 shows a 100% compliant evolved asymmetric bandpass filter.

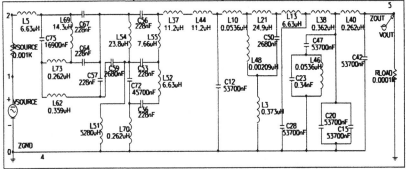

Figure 15   Genetically evolved asymmetric bandpass filter.

6.3. A Crossover (Woofer and Tweeter) Filter

The design for a crossover (woofer and tweeter) filter with a crossover frequency of 2,512 Hz (Koza, Bennett, Andre, and Keane 1996b) was evolved. This problem requires a one-input, two-output embryonic circuit and requires that the fitness be measured at two probe points. The lowpass part of the genetically evolved best-of-run circuit (figure 16) has the Butterworth topology. Except for additional capacitor **C36**, the highpass part of this circuit also has the Butterworth topology. This circuit is slightly better than the combination of lowpass and highpass Butterworth filters of order 7.

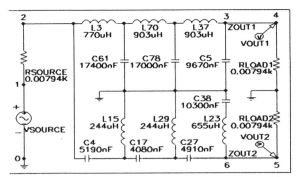

Figure 16   Genetically evolved crossover filter.

7. Conclusion

Genetic programming successfully evolved a 22-transistor amplifier that delivers a DC gain of 60 dB amplification and that has almost no bias or distortion. It generalizes in the frequency domain with a 3 dB bandwidth of 79,433 Hz. Moreover, this genetically evolved 60 dB amplifier generalizes in such a way as to deliver 80 dB of amplification (as measured by the DC sweep) when it is embedded in a test harness that allows 80 dB of amplification.

7.1. Related Paper in this Volume

See also Koza, Bennett, Andre, and Keane (1996e) in this volume.

Acknowledgments

Jason Lohn and Simon Handley made helpful comments on drafts of this paper.

References

Andre, David and Koza, John R. 1996. Parallel genetic programming: A scalable implementation using the transputer architecture. In Angeline, P. J. and Kinnear, K. E. Jr. (editors). 1996. *Advances in Genetic Programming 2*. Cambridge: MIT Press.

Angeline, Peter J. and Kinnear, Kenneth E. Jr. (editors). 1996. *Advances in Genetic Programming 2*. Cambridge, MA: The MIT Press.

Gruau, Frederic. 1996. Artificial cellular development in optimization and compilation. In Sanchez, Eduardo and Tomassini, Marco (editors). 1996. *Towards Evolvable Hardware*. Lecture Notes in Computer Science, Volume 1062. Berlin: Springer-Verlag. Pages 48 – 75.

Hemmi, Hitoshi, Mizoguchi, Jun'ichi, and Shimohara, Katsunori. 1994. Development and evolution of hardware behaviors. In Brooks, R. and Maes, P. (editors). *Artificial Life IV:* Cambridge, MA: MIT Press. Pages 371–376.

Higuchi, Tetsuya, Niwa, Tatsuya, Tanaka, Toshio, Iba, Hitoshi, de Garis, Hugo, and Furuya, Tatsumi. 1993a. In Meyer, Jean-Arcady, Roitblat, Herbert L. and Wilson, Stewart W. (editors). *From Animals to Animats 2: Proceedings of the Second International Conference on Simulation of Adaptive Behavior*. Cambridge, MA: The MIT Press. 1993. Pages 417 – 424.

Higuchi, Tetsuya, Niwa, Tatsuya, Tanaka, Toshio, Iba, Hitoshi, de Garis, Hugo, and Furuya, Tatsumi. 1993b. *Evolvable Hardware – Genetic-Based Generation of Electric Circuitry at Gate and Hardware Description Language (HDL) Levels*. Electrotechnical Laboratory technical report 93-4. Tsukuba, Japan: Electrotechnical Laboratory.

Holland, John H. 1975. *Adaptation in Natural and Artificial Systems*. Ann Arbor, MI: University of Michigan Press.

Kinnear, Kenneth E. Jr. (editor). 1994. *Advances in Genetic Programming*. Cambridge, MA: The MIT Press.

Koza, John R. 1992. *Genetic Programming: On the Programming of Computers by Means of Natural Selection*. Cambridge, MA: MIT Press.

Koza, John R. 1994a. *Genetic Programming II: Automatic Discovery of Reusable Programs*. Cambridge, MA: MIT Press.

Koza, John R. 1994b. *Genetic Programming II Videotape: The Next Generation*. Cambridge, MA: MIT Press.

Koza, John R., Andre, David, Bennett III, Forrest H, and Keane, Martin A. 1996. Use of automatically defined functions and architecture-altering operations in automated circuit synthesis using genetic programming. In Koza, John R., Goldberg, David E., Fogel, David B., and Riolo, Rick L. (editors). 1996. *Genetic Programming 1996: Proceedings of the First Annual Conference, July 28-31, 1996, Stanford University*. Cambridge, MA: The MIT Press.

Koza, John R., Bennett III, Forrest H, Andre, David, and Keane, Martin A. 1996a. Toward evolution of electronic animals using genetic programming. *Artificial Life V: Proceedings of the Fifth International Workshop on the Synthesis and Simulation of Living Systems.* Cambridge, MA: The MIT Press.

Koza, John R., Bennett III, Forrest H, Andre, David, and Keane, Martin A. 1996b. Four problems for which a computer program evolved by genetic programming is competitive with human performance. *Proceedings of the 1996 IEEE International Conference on Evolutionary Computation.* IEEE Press. Pages 1–10.

Koza, John R., Bennett III, Forrest H, Andre, David, and Keane, Martin A. 1996c. Automated design of both the topology and sizing of analog electrical circuits using genetic programming. In Gero, John S. and Sudweeks, Fay (editors). *Artificial Intelligence in Design '96.* Dordrecht: Kluwer. Pages 151-170.

Koza, John R., Bennett III, Forrest H, Andre, David, and Keane, Martin A. 1996d. Automated WYWIWYG design of both the topology and component values of analog electrical circuits using genetic programming. In Koza, John R., Goldberg, David E., Fogel, David B., and Riolo, Rick L. (editors). 1996. *Genetic Programming 1996: Proceedings of the First Annual Conference, July 28-31, 1996, Stanford University.* Cambridge, MA: The MIT Press.

Koza, John R., Bennett III, Forrest H, Andre, David, and Keane, Martin A. 1996e. Reuse, parameterized reuse, and hierarchical reuse of substructures in evolving electrical circuits using genetic programming. In this volume.

Koza, John R., Goldberg, David E., Fogel, David B., and Riolo, Rick L. (editors). 1996. *Genetic Programming 1996: Proceedings of the First Annual Conference, July 28-31, 1996, Stanford University.* Cambridge, MA: The MIT Press.

Koza, John R., and Rice, James P. 1992. *Genetic Programming: The Movie.* Cambridge, MA: MIT Press.

Kruiskamp, Marinum Wilhelmus. 1996. *Analog Design Automation using Genetic Algorithms and Polytopes.* Eindhoven, The Netherlands: Data Library Technische Universiteit Eindhoven.

Kruiskamp Marinum Wilhelmus and Leenaerts, Domine. 1995. DARWIN: CMOS opamp synthesis by means of a genetic algorithm. *Proceedings of the 32nd Design Automation Conference.* New York, NY: Association for Computing Machinery. Pages 433–438.

Mizoguchi, Junichi, Hemmi, Hitoshi, and Shimohara, Katsunori. 1994. Production genetic algorithms for automated hardware design through an evolutionary process. *Proceedings of the First IEEE Conference on Evolutionary Computation.* IEEE Press. Vol. I. 661-664.

Quarles, Thomas, Newton, A. R., Pederson, D. O., and Sangiovanni-Vincentelli, A. 1994. *SPICE 3 Version 3F5 User's Manual.* Department of Electrical Engineering and Computer Science, University of California, Berkeley, CA. March 1994.

Rutenbar, R. A. 1993. Analog design automation: Where are we? Where are we going? *Proceedings of the 15th IEEE CICC.* New York: IEEE. 13.1.1-13.1.8.

Sanchez, Eduardo and Tomassini, Marco (editors). 1996. *Towards Evolvable Hardware.* Lecture Notes in Computer Science, Vol 1062. Berlin: Springer-Verlag.

Thompson, Adrian. 1996. Silicon evolution. In Koza, John R., Goldberg, David E., Fogel, David B., and Riolo, Rick L. (editors). 1996. *Genetic Programming 1996: Proceedings of the First Annual Conference, July 28-31, 1996, Stanford University.* Cambridge, MA: MIT Press.

Evolution of Binary Decision Diagrams for Digital Circuit Design Using Genetic Programming

Hidenori Sakanashi [1], Tetsuya Higuchi [2], Hitoshi Iba [2] and Yukinori Kakazu [1]

[1] Autonomous Systems Eng., Complex Systems Eng., Hokkaido University
N13-W8, Kita-ku, Sapporo 060, Japan
{sakana, kakazu}@complex.hokudai.ac.jp
[2] Eloctrotechnical Laboratory, 1-1-4, Umezono, Tsukuba 304, Japan
{higuchi, iba}@etl.go.jp

Abstract. This paper proposes the methodology for hardware evolution by genetic programming (GP). By adopting Binary Decision Diagrams (BDDs) as hardware representation, larger circuits can be evolved, and they will be easily verified by utilizing commercial CAD software. The hardware descriptions specified in BDDs are improved by GP operators, to synthesize various combinatorial logical circuits.

From the viewpoint of GP, however, some constraints of BDD must be satisfied during its search process. In other words, GP must search not only in phenotype space, but also in genotype space. In order to resolve this problem, in this paper, we attempt two approaches. One concerns the operations to obtain BDDs satisfying the genotypical constraints, and the other is the method for balancing phenotypic and genotypic evaluations.

1. Introduction

In the last few years, there has been a growing interest in evolvable hardware (EHW) [Higuchi94]. By genetic learning, EHW allows the hardware to change its own hardware structure dynamically (on-line) and adaptively to attain better performance. Although current EHW research can evolve only circuits of small sizes, the verification of circuit design has to be taken into account for industrial EHW applications in future. The verification of hardware design has close relationship with the hardware representation schemes (e.g. truth table, finite-sate diagram, binary decision diagram, and hardware description language, etc).

In recent commercial CAD software, Binary Decision Diagram (BDD) [Bryant86 and Bryant95] is becoming a standard representation. This is because BDD can handle larger circuits with smaller memory capacities than other representation schemes. Moreover, if BDDs satisfy some constraints, they can represent any logical structures in unique and compact forms, then the verification can be performed more quickly. Therefore, this paper proposes the methodology for the hardware evolution of BDD descriptions. That is, we attempt to synthesize or improve BDD by means of the genetic learning techniques.

It is no guarantee that the above mentioned constraints of BDD continues to be satisfied, however, since the genetic learning basically adopts stochastic search procedures. Therefore, the genetic learner is required

- to improve BDDs through evolution, and
- to satisfy above mentioned constraints (those are explained in later section).

The former and latter requirements respectively corresponds the phenotypical and genotypical search problems of BDD. Moreover, the requirements about BDD may conflict with each other. The genotypical constraints force BDD to become small and simple, and BDD generally tends to be complex to improve its phenotypical evaluation. It is serious contradiction that they have quite different tendency in their improvements.

In this paper, for resolving this problem, we attempt to construct the methodology of improving the phenotype of BDD and satisfying its genotypical constraints, simultaneously. There are many learning methods treating graphs, such as Adaptive Logic Network (ALN) [Armstrong79], Genetic Programming (GP) [Koza94], and so on. Among them, considering that this methodology will implemented to EHW in future, we adopt GP as a search engine for improving BDDs. In particular, we use and modify Koza's canonical GP, instead of the extended GP like Genetic L-system Programming [Jacob94] or GP with subroutines [Kinnear94 and Rosca94], because of convenience to treat genotype of graphs.

Based on the canonical GP, our approach consists of two extensions: The first one is the method of evaluation which can measure the phenotypical and genotypical characteristics of BDDs. The second one is the operation to improve the genotype without obstructing evolution of phenotype of BDDs.

In section 2 and 3, we explain BDD and GP briefly. Then, section 4 gives our approach, and shows how BDD should be evolved and improved in GP. We improve GP according to our approach in section 5. In section 6, we execute some computer simulations, and their results let us know the efficiency and problems of our method.

2. Binary Decision Diagram

Binary Decision Diagram (BDD) is one of methods for representing logical structures, just like truth table and logical expression [Bryant86]. There are many types of BDDs and they have superior features [Bryant95], but, in this paper, I adopt the standard BDD as shown in **Fig.1**. In the figure, rectangular and circular vertexes are *terminal* and *nonterminal vertexes*, respectively. A nonterminal vertex v has an *argument index*, $index(v) \in \{1, \cdots, n\}$, and two children, $low(v)$ and $high(v)$, which are illustrated as left and right children in figures. A terminal vertex v has a value $value(v) \in \{0, 1\}$. Every vertex v in BDD denotes a logical function f_v, which is defined recurrently as:

$$f_v = \begin{cases} value(v); & \text{if } v \text{ is a terminal vertex,} \\ \overline{index(v)} \cdot f_{low(v)} + index(v) \cdot f_{high(v)}; & \text{otherwise.} \end{cases} \quad (1)$$

In other words, a vector of arguments $(x_1, \cdots, x_n)$ generates a path in the graph, when it is applied to BDD. The path starts from a root vertex, and, when some nonterminal vertexes along the path have $index(v) = i$, it continues to $low(v)$ if $x_i = 0$, and $high(v)$

if $x_i = 1$. The value of the function for the vector equals the value of the terminal vertex at the end of the path.

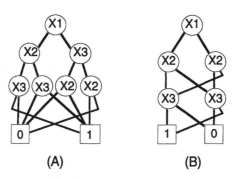

(A)   (B)

Fig.1: Examples of BDD.

Some BDDs of different structures may represent the same logical function, but a function is expressed by a unique Reduced Ordered BDD (RO-BDD). RO-BDD is a type of BDD satisfying the following (genotypical) conditions:

1. In any paths started from a root vertex of the RO-BDD, the order of argument indices assigned to appearing nonterminal vertexes must follow a unique total order.
2. The RO-BDD doesn't have any verbose vertexes and odd pairs of equivalent vertexes.

The verbose vertex u and a pair of equivalent vertexes, v and w, satisfy **equation 2** and **3**, respectively (also see **Fig.2**).

$$low(u) = high(u) \tag{2}$$

$$\begin{cases} high(v) = high(w) \\ low(v) = low(w) \end{cases} \tag{3}$$

Among two BDDs in **Fig.1**, (A) is not RO-BDD because it doesn't the first condition. The RO-BDD with these conditions has three significant advantages; (1) it requires small memory-size, (2) it assists fast computation, and (3) it can stores any logical structures as unique forms. If the order of argument indices assigned to nonterminal vertexes is not proper, however, a RO-BDDs satisfying the conditions may have a lot of vertexes, and it loses the first and second advantages. This is a serious problem, therefore, we must discover the proper order of argument indices before achieving RO-BDDs, and the method for resolving the problem is mentioned in the later section.

3. Genetic Programming

Genetic Programming (GP) is a search procedure developed based on Genetic Algorithm (GA) which is inspired by natural evolution [Holland75 and Goldberg89]. Both of traditional GA and GP have a population consisting of some members (solutions). For achieving members which works well in the environment (good or optimal solutions), they manipulates those members by genetic operators. Compared with GA, the most remarkable characteristic of GP is that it adopts tree-expression (S-expression of LISP language) for representing members, instead of binary or non-binary strings of characters, then GP has more powerful ability for representing solutions.

As genetic operators, GP uses the selection, the crossover, and the mutation. They are also used by GA, but are modified to treat tree-structure. The selection increases the number of good members which have large evaluation values, and removes bad members from the population. When the crossover occurs, two trees are randomly selected out from the population, one node is chosen at random in each tree, and they exchange sub-trees rooted from chosen nodes. The mutation is an operator to replace a sub-tree with a randomly produced sub-tree.

GP can be applied not only to the search problems, but also to the problems occurred in the domain of machine learning, because of its powerful ability for representation. On the other hand, GP requires heavy cost of memory and time for computation, and this is the most serious fault. This defect is caused because the generic operators of GP aren't radically improved from those of GA, although it has extremely superior representational ability.

In our approach which will be shown in next section, instead of S-expression, we attempt to adopt BDD for representing members in the population. Since BDD has some representational advantages mentioned in previous section, it is expected that our approach can overcome inefficiency in genetic operations of GP improve its performance.

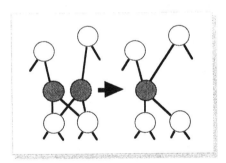
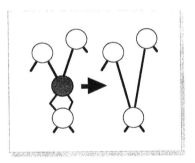

(a) Equivalent Function Node.   (b) Redundant Function Node.

Fig.2: Reduction

4. Approach

At the beginning of the discussion about our approach for evolution of BDD on GP, we suppose that the objective of GP in this paper is to learn a specific logical function. In other words, our GP must discover BDD representing a logical structure which can returns the same answers as the target logical function given as a problem.

For constructing our GP evolving BDD, we can imagine two technical specifications;

- The order of argument indices assigned to nonterminal vertexes must be protected from the genetic operators.
- The genetic operators must be extended to treat graphs containing self loops.

The former must be satisfied to obtain RO-BDDs of BDDs evolved by GP, as explained in the section 2. The latter indicates that the genetic operators of canonical GP are developed to treat tree-structure without self loops, then they can not deal with BDD without any extensions.

In this paper, to consider the former specification and to satisfy it, instead of BDD, we adopt Binary Decision Tree (BDT) for representing solutions. BDT may be regarded as the most specialized type of BDD, and never contains self loops of nonterminal vertexes.

Since we will adopt (RO-)BDD in future, optimal BDT must have the following features;

1. It completely expresses a logical structure of target function,
2. It doesn't have any verbose vertexes and any pairs of equivalent vertexes, and
3. All of the nonterminal vertexes don't violate a specific order of argument indices.

The first feature is a phenotypical requirement, and the second and third ones are genotypical constraints. In other words, a BDT with the first feature is the *correct BDT*, and a correct BDT with the second and third features is the *optimal BDT*. Among them, the first feature is the most important one, then every optimal BDT must possess the first feature, and a BDT without it is never optimal. The second feature is not so significant one, but the BDTs with it are more desirable than those without it, because this feature means that BDTs don't have any useless sub-tree. The third feature is necessary to obtain RO-BDD of BDT, and this seems the most serious problem as mentioned in section 2.

In this paper, in order to discover optimal BDT, we extend the architecture of GP as follows;

(a) Introducing an evaluation function considering the order of argument indices assigned to nonterminal vertexes in BDT, and
(b) Developing reduction-operator to remove all verbose vertexes and odd pairs of equivalent vertexes.

Namely, we expect that GP with a brand-new evaluation function discovers many BDTs with above features (the features 1 and 3, especially), and that the extension-b will help to acquire the BDTs with the features 2.

5. Proposed Method

In this section, we attempt to extend the architecture of GP to discover BDTs which possess the features mentioned in the previous section. Our extended GP adopts genetic operators ordinarily used in canonical GP, such as the selection, the crossover and the mutation. In this section, we explain only about two extended procedures, evaluation and reduction operator.

5.1 Evaluation

According to the discussion about the features which optimal BDT must possess, we settle the following evaluation function to be maximized;

$$F(T_i) = K \frac{G_1(T_i) - \min_k\{G_1(T_k)\}}{\max_k\{G_1(T_k)\} - \min_k\{G_1(T_k)\}} + (1-K) \frac{\max_k\{G_2(T_k)\} - G_2(T_i)}{\max_k\{G_2(T_k)\} - \min_k\{G_2(T_k)\}}$$

$$\text{(4)}$$

where $K = 0.1$, T_i is the ith member of tree in the population of GP, $G_1(T_i)$ and $G_2(T_i)$ are the phenotypical and genotypical evaluation functions to be minimized. $G_1(T_i)$ denotes the ratio how T_i returns wrong answers per all possible imputed vectors of arguments. $G_2(T_i)$ is the number of nonterminal vertexes which violate the *temporal order* of argument indices, the order which is defined below. This evaluation function concerns only the previous features 1 and 3, then even if a BDT receives the largest evaluation value, it is not always optimal one which have all three features. About second feature, however, the number of verbose vertexes and pairs of equivalent vertexes is expected to be small, because BDTs possessing the third features don't need so many vertexes to express any logical functions. Moreover, almost all of such vertexes are removed by the reduction operator defined later.

The temporal order of argument indices assigned to nonterminal vertexes is obtained in each tree as following steps;

1. Allocate smaller numbers to argument indices which is assigned to closer vertexes to the root of the BDT.
2. When some vertexes with different competing indices are equally distant from the root, among all competing argument indices, smaller numbers are assigned to indices which satisfies following conditions;
   2-a. The number of the index assigned to vertexes which is a child of vertexes with other competing indices is smaller than others.
   2-b. The number of the vertexes with the index is larger than the number of vertexes with others.
3. If the competition is not resolved during step 1 and 2, they are assigned smaller number in order of random selection.

We explain this procedure by using BDT containing 7 argument indices $\{X1, ..., X7\}$ shown in **Fig.3**.

1. *X3* is selected as first index in the temporal order, because it is possessed by root vertex A which is the *nearest* vertex from the root (step 1).
2. *X1* and *X2* compete each other at vertexes B and C (procedure 1). The two vertexes F and I, which possess *X1*, are children of vertexes with

$X2$, and only a vertex D with $X2$ is a child of vertex with $X1$, therefore $X2$ is assigned smaller number than $X1$ (step 2-a).

3. The competition between $X4$ at E and $X5$ at G can not be resolved by step 2-a. $X4$ appears at two nonterminal vertexes G and H, although $X5$ appears only at E, then smaller number is assigned to $X4$ (step 2-b).

4. Because there is no difference between $X6$ and $X7$ at step 1 and 2, one of them is chosen at random, and assigned smaller number than the other (step 3).

As a result of this example, the temporal order of the BDT in **Fig.3** is determined as [3214567] or [3214576]. Supposing that [3214567] is stochastically adopted as the temporal order, there is only one nonterminal vertex D (At the nonterminal vertex D, $X2$ belongs to the sub-tree belonging to $X1$, and there is no nonterminal vertex violating the temporal order). Therefore, in the case of the BDT in **Fig.3**, G_2 becomes equal to 1.

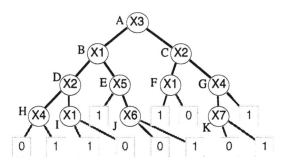

Fig.3: Example of BDT.

5.2 Reduction Operator

Just like the procedure to obtain RO-BDD of BDD, it is necessary for reducing BDT to remove all verbose vertexes and odd pairs of equivalent vertexes. We don't concern about pairs of equivalent vertexes, however, because of two reasons. One is because, compared with verbose nonterminal vertexes, discovering all of such pares in every BDT contained in the population of GP requires heavier computational cost. The other reason is because such pairs are expected to disappear during the search of GP with **equation 4**.

A reduced BDT is achieved by repeating the following procedure.

1. Remove all verbose vertexes.
2. Remove all *ineffectual sub-trees* defined below.

It is obvious the way of removing verbose vertexes, as shown in **Fig.4-A**. In **Fig.4-B**, we illustrate an example of ineffectual sub-tree. We can see that a nonterminal vertex B belongs to the sub-tree *high*(A), and has argument index $X1$. Namely, reference of vertex B means that $X1 = 1$, then *low*(B) is never referred by any argument vector.

Therefore, the logical function of this BDT is never changed, even if vertex B is replaced by *high*(B) (terminal vertex D, in the figure). In other words, ineffectual subtrees are parts which are never referred by any argument vectors.

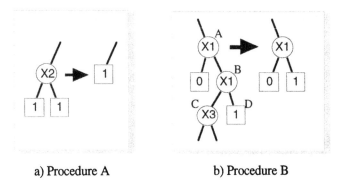

a) Procedure A   b) Procedure B

Fig.4: Two procedures of Reduction Operation

As explained above, the procedure of reduction operation is very simple, but it has a serious problem to determine which BDTs should be reduced and when the operator is executed. The reason of this difficulty is because the reduction may decrease the variety in the population of GP, although the variety is necessary for efficient search. On the contrary, small and simple BDTs achieved by the reduction may save the computational costs for genetic operations of GP.

The optimal BDT can be produced not only by the search of GP, but also by reducing correct BDTs discovered in the search. On the contrary, the possibility that the reduction produces the optimal BDT seems much larger than the possibility of discovering it in the search of GP. Namely, the reduction operation has important role to produce the optimal BDT from the correct BDTs, and correct BDTs should be reduced by the operator. In the computer simulation, therefore, the following BDTs in the populations are reduced by the operation;

1. All copies of BDTs whose evaluation values $G_1 = 0$.
2. All BDTs whose evaluation values F are smaller than the middle of value F in the population.

The former condition never causes any effects to the search of GP, and by the latter condition, about half number of BDTs in the population are reduced.

6. Computer Simulation

In this section, through some computer simulation, we examine whether our extended GP can learn the famous logical functions shown in the **Table 1**. That is, the objective of GP is to discover optimal BDTs possessing three features denoted in the section 4. Concerning parameter setting, In all of problems, it used tournament selection whose tournament size is 6, its population consisted of 500 members, and other parameters followed the setting in Koza's textbook [Koza94].

The results of experiments are shown in **Table 2**. In this table, there are two large rows which show the results derived from our GP and those of the canonical GP. The first large row have three medium rows which consist of two small rows, and first two medium rows concern with the performance of discovering *correct BDTs* and *optimal BDTs*, respectively. The correct BDT possess only the first feature denoted in the section 4, and the optimal BDT have all features. In other words, we can easily transform the optimal BDT to RO-BDD, but it is difficult or impossible to get RO-BDDs from correct BDTs which are not optimal. The first small rows show the ratios that our GP can discover correct or optimal BDTs, and, if those ratios are 100%, the second rows indicate the worst generations when such BDTs are discovered. The third medium row shows the number of trial and the size of the population.

Table 1: Problems.

Problem No.	Problem Name
1, 2, 3, 4	Even {3,4,5,6}-Parity
5, 6	{6-11}-Multiplexer
7, 8, 9	{4,5,6}-Symmetry
10	6 bit Random

Table 2: Simulation Results.

Problem No.		1	2	3	4	5	6	7	8	9	10
Proposed correct	Ratio	1.0	1.0	1.0	1.0	1.0	1.0	1.0	1.0	0.0	0.5
Method	Gen.	6	17	51	170	13	159.2	19	10		
optimal	Ratio	10/10	6/10	2/10	2/10	10/10	2/10	10/10	10/10	0/0	0/5
	Gen.	7				35		20.6	72		
	Trial	10	10	10	10	10	10	10	10	10	10
	size	500	500	500	500	500	500	500	500	500	500
Koza's GP	Ratio	1.0	1.0	1.0	0.9	1.0	1.0	0.97	0.99	0.54	
with ADF	Gen.	3	10	28	< 50	9	15	< 50	< 50	< 50	
	Trial	33	18	19	21	1	21	62	375	28	
	size	16k	16k	16k	16k	4k	4k	4k	4k	4k	

The second large row shows the results of the canonical GP quoted from Koza's textbook. Each of small rows, average ratio of correct answers of the best members, the generation when the average ratio is obtained, the number of trial, and the population size.

In many articles, for evaluating the search performance of GP (or other procedures), the average trials among several executions. In this paper, however, we think that it is important whether the search procedure can solve the given problems or not, rather than how fast it can solve them. GP is never a efficient and speedy procedures, and it exhausts a large amount of memory and time. Then, we must understand how much costs GP (or GP-based system) spends in the worst case. This is the reason why we adopted the worst generation as performance indices of our GP and the canonical GP, instead of the average generation.

Compared with the canonical GP, our GP exhibited fairly good results although it had a much smaller sized population than the canonical GP. Even in the problems which the canonical GP can not solve correctly, our GP could discover correct or optimal BDTs. Except in the problem 9 and 10, our GP could completely discover the correct BDTs. In the problem 1, 5 and 7, it could find all of the optimal BDTs, while there is no guarantee that the canonical GP could discover the optimal ones.

Fig.5 presents the typical search process of our GP, and is obtained by averaging the evaluation value of best BDT at each generation in 10 simulations. In the graph, the black and gray lines shows the phenotypical evaluation (G_1) and the genotypical evaluation (G_2), respectively. The story of the search can be described as follows;

- At initial state, the randomly generated BDTs are genotypically simple, and don't receive good evaluation values for their phenotypes.
- As generations go on, the population is phenotypically improved, but it increases the number of violation of the genotypical constraints. The reason of this phenomenon seems that the genotypical information is easily broken by the genetic operators, and that the phenotypical information is robust because plural BDTs can denote an equal logic structure (phenotype).
- If the phenotype improvement stick into local optima, the genotypical constraints begin to be satisfied. This process continues to progress evolution in the population, and this genotypical evolution helps the population to escape from phenotypical local optima.
- After the discovery of the correct BDTs (completion of the phenotypical evolution), the genetic operators and the reduction operator discover the optimal BDT which satisfies all of the constraints itemized in section 4.

Table 2 shows the improved performance of our GP, but it also exposes the faults of our GP. **Fig.6** and **7** show the search process of our GP in the problem 9 and 10 in which our GP could not discover correct BDTs completely. Especially, our GP could not solve the problem 9 any more during 10 simulations. In this problem, the BDT which consists of only one terminal vertex with value 0 can receive relatively good evaluation value, because $G_1 = 0.125$ and $G_2 = 0$ (**equation 4**). This BDT is easily discovered by GP in the early generations because of its simpleness and its good evaluation value, and the population converged to BDT in the early generations.

In the case of the problem 10, the population could not evolve genotypically any more. This problem has complicated logical structure because it is generated at random, then it seems difficult to represent the structure by a simple and small BDT. That is, the reason of this failure is that the reduction operator didn't work well in this problem.

These two failure tell us the insufficiency in the evaluation function and the reduction operation. In the concrete, we need adaptive mechanisms for changing the parameter K which we fixed in the **equation 4**, and for determining what BDTs in the population should be reduced. Even by the fixed schemes in reduction and evaluation, our GP showed remarkable performance, our approach will exhibit noteworthy improvement if we achieve those adaptive mechanisms.

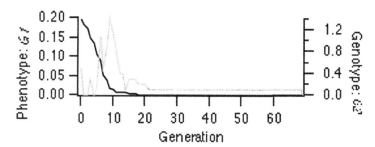

Fig.5: Result of proposed method on 5-Symmetry (problem 8).

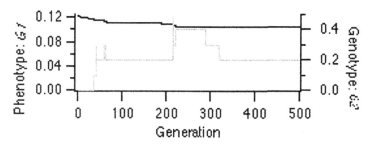

Fig.6: Result of proposed method on 6-Symmetry (problem 9).

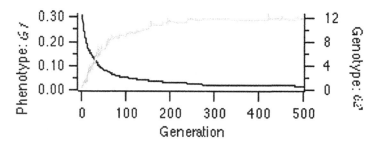

Fig.7: Result of proposed method on 6 bit Random (problem 10).

7. Conclusion

This paper proposed the methodology for evolving hardware, whose descriptions are specified in BDDs, by genetic programming (GP). To realize this methodology, GP must search in both of phenotype and genotype spaces, therefore we attempted two approaches; the operations to obtain BDDs satisfying the genotypical constraints and the method for balancing phenotypic and genotypic evaluations.

In this paper, we explained BDD and GP briefly, and discussed their characteristics and constraints. Then, we proposed the evaluation method and introduced brand-new operator, for improving phenotype and for satisfying genotypical constraints.

In the computer simulations, our GP exhibited superior results than the canonical GP with ADF, even by adopting much smaller-sized population. The search process showed the internal dynamics of the phenotypical and genotypical evolution. Unfortunately, our GP could not completely discover the correct solutions in two problems among 10, and these failure told us the insufficient features of our proposed method and gave us suggestions of additional mechanisms for more improved performance.

Reference

[Armstrong79] W. W. Armstrong and J. Gecsei: *Adaptation Algorithms for Binary Tree Networks*, IEEE Trans. on SMC, vol.SMC-9, No.5, pp.276-285, 1979.

[Bryant86] R. E. Bryant: *Graph-Based Algorithms for Boolean Function Manipulation*, IEEE Trans. on computers, Vol.C-35, No.8, pp.677-691, 1986.

[Bryant95] R. E. Bryant: *Binary Decision Diagrams and Beyond: Enabling Technologies for Formal Verification*, Embedded tutorial at International Conference on Computer-Aided Design November, 1995.

[Goldberg89] D. E. Goldberg: *Genetic Algorithms in Search, Optimization and Machine Learning*, p.412, Addison-Wesley, 1989.

[Holland75] J. H. Holland, *Adaptation in Natural and Artificial Systems*, University of Michigan Press, 1975.

[Higuchi94] T. Higuchi, H. Iba and B. Manderick: *Applying Evolvable Hardware to Autonomous Agents*, Parallel Problem Solving from Nature 3, pp.524-533, Springer, 1994.

[Jacob94] C. Jacob, *Genetic L-System Programming*, Parallel Problem Solving from Nature 3, pp.334-343, Springer, 1994.

[Kinnear94] K.E. Kinner, Jr., *Alternatives in Automatic Function Definition: A Comparison of Performance*, Advances in Genetic Programming (Edited by K. E. Kinnear, Jr.), MIT Press, 1994.

[Koza94] J. R. Koza: *Genetic Programming II*, p.746, MIT Press, 1994.

[Rosca94] J. P. Rosca and D. H. Ballard, *Hierarchical Self-Organization in Genetic Programming*, Machine Learning, Proc. of 11th Int. Conf., pp.251-258, 1994.

Author Index

Springer
and the
environment

At Springer we firmly believe that an international science publisher has a special obligation to the environment, and our corporate policies consistently reflect this conviction.

We also expect our business partners – paper mills, printers, packaging manufacturers, etc. – to commit themselves to using materials and production processes that do not harm the environment. The paper in this book is made from low- or no-chlorine pulp and is acid free, in conformance with international standards for paper permanency.

 Springer

Lecture Notes in Computer Science

For information about Vols. 1–1179

please contact your bookseller or Springer-Verlag

Vol. 1216: J. Dix, L. Moniz Pereira, T.C. Przymusinski (Eds.), Non-Monotonic Extensions of Logic Programming. Proceedings, 1996. XI, 224 pages. 1997. (Subseries LNAI).

Vol. 1217: E. Brinksma (Ed.), Tools and Algorithms for the Construction and Analysis of Systems. Proceedings, 1997. X, 433 pages. 1997.

Vol. 1218: G. Păun, A. Salomaa (Eds.), New Trends in Formal Languages. IX, 465 pages. 1997.

Vol. 1219: K. Rothermel, R. Popescu-Zeletin (Eds.), Mobile Agents. Proceedings, 1997. VIII, 223 pages. 1997.

Vol. 1220: P. Brezany, Input/Output Intensive Massively Parallel Computing. XIV, 288 pages. 1997.

Vol. 1221: G. Weiß (Ed.), Distributed Artificial Intelligence Meets Machine Learning. Proceedings, 1996. X, 294 pages. 1997. (Subseries LNAI).

Vol. 1222: J. Vitek, C. Tschudin (Eds.), Mobile Object Systems. Proceedings, 1996. X, 319 pages. 1997.

Vol. 1223: M. Pelillo, E.R. Hancock (Eds.), Energy Minimization Methods in Computer Vision and Pattern Recognition. Proceedings, 1997. XII, 549 pages. 1997.

Vol. 1224: M. van Someren, G. Widmer (Eds.), Machine Learning: ECML-97. Proceedings, 1997. XI, 361 pages. 1997. (Subseries LNAI).

Vol. 1225: B. Hertzberger, P. Sloot (Eds.), High-Performance Computing and Networking. Proceedings, 1997. XXI, 1066 pages. 1997.

Vol. 1226: B. Reusch (Ed.), Computational Intelligence. Proceedings, 1997. XIII, 609 pages. 1997.

Vol. 1227: D. Galmiche (Ed.), Automated Reasoning with Analytic Tableaux and Related Methods. Proceedings, 1997. XI, 373 pages. 1997. (Subseries LNAI).

Vol. 1228: S.-H. Nienhuys-Cheng, R. de Wolf, Foundations of Inductive Logic Programming. XVII, 404 pages. 1997. (Subseries LNAI).

Vol. 1230: J. Duncan, G. Gindi (Eds.), Information Processing in Medical Imaging. Proceedings, 1997. XVI, 557 pages. 1997.

Vol. 1231: M. Bertran, T. Rus (Eds.), Transformation-Based Reactive Systems Development. Proceedings, 1997. XI, 431 pages. 1997.

Vol. 1232: H. Comon (Ed.), Rewriting Techniques and Applications. Proceedings, 1997. XI, 339 pages. 1997.

Vol. 1233: W. Fumy (Ed.), Advances in Cryptology — EUROCRYPT '97. Proceedings, 1997. XI, 509 pages. 1997.

Vol 1234: S. Adian, A. Nerode (Eds.), Logical Foundations of Computer Science. Proceedings, 1997. IX, 431 pages. 1997.

Vol. 1235: R. Conradi (Ed.), Software Configuration Management. Proceedings, 1997. VIII, 234 pages. 1997.

Vol. 1236: E. Maier, M. Mast, S. LuperFoy (Eds.), Dialogue Processing in Spoken Language Systems. Proceedings, 1996. VIII, 220 pages. 1997. (Subseries LNAI).

Vol. 1238: A. Mullery, M. Besson, M. Campolargo, R. Gobbi, R. Reed (Eds.), Intelligence in Services and Networks: Technology for Cooperative Competition. Proceedings, 1997. XII, 480 pages. 1997.

Vol. 1239: D. Sehr, U. Banerjee, D. Gelernter, A. Nicolau, D. Padua (Eds.), Languages and Compilers for Parallel Computing. Proceedings, 1996. XIII, 612 pages. 1997.

Vol. 1240: J. Mira, R. Moreno-Díaz, J. Cabestany (Eds.), Biological and Artificial Computation: From Neuroscience to Technology. Proceedings, 1997. XXI, 1401 pages. 1997.

Vol. 1241: M. Akşit, S. Matsuoka (Eds.), ECOOP'97 – Object-Oriented Programming. Proceedings, 1997. XI, 531 pages. 1997.

Vol. 1242: S. Fdida, M. Morganti (Eds.), Multimedia Applications, Services and Techniques – ECMAST '97. Proceedings, 1997. XIV, 772 pages. 1997.

Vol. 1243: A. Mazurkiewicz, J. Winkowski (Eds.), CONCUR'97: Concurrency Theory. Proceedings, 1997. VIII, 421 pages. 1997.

Vol. 1244: D. M. Gabbay, R. Kruse, A. Nonnengart, H.J. Ohlbach (Eds.), Qualitative and Quantitative Practical Reasoning. Proceedings, 1997. X, 621 pages. 1997. (Subseries LNAI).

Vol. 1245: M. Calzarossa, R. Marie, B. Plateau, G. Rubino (Eds.), Computer Performance Evaluation. Proceedings, 1997. VIII, 231 pages. 1997.

Vol. 1246: S. Tucker Taft, R. A. Duff (Eds.), Ada 95 Reference Manual. XXII, 526 pages. 1997.

Vol. 1247: J. Barnes (Ed.), Ada 95 Rationale. XVI, 458 pages. 1997.

Vol. 1248: P. Azéma, G. Balbo (Eds.), Application and Theory of Petri Nets 1997. Proceedings, 1997. VIII, 467 pages. 1997.

Vol. 1249: W. McCune (Ed.), Automated Deduction – CADE-14. Proceedings, 1997. XIV, 462 pages. 1997. (Subseries LNAI).

Vol. 1250: A. Olivé, J.A. Pastor (Eds.), Advanced Information Systems Engineering. Proceedings, 1997. XI, 451 pages. 1997.

Vol. 1251: K. Hardy, J. Briggs (Eds.), Reliable Software Technologies – Ada-Europe '97. Proceedings, 1997. VIII, 293 pages. 1997.

Vol. 1252: B. ter Haar Romeny, L. Florack, J. Koenderink, M. Viergever (Eds.), Scale-Space Theory in Computer Vision. Proceedings, 1997. IX, 365 pages. 1997.

Vol. 1253: G. Bilardi, A. Ferreira, R. Lüling, J. Rolim (Eds.), Solving Irregularly Structured Problems in Parallel. Proceedings, 1997. X, 287 pages. 1997.

Vol. 1254: O. Grumberg (Ed.), Computer Aided Verification. Proceedings, 1997. XI, 486 pages. 1997.

Vol. 1255: T. Mora, H. Mattson (Eds.), Applied Algebra, Algebraic Algorithms and Error-Correcting Codes. Proceedings, 1997. X, 353 pages. 1997.

Vol. 1256: P. Degano, R. Gorrieri, A. Marchetti-Spaccamela (Eds.), Automata, Languages and Programming. Proceedings, 1997. XIV, 862 pages. 1997.

Vol. 1258: D. van Dalen, M. Bezem (Eds.), Computer Science Logic. Proceedings, 1996. VIII, 473 pages. 1997.

Vol. 1259: T. Higuchi, M. Iwata, W. Liu (Eds.), Evolvable Systems: From Biology to Hardware. Proceedings, 1996. XI, 484 pages. 1997.

Vol. 1260: D. Raymond, D. Wood, S. Yu (Eds.), Automata Implementation. Proceedings, 1996. VIII, 189 pages. 1997.